"Allan Chapman is a polymath, celebrated for his superb lectures on astronomical history. This engrossing book contains an immense amount of recondite information. His lively writing retains the flavour of his lectures, and will enlighten, fascinate and entertain anyone interested in science and its social context."

Lord Martin Rees, Astronomer Royal

"A fascinating narrative, full of delightful anecdotes, giving a very readable overview of astronomy and our understanding of the universe."

Martin Grossel, Emeritus Student in Organic Chemistry, Christ Church, Oxford, and Emeritus Fellow in Organic Chemistry, University of Southampton

"Allan Chapman writes with clarity and energy in a manner designed to both inform the general reader and stimulate thought. Engagingly written, and with great authority, he combines a manageable level of detail regarding this vast subject, with his own personal insights and experiences.

His work enables the reader to both grapple with the complex historical 'big picture' of unfolding ideas over the centuries, while also appreciating the significant impact and discoveries of individual pioneers in the field. Allan is not afraid to offer challenging personal insights and raises important questions for the reader to consider. This is an engaging, detailed, informative and thought-provoking book."

Martyn Whittock, historian, teacher, and writer

Also by Allan Chapman:

Slaying the Dragons: Destroying Myths in the History of Science and Faith

Stargazers: Copernicus, Galileo, the Telescope and the Church

Physicians, Plagues and Progress: The History of Western Medicine from Antiquity to Antibiotics

ALLAN CHAPMAN

Comets, Cosmology and The Big Bang

A HISTORY *of* ASTRONOMY *from* EDMOND HALLEY *to* EDWIN HUBBLE 1700 – 2000

Published by
Lion Hudson Limited
Wilkinson House, Jordan Hill Business Park
Banbury Road, Oxford OX2 8DR, England
www.lionhudson.com

ISBN 9780 7459 8031 7
e-ISBN 9780 7459 8030 0

First edition 2018

A catalogue record for this book is available from the British Library

Printed and bound in the UK, September 2018, LH29

To Rachel: Wife, Scholar, and Best Friend

CONTENTS

Acknowledgments

As with my *Stargazers* of 2014, there are many people who have, in various ways, encouraged and assisted me over several decades with the research that lies behind this book; they are too numerous to list here, but I am extremely grateful to them all. Especial thanks, however, are due to my late friend Peter Hingley, Royal Astronomical Society Librarian, and my still very active friend Tony Simcock, Emeritus Archivist of the Museum of the History of Science, Oxford, for their unfailingly generous help with all the queries I have thrown in their direction; Kevin Kilburn, of the Manchester Astronomical Society, for his planetary computing skills; my late good friend Sir Patrick Moore, CBE, FRS, for his inspiration from childhood onwards; Jane Fletcher and the BBC *Sky at Night* team; and Martin Durkin, Director of WAG TV. Nor can I forget the inspiration I have received from the late Colin Ronan and Andrew Murray, who (along with Patrick Moore), at RAS Club dinners, knew how to combine erudition and stimulating conversation with good fellowship: a tradition carried on by Professor Mike Edmunds, Charlie Barclay, and many present-day Club members.

Institutionally, I am grateful for the skills, assistance, and great kindness of librarian friends in Wadham College (Tim Kirtley and Francesca Heaney), Christ Church (Dr Cristina Neagu, Alina Nachescu, Dr Judith Curthoys, Angela Edward, and the late Janet McMullin), the Museum of the History of Science, (Dr Lee Macdonald), in the History Faculty Library, and in the Bodleian Library, all in Oxford; and in the Royal Society, the Royal Astronomical Society (Dr Sian Prosser), the National Maritime Museum, Greenwich, Manchester Central and Salford City Libraries, and numerous other institutions. In addition, I would like to express my gratitude to the Curators of the Collections of the Museum of the History of Science, Oxford (especially the late Francis R. Maddison, Gerard L'Estrange Turner, and John D. North, who taught me a great deal about how to examine and to learn from historical scientific artefacts), and of the National

Maritime Museum (in particular the late Commander Derek Howse RN). And I would also like to acknowledge my debt to Lancaster University, where I took my first degree, in History, and in particular to Professors Robert Fox and John Hedley Brooke, my inspiring teachers.

I am, too, greatly indebted to the Warden and Fellows, and the Chapel, of Wadham College, Oxford, and to the Dean, Chapter, and Governing Body of Christ Church, Oxford, for their friendship, encouragement, intellectual and spiritual support, and provision of academic sanctuary. I owe a special debt to my scientist friend Dr Martin Grossel of Christ Church, Oxford, for kindly taking the time in his busy life to read through and comment on my manuscript, although any remaining errors are, I emphasize, entirely my own. Likewise, I am grateful to friends and clergy at my "native" parish churches of St Thomas and St Anne, Clifton, Salford, Lancashire, for their interest and prayers over most of my life.

As I always learn so much from my research students, I would like to express my heartfelt thanks to them; and also to the members of the numerous astronomical and scientific societies to which I am invited to lecture, especially the Astronomical Societies of Salford, Lancashire (in particular Ken Irving), the Preston and District, Lancashire, and the Mexborough and Swinton, Yorkshire; the Society for the History of Astronomy, and the William Herschel Society, all of which I have the honour of serving as Honorary President. I always maintain that there is nothing like teaching and lecturing before live audiences to oblige one to master one's ideas thoroughly and present them with lucidity, and to defend oneself when challenged.

My sincere thanks go to Ali Hull, formerly of Lion Hudson, who commissioned this book, for all her advice and encouragement over the years we have worked together, and for her sensitive efficiency in weeding out the inevitable errors and infelicities of style. At Lion Hudson thanks are also due to Suzanne Wilson-Higgins and Jessica Tinker for their help and encouragement, and to Joy Tibbs, Kirsten Etheridge, Jacqui Crawford, and Clair Lansley for their skill in turning my manuscript into a beautiful book. I am indebted to my friend Bob Marriott, who compiled the index and also drew my attention to several small errors of fact, which were subsequently corrected.

But most of all, I am hugely indebted to my wife, Rachel, who has worked alongside me at every stage of this book, for her manifold skills, judgment, encouragement, efficiency, and infinite patience in coping with a husband whose office and administrative skills come a poor second to those of a confused chimpanzee. Rachel routinely rescues me from electronic muddles and helps search for misplaced books and research notes with great forbearance. While I possess an excellent memory for facts and information, I can never remember where I have put something. Once again, she has typed this book from my original fountain-pen-written manuscript, checked and edited it, and has made the whole process of writing this work immeasurably easier than it would otherwise have been.

Allan Chapman

January 2018

Preface

In 2014, I published *Stargazers: Copernicus, Galileo, the Telescope and the Church: The Astronomical Renaissance, 1500–1700*, which presented a revised interpretation of the Galileo story, setting it within the wider scientific, cultural, and religious context of the European Renaissance. This book aroused considerable interest, following which I was encouraged to continue the story from 1700 down to the present day.

Stargazers crept beyond 1700 and came to a close in 1728, when the Revd Dr James Bradley, Savilian Professor of Astronomy at Oxford and later Astronomer Royal, discovered the aberration of light, which supplied the first clear evidence that the earth really did move, and that Nicholas Copernicus had been right. He did this *not* by rhetoric or bluster, but by patient, exact measurement with instruments capable of new levels of accuracy, followed by meticulous calculation. A scientific tradition of precision that was inaugurated by Tycho Brahe and developed by Robert Hooke, John Flamsteed, and others in the seventeenth century, before coming to maturity in the eighteenth. It was Edmond Halley who furthered the tradition during the first forty years of the eighteenth century.

Yet who, in 1728, could have dreamed how astronomy would have developed by AD 2000?

Our story starts with Edmond Halley, who was forty-four years old in 1700, with an already illustrious career behind him. He had also been something of a mentor to Bradley in Oxford. By any standards, Halley was colourful, his career embracing achievements as a scientist, diplomat, Royal Navy sea captain, and explorer, and he was also good company. Edmond Halley epitomizes several of the key themes that run through this book, for he was a pioneer of cometary science and a deep-space cosmologist. He studied the "lucid spots" or nebulae and stellar distribution, and even asked why, if the stars extended to infinity, the sky went dark at night. Halley also pondered what forces from space might have moulded the earth's surface, extending from cometary impacts in remote, prehuman times, and speculated about the possible cause of the spectacular aurora borealis of 1716. He was, too, an instinctive astronomical historian, using ancient and modern star positions to discover the "proper motions" of stars: motions that would prove essential to measuring stellar distances in the nineteenth century.

Comets, Cosmology, and the Big Bang also documents the growing public fascination with astronomy, from the paid "box office" lectures of James Short and his eighteenth-century contemporaries to the "astronomical mania" of early Victorian Britain. It then looks at the founding of the large civic and national amateur astronomical societies after 1859, and the impact of radio, TV, and astronomical personalities, such as Sir Patrick Moore and Carl Sagan. Growing prosperity and good quality yet cheap books, magazines, and instruments turned astronomy into something of a national passion by the twentieth century: a passion further fuelled by the "Space Race" of the 1950s.

This popular passion drew endless inspiration from the rapid technical progress of the science, and from the fact that so many British astronomers did not hold scientific jobs, but were self-funded Grand Amateurs. And what a harvest of transformative discoveries they would gather, starting with Sir William Herschel and his sister Caroline, who, from the last twenty years of the eighteenth century, would design and build great reflecting telescopes that would create observation-based "deep space" cosmology. The Herschels, who had originally funded their researches from William's successful musical career, won the plaudits of the scientific community, and bedazzled the Romantic age – that age of awe, imagination, and sensitivity to artistic and literary beauty – with the wonders of deep space that their telescopes revealed.

The book goes on to explore the role of women in nineteenth- and twentieth-century astronomy, beginning with Caroline Herschel and Mary Somerville, and on to modern women astronomers such as Dame Jocelyn Bell Burnell, who discovered pulsars in 1967.

The mid-nineteenth century witnessed a series of sea changes in physical science that would transform not only astronomy but also physics, chemistry, and even medicine, all through rapid advancements in technology. First there was photography, which, within forty years after 1840, would go on to fundamentally change all manner of data recording and storage, making it possible to build up vast archives of astronomical and other scientific images. Then, after 1860, came spectroscopy, which revealed both the chemistry and the physics of the sun and stars, transforming our knowledge of deep space in a way that Sir William Herschel could never have imagined, but which his "universal philosopher" son, Sir John, would live long enough to witness.

It was in the 1890s, however, that physical science would enter a world of topsy-turvy, where one ancient truism after another bit the dust. Firstly, there were the new-found marvels of the electromagnetic spectrum, replete with all kinds of invisible yet measurable energies such as X-ray and radio waves. Secondly, there was radiation physics, which began life in the Parisian laboratories of Antoine Henri Becquerel and Marie and Pierre Curie, and would, within a couple of decades, explain how the stars were born, radiated, and died. Perhaps strangest of all, between J. J. Thomson's discovery of the electron in 1897 and James Chadwick's elucidation of the neutron in 1932–34 came a fundamental reappraisal of what matter was made of, through the developing science of nuclear physics.

By the end of the second decade of the twentieth century, so many of the old physical certainties seemed to have gone, replaced by one cosmological puzzle after another. Was there just one big finite galaxy, or were the nebulae actually vast "island universes" of stars scattered to infinity, appearing dim only because of their vast distances? How could stellar nuclei generate such prodigious energy, then blow up or shrink into tiny white dwarves? Stars, far from being eternal, were now understood to have life cycles, like humans.

This bizarre new universe of the "pre-watershed" decades before 1925 required new laws of matter, space, time, and motion that went beyond the perfectly true yet local cosmology of Sir Isaac Newton of 1687. The breathtaking universe emerging with the twentieth century needed Albert Einstein, Niels Bohr, Max Planck, and Sir Arthur Eddington to make sense of it, and to see great mathematical truths underlying the new precision-instrument-based primary data harvested from both the sky and the physics laboratory.

The watershed was eventually crossed after 1924, when the astronomical celebrity and Anglophile Edwin Hubble studied Cepheid and other stars in the Andromeda nebula, whose optical physical characteristics suggested they were vastly beyond the stars of our Milky Way galaxy. Hubble's reputation was immortalized in the Hubble Space Telescope in the 1990s, the digitized images of which would entrance the world and advance astronomy yet further: just as Hubble did himself at Mount Wilson in the 1920s and 30s.

The whole cosmological scale of reference had changed by 1930 to suggest a universe that seemed to be expanding rather than remaining static, as astronomers had believed since antiquity. Crucial to this new interpretation was Father Georges Lemaître's hypothesis of a cosmic atom, which had exploded and sent out its contents to expand for ever through empty space, to create the visible universe: a primal explosion later derided by Sir Fred Hoyle as a "big bang". Yet without the new post-1860 technologies and the great international archive of data recorded on glass plates, none of this would have been possible. Then in the 1930s two new technologies were born that would, within thirty years, transform astronomy yet again: radio astronomy and rocket-powered space flight, both of which we will meet later. Then at the end of the book, we will return to the self-funded amateur astronomers, and their own continuing contribution to the advancement of the science.

Running through it all is the perennial question: are we alone in the universe? I showed in *Stargazers* that an inhabited universe was being seriously discussed in the seventeenth century, while Sir William Herschel even wondered whether there might be cool habitable zones beneath the sun's fiery exterior. But it was the Georgians and Victorians who truly took "aliens" to heart, with Thomas Dick's bustlingly populated cosmos, the proclamation of cities discovered on the moon and "canals" on Mars, and the invention of the science fiction space journey, starting with Jules Verne and H. G. Wells, and moving on to the twentieth-century cinema and TV industry. We will end Chapter 29 with questions about life on other worlds.

Comets, Cosmology, and the Big Bang covers a vast sweep of human endeavour and achievement, stretching across three centuries. Yet it is not just about technology and physics; it is about people, individual lives, and why we are motivated to discover new things. When we ponder the universe, it is impossible to avoid the big questions pertaining to origins, meaning, symmetry, beauty, and human curiosity. These question take on a metaphysical or religious dimension, such as "Why are we here?", "Why does the universe exist?", and "Why do we humans care?"

It will be a fascinating ride, I can assure you. So sit comfortably, fasten your seat belt – and read on!

Chapter 1

From the Beginning to 1700: The Origins of Astronomy

It was in the sixteenth and seventeenth centuries that modern astronomy was born: the astronomy of cosmological vastness, of planets, not as lights in the sky but as spherical worlds encircling the sun, of a Milky Way composed of millions of individual stars, and of strange, feebly glowing "nebulae" or cloudy patches seen against a black sky. This "new" universe came about, however, not because some enlightened intellectual crusaders dared to think for themselves, or challenge the blind Church dogma of mythology, but because of new unexpected discoveries, made with instruments of ever-increasing precision, and, after 1609, with telescopes. The "new" astronomy was created by a progressive technology, developed first in the Greco-Roman world, and then the Middle Ages, as Western Europeans in particular became enchanted by precision devices such as complex mechanical clocks, astrolabes, and sundials, and by sophisticated musical organs, magnetic compasses, Gothic cathedral technology, guns, experimental optics, naturalistic oil painting, and, by 1450, the printing press. All this was driven by an increasingly prosperous Europe-wide market economy. But if these were the essential preconditions for the Renaissance watershed, how had astronomy originated in the first place?

The origins of astronomy

It is generally agreed that astronomy is the oldest of the sciences. By a "science", I mean a body of ideas founded upon a logical structure and verifiable predictability. Astronomy's logical and demonstrable roots were established long before medicine, botany, chemistry, and biology became precise sciences, having graduated from being observation-based, speculative arts, based on hunches and accumulated experiences.

A Babylonian *magus* of 2000 BC, for example, could confidently predict how many days must come before the next new moon, while the Judean vineyard owner knew how many weeks or months must run their course before his grapes were ready for harvest. The heavens provided a permanent and unchanging backcloth against which life could be measured, for people knew the seasons and constellations of their ancestors' days would still be there when their great-grandchildren were ancient.

Everything else was a lottery. A virulent epidemic would kill one man, yet his wife would escape unscathed. A herbal concoction would cure one person, and kill his neighbour. One wolf might run away howling when you threw a stone at it, another would turn and maul you. Snakebites sometimes killed, yet occasionally did not. A lump of copper stone heated in a crucible might yield an abundance of useful metal, or might not. And why did milk turn sour and wine become vinegar? Yet the times, seasons, planets, and constellations ran through their eternal and predictable courses – world without end.

The earliest astronomers

The oldest human cultures for which we possess written records, such as those of Egypt, Babylon, India, China, and Israel, all left astronomical records of some sort. They might, like the Chinese and Babylonian, be meticulous records of eclipses, comets, and maximum elongations of Venus from the sun (the Babylonian goddess Ishtar, the curiously horned star of legend). From ancient times, the Chinese imperial sky-watchers recorded the appearance of guest stars, now known to be supernovae exploding in deep space, and this provides valuable pieces of data for present-day astronomers and astrophysicists, studying the cosmological detritus of "supernovae remnants".

It is fascinating to speculate how high-civilization cultures, from Egypt to China, developed coherent cosmologies that had much in common. These cultures often had trade relations with each other, as the archaeological record shows. The Old Testament tells of active relationships of trade, diplomacy, and war between

Egypt and Babylonia, often with little Israel – "the (defenceless) Belgium of the ancient world"[1] – acting as a reluctant highway between them. When you were in Nineveh or Babylon, you had only to sail down the Euphrates for 800 miles, and you reached Ur of the Chaldees, Abraham's supposed birthplace; go a bit further, you were in the Persian Gulf. Hug the coast for another 1,300 miles around the Indian Ocean, and you would arrive at the Indus, with a culture extending back to at least 3000 BC. Unsurprisingly then, nineteenth-century European Sanskrit scholars found ancient Indian constellation names that may have been the originals behind Perseus, Andromeda, and Orion.[2] Spices, gold, and exotic artefacts travel across trade routes; ideas of all kinds may do likewise.

In many ways, Chinese astronomy, ancient as it was, seems to have developed within a different cultural orbit, extending primarily to Indonesia and Japan. But the presence of jade artefacts in Indian, Babylonian, and even European archaeological digs reveals that some kind of trade with China passed through Indonesia to India, before crossing the great Asiatic steppe, the future fabled Silk Road to the East.

Whether one were an Egyptian, a Babylonian, or a Hebrew in *c.* 1500 BC, the abundance of surviving records reveals a common-sense cosmology that looked something like the following. The earth was flat: a perfectly reasonable conclusion to draw if, no matter how far you walked or rode your donkey, a distant flat horizon *always* unfolded before you. Also, *your* land was special, for a great, life-giving, north- or south-flowing river irrigated it, be it the Nile, the Euphrates, or the Jordan: Middle Eastern rivers bounded east and west by arid desert, or the Mediterranean "Great Sea". Your cultural specialness was apparent daily, when the sun rose to its highest noonday point above your river, as if conferring a blessing.

The sky above was envisaged as a vault, or tent, or for the ancient Hebrews, a "tabernacle", probably supported by four distant pillars. Beneath its canopy, the sun, moon, stars, and planets described their exact and eternal celestial journeys. Generally speaking, these astronomical bodies were believed to pass under the earth at night, to be born afresh at dawn; the ancient Egyptians maintained nocturnal services in their temples to ensure the sun's (or Ra's) safe passage through the twelve subterranean gates. The elaborately painted tomb of Pharaoh Rameses VI (1144–1137 BC) in the Egyptian Valley of the Kings makes this cosmology explicit, as the body of the sky-goddess Nut encircles the heavens like a slender dancing-girl, arched and balancing on her toes and fingers above a flat world. Twelve maidens represent the hours of the night – perhaps the origin of our division of the day into twelve hours of light and twelve hours of dark. All deeply astronomical.

The Greeks of post-*c.* 650 BC, however, would develop a very different cosmology: one that would underpin European (and Arabic) astronomy until *c.* AD 1600, before bequeathing an observational, mathematical, and instrument-based approach to astronomy, starting with the bronze armillary sphere graduated circles and continuing to the Hubble Space Telescope and the International Space Station.

What made the "Greek experience" central to Western thought?

Wherever you come from, you are a child of classical Greece. As Greek science was absorbed in translation into the Arab world after *c.* AD 900, and even into Japan by the nineteenth century, Greek thinking came to constitute the quintessence of Western civilization. By AD 50, Greek became the language through which Christianity would spread, creating a whole new moral, spiritual, and humanitarian framework for the world. Yet why was it *Greek* thinking, rather than the much older Egyptian or Babylonian, that exerted such a profound influence?

In *c.* 1200 BC, when Agamemnon, Helen of Troy, Odysseus, and the other characters later immortalized by Homer were supposedly around, Greece was a war-torn, chaotic place where various heroes constantly had fun knocking the stuffing out of each other. Yet by the eighth century BC, something very significant was happening; and the reasons behind it have fascinated scholars for several centuries. First there was Hesiod, *c.* 750 BC, whose *Theogony* in many ways invented theology, or the study of the origins, nature, and relationships of the gods. Hesiod's *Works and Days* then praises not the superhero soldier, but the farmer, whose careful tillage peacefully fed the population, while making references to the heavenly bodies and their relationship to the seasons. After *c.* 640 BC, one meets a crucial innovation: philosophy, mathematics, geometry, and astronomy. Thales and Pythagoras exercise their brains upon the properties of circles and triangles, and the exact mathematical sequences of the astronomical bodies. How much they drew upon practical aspects of Babylonian astronomy – such as the sun moving through a single digit (or degree) of its annual course each day of the assumed 360-day year – we cannot be sure. But most likely their basic sky geography of the twelve signs of the zodiac and constellation boundaries were inherited from Babylon.

We cannot be sure whether Thales and Pythagoras ever wrote their ideas down or passed them on through an oral teaching tradition, to be written down by later disciples. But we do know that they, and their ideas, made a formative and indelible impression on the Western mind: and most of all, on astronomy, with the proposition that mathematics contained eternal truths that transcended mere numerical reckoning. The Babylonians had been experts in reckoning and recording numerical data, such as those used in tax collection or astrology, while Egyptian surveyors were masters of lines and triangles, and able to delineate who owned what piece of ground when the annual Nile flood receded, as well as laying out monumental architecture such as pyramids and ziggurats. What the Greeks did, however, transcended the practice of mathematics as just a useful form of social technology: they saw it, rather, as possessing a compelling philosophical power.

As far as we can tell from surviving clay tablets and papyri, the Babylonians and Egyptians did not ask *why* two plus two cows always equalled four cows, or why circles and triangles had useful properties when it came to surveying a field. Yet the Greeks did; and by the time of Thales and Pythagoras after 600 BC, these questions were being considered in a big way. Was it not wonderful that the radius of a circle unfailingly divided into the circumference 6.8 times? And that the areas of two squares erected on the shorter sides of a right-angled triangle always added up to the area of a square on the longest, hypotenuse, side – Pythagoras' famous

Theorem? It was the Greeks who turned geometry and arithmetic from useful skills into precise intellectually coherent *sciences*, and when they were applied to the celestial bodies, astronomy became a science as well.

Just think, blasé as we may be about it today, of the sheer intellectual force behind the realization that the same numerical reckoning tools used to count a flock of sheep or mark out a field boundary also applied to the great and small cycles of the sun, moon, and planets, and with stunning and repeatable accuracy. These techniques, moreover, could be developed and taught to others, leading on to a precise accumulation of systematic knowledge. Mathematics and astronomy were not the only *sciences*, or publicly demonstrable intellectual disciplines, to which the Greeks paid attention. There were also philosophy and linguistics, the arts of thinking, speech, and coherent communication, the politics of the negotiated public space as opposed to the tribal hierarchy, economics, music as a mathematical discipline, rational medicine, and even organized sports, such as the Olympic Games. The Greeks, too, invented public theatre and plays that pursued a particular theme, encompassing the classical tragedies of Aeschylus and the bawdy comedies of Aristophanes, which still make us alternately cry and laugh.

Over the centuries, many scholars have wrestled with what caused the Greek experience, and why it was so formative in the development of subsequent Roman and Christian medieval European civilizations. I have made my own suggestions in my previous writings.[3] Perhaps the independent-minded commercial and intellectual initiatives that developed with the city-state system after *c.* 700 BC played a significant part, along with Greek philosophical ideas of the world embodying a coherent unifying intelligence, referred to variously as the *logos* or the *nous*. This world was not the plaything of warring gods and spirits, but embodied a rationality of which the human race was a part. Greek astronomy cannot be seen in isolation; to make sense of its formidable influence, it must be viewed as a part of the wider cultural and philosophical package. By *c.* 600 BC, the Greeks had come to realize, perhaps as a result of their sea-voyaging and travel, that the earth was a sphere and *not* a flat plane; by 300 BC Eratosthenes had even made a remarkably good reckoning of its size, based upon shadows and geometrical projections. By 125 BC, Hipparchus had made a list of constellations, no doubt based upon Babylonian predecessors, and from a mathematical analysis of already ancient Egyptian in comparison with modern data, deduced the existence of the precession of the equinoxes: a tiny annual slipping back of the equinoxes along the celestial equator. By AD 150, Ptolemy – sometimes cited as Claudius – had produced a star-map, whose basics we still use today, as well as formalized the classical geocentric cosmos with its planetary and stellar spheres: all of which accorded elegantly with the best knowledge of the time. As far back as 350 BC, Aristotle had produced a coherent *physics*, or logical explanation of how matter was formed, behaved, and changed, and how the heavenly bodies might relate to terrestrial bodies. All this was with exquisite rationality from the then best-known common-sense logical knowledge, a million miles from superstitious tales and myths.

Medieval consolidation

For 1,000 years after the slow decline of classical Greco-Roman civilization, by the sixth century AD, the cosmology and physics outlined above (just like its medicine and physiology) formed the bedrock of Christian Europe's explanations of the heavens and most natural phenomena. Between *c.* AD 900 and 1400, this system was absorbed into the intellectual traditions of the Arab world, from southern Spain to Iran, largely at the hands of Greek, Syriac, and Mesopotamian Christian and Jewish translators who found themselves living under new Islamic overlords, following Muhammad's and his disciples' lightning conquests of their lands between AD 622 and *c.* 720.

The Islamic world would produce some astronomers of genius between *c.* AD 950 and 1449. These included great observational astronomers and table calculators, such as Al-Battani, Thabit ibn Qurra, Al-Zarqali, and Ulugh Beg, along with planetary dynamic mathematical theorists and calculators such as Nasir al-Din al-Tusi of the Maragha Observatory in Persia. The observatories at Maragha,

Samarkand, and elsewhere used enormous Greek-derived 90° quadrants and circles to measure tiny angles between astronomical bodies, so that this data could be used to compute refinements to the geocentric cosmology of the classical Greeks, and to work out precise corrections that could be applied when comparing Greek observations of 125 BC or AD 150 with those made in AD 1000.

Like the Europeans, the Arabs fell in love with mathematics and geometry, and began to devise ingenious angle-measuring instruments, often inspired by Ptolemy's *Almagest* of *c.* AD 150, such as the astrolabe. Consisting of a set of angle-graduated brass plates, usually between 5 and 10 inches in diameter, incorporating a rotating star map and engraved tables, this would become the quintessential medieval astronomical instrument. It could be used to measure celestial angles, find the local time, or compute the rising and setting times of the sun, moon, planets, and stars. Astrolabes became commonplace across Europe from the twelfth century onwards, and the poet Geoffrey Chaucer's *A Treatise on the Astrolabe* (*c.* 1391), based upon Latin and Arabic predecessors, probably ranks as the first book on a science-based technology to be written in the English language.

1.1 Claudius Ptolemy, holding a set of the angle-measuring "rulers" that he is said to have invented. No authentic portrait of Ptolemy exists, and this is an imaginary one of c. 1600. (R. S. Ball, *Great Astronomers* (1895), p. 9. Author's collection.)

While both observational and mathematical astronomy were advanced in the medieval Arab world, Europe was also astronomically and scientifically active. The Venerable Bede's *De Temporibus* (*On the Times*, or calendar) and *De Arte Metrica* (*On the Art of Measurement*) were written, impeccably within the classical tradition, at Bede's Northumbrian monastery at Jarrow (*c.* AD 710-730). When the second millennium arrived in AD 1000, the reigning Pope Sylvester II (a Frenchman, formerly Gerbert of Aurillac) was a renowned astronomer and mathematician.

When Europe's first universities came into being in the twelfth century, astronomy, geography, arithmetic, and music (the planetary harmonies) formed a cornerstone of the undergraduate Arts curriculum, and most students would have learned how to observe and calculate with astrolabes. By the 1460s, astronomy was taking off in Europe, with groups of serious observers, instrument-users, and calculators in cities such as Nuremberg, Augsburg, Bologna, Oxford, and elsewhere. These men not only observed the heavens, but also pondered the nature of motion, physics, infinity, and cosmology. Medieval Europe's great single contribution to astronomical technology, however, was the mechanical clock: one of the truly world-changing inventions of all time. Richard of Wallingford's great St Albans Abbey clock of 1326 even incorporated a mechanized astrolabe into its complex weight-driven mechanism, while the Exeter Cathedral clock incorporated a – still working – lunar calendar. All of this puts paid to the myth that medieval Europe was scientifically backward.

Why was all of this astronomical achievement pursued firmly in the context of the geocentric, Ptolemaic cosmology, in which the sun stood at the centre of nine transparent spheres that between them carried the moon, sun, planets, and stars around us? It had nothing to do with the church suppressing novel ideas, or with Dark Ages superstition. The classical cosmos survived unchallenged in both Europe and Arabia for one simple reason: it accorded with the best observed facts then available. Do we feel that we are spinning through space at 19 miles per second, as, in fact, we are? Shouldn't objects get flung into space rather than remaining where they are, if the earth is in motion? While there were several long-term incongruities in Ptolemy's classical geocentric cosmos, such as strange irregularities in the orbits of certain planets, they could be compensated for in practical terms by a variety of computational models. When all things were considered, therefore, the classical geocentric cosmology made most sense and accorded best with human experience. And that is why it survived so long.

Europe's astronomical Renaissance

As I have shown in my *Stargazers*,[4] the formulation, testing, and eventual physical geometric demonstration of a *helio-* or sun-centric system between 1500 and 1728 was *not* the work of a few revolutionary geniuses who brought "superstition" crashing down. It was, to the contrary, the product of a long and meticulously conducted *quest*; and while Galileo, in spite of the brilliance of his telescopic discoveries, had a misplaced penchant for bombast, the real progress was made by meticulous observers, designers and builders of new instruments, experimentalists, and painstaking calculators.

The real-life Copernicus was not the "timid canon" of legend who for forty years kept his revolutionary idea of a moving earth to himself, only daring to publish on his deathbed. He was, instead, an eminent ecclesiastical lawyer, a renowned medical doctor, and a Polish cathedral dignitary. He was a man of high professional status, understandably cautious about publishing a theory for which he knew he could not supply an atom of proof as things stood in 1543 – the year in which his *De Revolutionibus Orbium Coelestium* (*On the Revolutions of the Heavenly Spheres*) eventually came off the presses in Nuremberg, shortly before his sudden death, probably from a stroke. A lack of demonstrable evidence was a notable weakness within the world of Europe's intellectual rigour-driven universities. As Copernicus fully understood, the physical proof of the earth's motion could only come about through meticulous geometry and the detection of tiny angles in space. In 1543, there was not an instrument in Europe sufficiently accurate to make the measurement of that angle possible, equivalent as it was to the angle subtended by a British 10 pence coin (diameter 0.95 inches, or 28 mm) at about 3.25 miles. Nor would there be one until 1728, when the Revd Dr James Bradley, using a telescopic instrument of such an accuracy as Copernicus could never have imagined, chanced to discover the aberration of light, or a six-monthly star displacement that could only be explained on the assumption that the earth moved in space. The key proof of the earth's rotation, the measurement of a stellar parallax, had to wait until 1838, when Friedrich Wilhelm Bessel in Germany successfully measured a six-monthly parallax angle for the fifth-magnitude star 61 Cygni in the constellation of Cygnus, the Swan.

What had made this tiny angular measurement possible was a true revolution in precision engineering and optical technology. This technological transformation had begun in late sixteenth-century Denmark, with the Lutheran Tycho Brahe, a staunch admirer of the Roman Catholic astronomer Copernicus. Expending his own and royal revenue wealth, Tycho began that technological quest that still lies at the heart of modern science. Employing the finest craftsmen of northern Europe, he evolved a sequence of engineering designs that, by 1585, enabled him to measure celestial angles ten times more accurate than those of Copernicus fifty years before. Yet still no six-monthly star displacement, or stellar parallax, could be detected. Profound admirer of Copernicus as he was, Tycho was to conclude, on solid physical grounds, that he did not believe that the earth moved.

Tycho was impeccably correct in his logic and approach, for the demonstration of the heliocentric system was not about rhetoric or philosophy; it was about precision technology and angular measurement. The whole of modern science has Tycho to thank for this crucial realization, for every aspect of it hinges upon making our technologies ever more sensitive, from digitally controlled rover vehicles on Mars to body scans in our hospitals that can detect cancerous growths.

Less than a decade after Tycho's own sudden death in November 1601, Europe was to come upon a new piece of technology that would also play a transformative role in astronomical progress: the telescope. By 1608, lenses were commonplace across Europe. They were used as magnifying glasses, burning-glasses, and most commonly, for spectacles. By the seventeenth century, this late-thirteenth-century Italian invention had already given clear vision to countless millions of people across Europe – one of several great medieval inventions that would create the modern world. Then in 1608, a Dutch spectacle-maker of Middelburg, one Hans Lippershey (pronounced Lipper*hay*), stumbled across a novel use for the familiar lenses. Whether Lippershey made the key discovery or – according to one version – some children playing with lenses in his workshop did so is unknown. Yet Lippershey quickly realized that when two lenses of the correct curvature, one ground to a convex curve (held away from the eye), the other ground to a concave curve (positioned close to the eye), were held up at their mutual focal point, they made distant objects appear much closer, and beautifully detailed. Not a man to miss out on a guilder, Lippershey, known as the clever burgher, mounted the lenses in a tube and tried to secure a patent, for in 1608 Holland (the Spanish Netherlands) was winning a war of liberation from Spanish rule, and being able to see what the distant enemy were up to could confer distinct strategic advantages. As some fellow Dutchmen challenged Lippershey's priority of invention, however, no patent was forthcoming, only a reward and a commission to supply the States General with duplicates of his "truncke" or box. Consequently, "perspectives", "cylinders", "glasses", or "Dutch truncke" were soon being made in Amsterdam, Paris, and across Europe. The Latin word *telescopium*, or telescope, probably dates from 1613, and derives from the classical Greek words *tēle* ("far off") and *skopein* ("to look at").

It appears that the forty-eight-year-old Thomas Harriot, an Oxford graduate, philosopher, mathematician, and friend of Sir Walter Raleigh, came by such a "truncke" in the spring of 1609. Harriot made his first, carefully dated telescopic moon drawing on 26 July 1609, at 9 p.m., of the five-day-old crescent moon. The drawing, on a fine sheet of foolscap paper, is still preserved in the West Sussex Record Office, Chichester, England.[5] Soon after, according to their surviving correspondence, Harriot was comparing telescopic observations with his friend Sir William Lower in South Wales. He also made some 200 observations of sunspots, all independently of Galileo, who, as far as we can tell, did not use the telescope for astronomical purposes until the very end of November 1609, nearly four months after Harriot.

Although a famous mathematician, pioneer of algebra and binomial mathematics, and correspondent of Johannes Kepler, Harriot never published his telescopic astronomical work, and it only began to come to light after 1784. Harriot, a convinced follower of Copernicus, was a comfortably off bachelor, whereas Galileo was a hard-up, middle-aged Paduan professor, longing for wealth and fame. The telescope provided Galileo with both after 1610. Yet Harriot, and his Welsh friends, did not resent Galileo's success, openly admiring him.

As Galileo hammered home in his *Sidereus Nuncius* (*Starry Messenger*), 1610, the telescope revealed a universe radically different from the naked-eye cosmos familiar since antiquity. Not only was the moon, with its newly discovered craters, mountains, and seas, a world in its own right, but the planets appeared to be worlds as well, when observed through the telescope. The stars, instead of appearing to be fixed inside a great black sphere, all at the same distance, now seemed to recede to a three-dimensional deep-space eternity, as each improvement in telescope construction revealed yet more myriads of hitherto unimagined stars. So was the universe infinite?

None of this proved that Copernicus's sun-centred universe was correct, but it fundamentally challenged and undermined the ancestral truisms of the earth-centred cosmos of Ptolemy. Then other discoveries flew thick and fast. Kepler's Three Laws of Planetary Motion, based upon the idea of elliptical rather than circular orbits, provided a whole new way of thinking of the proportions and invisible forces that bound the solar system with perfect laws. Fathoming these laws occupied men across Europe, such as Jeremiah Horrocks, Christiaan Huygens, Jacob Bernoulli, Gottfried Leibniz, and Sir Christopher Wren. In 1674, Robert Hooke would begin to get close, but it was Sir Isaac Newton who – in his *Principia Mathematica* (1687) – would provide a cogent and comprehensive mathematical analysis of the strange invisible force that bound creation: gravity.

Gravity, however, significant as it was, must be seen in context, and not elevated to the status of the sole golden key that unlocked science. Seventeenth-century Europe saw a veritable cascade of new precision technologies that made new scientific discoveries possible, including gravity. In astronomy alone, these technologies included precision clock- and watch-making, new vernier scales, and micrometers, which, when combined with telescopes and fine telescopic optics, enabled the measurement of those tiny celestial angles that would prove decisive in confirming Copernicus's heliocentric solar system and Newtonian gravitation as physical truths as opposed to theories. Theories are all well and good, but if they are to become demonstrated facts, verifiable precision measurements are necessary.

Outside astronomy, William Harvey's 1628 theory of blood circulation would be factually confirmed by experiments, with precision measurements of blood flow, and, crucially, by Marcello Malpighi's use of the microscope to observe the capillary vessels in 1661. Similarly, after 1659, Robert Hooke's precision volumetric air pump for Robert Boyle would facilitate a whole raft of discoveries in respiratory physiology, gas chemistry, and combustion. By 1700, precision scientific instrument-making was being developed across Europe from Poland to Paris, but perhaps nowhere saw the wholesale manufacture of precision tools of scientific research more than did London, which dominated precision scientific instrument-making until the late nineteenth century and beyond. The men who devised and built these new research instruments were highly intelligent craftsmen – scientists in their own right, who fully understood the problems the formally acknowledged scientists were trying to solve, and worked closely alongside them. The clockmaker Thomas Tompion's close working relationship with Robert Hooke is a case in point, while Tompion's successor, George Graham FRS, designed, built, and refurbished the instruments for the Greenwich Royal Observatory after Edmond Halley became Astronomer Royal in 1720. Graham also built the instruments that enabled the Revd Dr James Bradley to finally demonstrate the earth's motion in space from the newly discovered aberration of light over 1727–28.

George Graham, like so many of his colleagues, was far more than just clever with his hands; he undertook fundamental scientific researches in his own right, and by the early nineteenth century, Graham, James Short, John Dollond, Jesse Ramsden, Edward Troughton, and others had been elected full Fellows of the Royal Society on the strength of their published physical researches. The chronometer-maker John Harrison was awarded the Society's prestigious Copley Medal in 1749. Science, as we shall see, was by its very nature a meritocracy, and as such, always open to talent, irrespective of the scientist's background, daily occupation, or nationality.

So this is where astronomy stood by 1700: on the eve of one of the greatest odysseys in the life of the human mind, as humanity was all set to begin what Sir William Herschel in 1817 would style the fathoming of "the length, breadth, and depth, or latitude, longitude, and profundity" of space,[6] to hopefully reveal the "Interior Construction of the Universe" itself. So read on!

Chapter 2

Cosmology Begins at Home: Captain Edmond Halley, FRS, RN, Astronomer, Geophysicist, and Adventurer

When Edmond Halley was born in October 1656, astronomy, physics, geomagnetism, meteorology, chemistry, and experimental physiology, as well as some other sciences, were all advancing rapidly across Europe, in particular in Leiden, Paris, Bologna, Poland, and Scandinavia. Roman Catholic Jesuit missionary scientists were also making discoveries, drawing maps, and teaching the local peoples, in Latin America, India, and China, for the new sciences of experiment and instrumental measurement were truly international.

In England, Oxford and London were at the forefront, with the Revd Dr John Wilkins's Oxford "Philosophical" or scientific Club, centred upon Wadham College, advancing half a dozen sciences, along with Gresham College in the City of London. The Oxford and Gresham groups had a fairly fluid membership, as Wilkins, the young (Sir) Christopher Wren, Robert Hooke, Robert Boyle, Thomas Willis, and others moved between the two cities. The basis of the Oxford and Gresham groups was friendship and a mutual curiosity. None of these men did science as a paid job, but were independent *gentlemen* – clergymen, doctors, lawyers, and academics – in love with the new learning, and they financed their own researches out of their own pockets. They corresponded with colleagues in Continental Europe and tested each other's experiments, be they reported from Wiltshire, Bologna, Paris, or Leiden.

When Halley was four, in 1660, the Stuart monarchy was restored. Eighteen years of civil war and political and religious disruption ended, as King Charles II re-entered his executed father's kingdom. The Oxford and Gresham scientists, men of clout in English society, approached His Majesty for support and formal recognition. The bankrupt king could give them no money, but he gave them a charter, some ceremonial regalia, and a title: "The Royal Society of London". This became a self-electing international club of scientific gentlemen, which, within a decade, would also elect distinguished Frenchmen, Italians, Poles, Dutchmen, and Germans to join its ranks. Their intellectual global descendants are still to be found there.

Edmond Halley would be elected to this gilded circle at the age of twenty-two, later serving as its Secretary, and remaining an active Fellow until his death in 1742. But who was Edmond Halley?

2.1 Edmond Halley. (R. S. Ball, *Great Astronomers* (1895), p. 167. Author's collection.)

The schoolboy scientist

It is uncertain precisely where Edmond Halley was born or when, primarily because of the loss of parish registers in the Great Fire of London in 1666. One early memoir states that he was born on 8 November; another puts the date as 29 October 1656, at Shoreditch, then a London suburb. The Halley family home, however, was in Winchester Street, a well-to-do district in the City of London. The astronomer's father, Edmond senior, was a prosperous merchant who owned various properties in the City and suburbs, including the Dog Tavern in Billingsgate, while the family also owned land in the county of Huntingdonshire.

His parents' marriage may have been a hasty affair, for if the Anne Robinson who married an Edmond Halley in the late summer of 1656 was the astronomer's mother (we know that his mother was named Anne), they had only been married a couple of months when Edmond junior came into the world. Being the son of a liveryman, or member, of the London Soap-Boilers' Company, he and his family would have been well off. Edmond senior is unlikely to have done the messy manual work of boiling fats and oils to make soap, having apprentices and employees to do so. He was an entrepreneur and held various commercially related appointments in the City, and as the rentals from the various family properties brought in several hundreds of pounds per annum (tens or hundreds of thousands in modern money), Edmond junior could have looked forward to the life of an independent gentleman with private means.

As a boy Edmond would have had only a short walk to school: that already prestigious academic institution hard by the vast Gothic structure of old St Paul's Cathedral – destined to burn down in the Great Fire in 1666 – and founded 150 years before by John Colet, the great Christian humanist scholar and Dean of St Paul's. In those days, St Paul's was not a boarding but a day school, educating the boys of the City families, and here Edmond began to show his intellectual precocity. Latin would have been the staple of the educational diet, followed by Greek. No Stuart schoolboy would have studied English, for you learned this at home or from a hired tutor, nor would Shakespeare's plays be on the curriculum. Instead, it would be the Latin classics, the poetry of Ovid and the plays of Terence, and then, perhaps, Sophocles or Aristophanes in the Greek. Such plays not only taught a youth his classical tongues, but also how to declaim and be confident at public speaking: useful skills for a young man who might later go into the church, the law, Parliament, or public life; and from what we know, it is clear that Halley never lacked confidence.

It is plain from his later remarks that even as a schoolboy, Halley was undertaking independent scientific investigations, such as monitoring deviations in the earth's magnetic field. It would be interesting to know how far his passion for science was encouraged by the headmaster of St Paul's, the Revd Dr Thomas Gale, who in 1677 would be elected into the Royal Society Fellowship. By the time he went up to The Queen's College, Oxford, in 1673, not yet seventeen, Edmond Halley took with him a set of state-of-the-art instruments said to be superior to those of the Savilian Professor of Astronomy. Few undergraduates go up to university so well prepared or equipped, and by all accounts he won his laurels early, with even greater ones in the offing.

EARLY ADVENTURES: ST HELENA, DANZIG, AND ACROSS EUROPE: THE MAKING OF A PHYSICAL SCIENTIST

In 1675 the Greenwich Royal Observatory had been established, primarily to produce star charts and navigational tables to help a sailor work out how he might fix his longitude at sea. The Revd John Flamsteed, the first Astronomer Royal, was aware that the star charts of the southern hemisphere were inadequate by modern instrumental standards, and Halley, the twenty-year-old undergraduate, leaped at the chance of doing something about it.

No doubt utilizing his father's City of London connections, he secured passage for himself and a friend, plus his own collection of instruments, to the remote island of St Helena on board the India-bound *Unity* in 1676. St Helena is 16 degrees south of the equator, between the African and South American continents and in the middle of nowhere. Nonetheless, it gave spectacular views of the southern skies, along with several latitudes of the northern constellations, which was useful for establishing mutual coordinates to dovetail the stars in both hemispheres together cartographically. The years spent on St Helena, 1676–78, were fundamental to the making of Halley the astronomer, for what he saw and measured on that island opened up many new lines of physical thinking. In addition to producing an excellent star chart, using measuring instruments fitted with the very latest telescopic sights and precision screw micrometers, he was struck by some curious physical phenomena.

Halley made observations near to sea level, and from high up the "mountain" on the island. When observing at sea level, everything was straightforward. Yet when observing up the "mountain", under a seemingly clear sky, he found that everything, including his telescope lenses, became damp, and paper and cardboard turned soggy. This made Halley one of the first scientists to study "dew point", or the moisture-bearing and depositing properties of the atmosphere: a topic he would return to back in Europe, in his subsequent meteorological studies.

Then, quite bafflingly, he found that when he set up his fine recently invented long-pendulum clock on the dockside, he could regulate it to keep perfect time by the daily meridian passage of the sun, whereas when the same clock was taken up to his "mountain" observatory, it ran slow. Could this have something to do with the fact that the earth's still imperfectly understood gravitational attraction was slightly weaker up the mountain, it being a few hundred feet further away from the earth's centre than sea level? Jean Richer had found that his Paris-regulated clocks lost 2 minutes per day when taken to the equatorial island of Cayenne, off Brazil, in 1672. This led to the idea that the earth might be an *oblate*, or orange-shaped sphere, with northerly Paris and London being closer to the gravitational centre of our planet than the more equatorial Cayenne or St Helena. We will meet Monsieur Richer again in Chapter 8.

Halley reported his pendulum observations back to Robert Hooke at the Royal Society, adding to his own subsequent reputation as a founding father of geophysics. Yet perhaps Halley's most significant realization on St Helena came from his suggested trigonometrical technique for measuring the distance between the earth and the sun, by using the planet Mercury as an intermediary. It occurred to him when observing Mercury's transit, the planet's passage across the sun's disc, in 1677. All this was the work of a man who was still officially an Oxford undergraduate.

When Halley returned to England in 1678, having not yet even reached his twenty-second birthday, he found himself a celebrity. Even the king, Charles II, wanted to meet him, and young Edmond astutely named one of his southern constellations *Robur Carolini*, or

"Charles's Oak", in honour of His Majesty – an allusion to the twenty-one-year-old Prince Charles's daring escape from Oliver Cromwell's troops by hiding up an oak tree following the battle of Worcester in 1651. King Charles ordered that the Vice-Chancellor of Oxford University should confer Halley's MA degree without examination: by *mandamus*, or royal command. And to cap it all, the Royal Society elected the daring and brilliant young man to its Fellowship. He would continue adding lustre to the Society for the next sixty-four years.

Scarcely had Halley returned home to his laurels from his St Helena expedition than the Royal Society dispatched him, in August 1679, on a delicate diplomatic mission to Danzig, Poland. An acrimonious dispute had broken out about the use of telescopes in making accurate angular measurements between Robert Hooke FRS and the "grand old man" of north European astronomy, Johannes Hevelius FRS, of Danzig. Halley spent several weeks in Danzig observing with Hevelius, and admitted to the superlative quality of his large, naked eye-sight quadrants and other angle-measuring instruments. The young astronomer's charm and gift for friendship went a long way to healing the dispute, and it was even hinted (which Halley subsequently denied) that he and the thirty-two-year-old Lady Elisabetha Hevelius had become very good friends. Hevelius was sixty-eight years old, and Elisabetha was his second wife. Following his return to England, Edmond obtained expensive silk dresses for the lady – not forgetting, however, to send her the bills.

Next, Halley set out on a triumphal tour of Europe, being fêted by the astronomers and *académiciens* of Paris, and then went on to Rome. For ever on the lookout for useful information, he made accurate measurements of some ancient public standard lengths, set up in the Campidoglio in Rome, to establish their equivalents in English inches. Then domestic tragedy struck. Edmond Halley senior went out as usual one day, and never came home. Some weeks later, his decomposed body was found in a ditch near Rochester.[1] His purse and watch were still in his pockets, and his cause of death was never established. Edmond identified the remains by his father's new shoes. Could Edmond senior have been abducted and killed because of a City or political dispute?

Tragedy then turned into lawsuits. Edmond senior had recently remarried (Edmond's mother Anne having died in 1672), and his new wife was not much older than her new son-in-law. The new Mrs Halley, née Joane Cleeter, made legal claims for a hefty chunk of the family property. This legal wrangle is the reason why we know so much about the Halley family property holdings: the Court of Chancery records of the case still survive, and were examined by the modern Halley scholar, the late Sir Alan Cook FRS[2]

As a result, Halley's expectation of a life of comfortable financial independence was badly dented, although his obvious commercial as well as scientific abilities would eventually enable him to break into clear financial waters once again. Then in 1682, he married Miss Mary Tooke, the daughter of another City family. Edmond also took a paid job, becoming one of the Secretaries of the Royal Society, which placed him in an excellent position to obtain all kinds of new physical data, from returning sea captains, fellow scientists, and researchers across Europe coming into the Royal Society. This was all grist to the mill of his endlessly inquisitive and highly organized information-driven intellect. By 1686–87, Halley was sufficiently well-off once more, probably from rental income and investments in overseas trade, to be able to finance the publication of his friend Isaac Newton's monumental *Principia Mathematica*, from a personal outlay of £60. (To give some idea of the purchasing power of this sum, the Warden, or Head of Wadham College, Oxford, received a stipend of £100 per annum at this time: itself a hefty sum.)

EDMOND HALLEY, THE FATHER OF METEOROLOGY AND GEOPHYSICS

By the time that Halley was a schoolboy at St Paul's in the late 1660s, accurate magnetic compasses and dip needles (measuring the vertical magnetic attraction of the earth) had come into being, enabling "philosophers", as scientists were then called, to map the globe's magnetic field. What a strange and bent thing this field was, snaking in three dimensions between the poles. Halley, the great physical-fact collector, had begun geomagnetic observations while still at school, and by his thirties was struck, as were other scientists, by how the magnetic field seemed to have a global shift to the west. Could this be caused by the earth's axial rotation and motion through space, at a time when the Copernican theory still lacked physical proof? Halley would later come to speculate that the earth might well possess a complex interior, consisting of several concentric magnetic shells, fitting snugly inside the outer shell upon which we live, rather like the internal skins of an onion. If these shells rotated at slightly different speeds under the influence of Newtonian gravitation, then they might result in that complex, *torqued* or twisted westward drift observed in the planet's global magnetic field.

We now know that Halley's model was wrong, yet modern geophysicists have discovered that our rotating earth's very complex, semi-fluid, interior does produce a range of energy systems that affect, among other things, our magnetic field. We shall return to Halley and terrestrial magnetism in the next section, when we look at his captaincy of the survey ship HMS *Paramore*, between 1698 and 1701, and his encounters with floating ice-mountains in the remote south Atlantic.

Running parallel in Halley's mind with magnetism and geophysics was meteorology. Shortly before his birth in 1656, four key meteorological measuring instruments had been invented: the thermometer, rain gauge, barometer, and wind gauge; and in the 1660s, Robert Hooke proposed that a register of the weather should be kept by the Royal Society and by scientists across Europe. If this could be successfully accomplished, two things might become possible: an understanding of the physical causes of climate and weather, and how to analyse the data and predict it. For commercial places such as the City of London – and Hamburg, Amsterdam, and other great European seagoing cities – navigation would be made safer and the risks involved in great voyages reduced, if the weather could be predicted.

In the 1680s in particular, Halley began to look into the mechanisms behind climate, and in particular, the famous trade winds that tended to blow in a westerly direction and sped Europe's great merchantmen to the Americas, India, and China. He began his climatological research with a laboratory experiment: by measuring the volume of water that was evaporated by the sun from a 9-inch-diameter pan on a temperate summer's day. From this, he computed how much water was "licked up" in a square mile, then a square degree (69 square miles) of ocean, and he informed the Royal Society that in a single day, each square degree of ocean in the temperate latitude put a staggering 33 million tons of water into the atmosphere. The tropics would have seen much more. Further calculations, such as measuring the rate of flow of the River Thames, enabled him to build up a quantified rain cycle, as vast volumes of moisture rose from the sea and condensed to become rain when hitting the colder rocky masses of the continents.

The keys to the whole process were heat and motion. As the sun moved above the equatorial regions each day, there was an inevitable global hot spot directly overhead. This hot spot caused a great up-draught of equatorial air from both hemispheres, north and south of the equator.

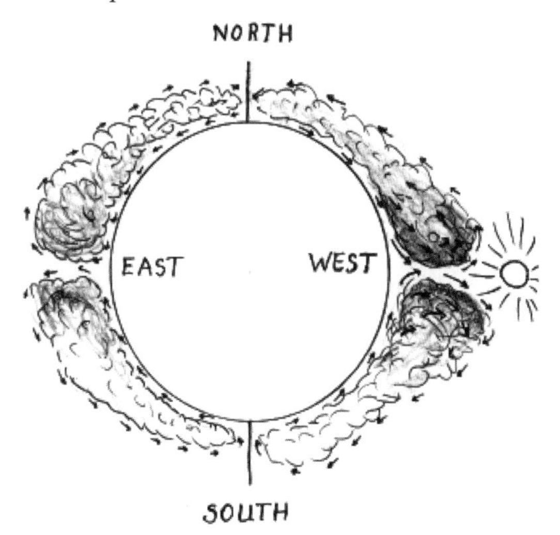

2.2 Halley's trade winds and climate model. See the text for an explanation. (Drawing by Allan Chapman.)

Then, as this atmospheric up-draught rose, bearing many millions of tons of water vapour, it cooled in the upper atmosphere and gradually descended to the temperate latitudes, depositing its heat and water, only to be sucked back to the equator in an endless cycle.

As the sun moved for ever westward, the sun-chasing equatorial winds tended to do likewise – caused by the earth's daily axial rotation, and further complicated by tidal drags, in both the sea and the atmosphere. Not all winds blow toward the west, and Halley suggested that ocean depth, shallow seas, archipelagos, and great "enclosed" bodies of water, like the Indian Ocean and the Caribbean, produced their own regional peculiarities – local ecosystems within a greater ecosystem – generating the Indian monsoon, the Far-Eastern typhoons, and the Caribbean hurricanes.

By the late 1680s, Halley was thinking along the right lines, both geophysically and meteorologically, and many of his models were basically correct. In pursuit of their further substantiation, he would command a series of voyages in a ship so small that most modern people would fear being beyond the sight of land in it, let alone venturing into the storm-tossed, uncharted frozen wastes of the south Atlantic.

LATER ADVENTURES: CAPTAIN HALLEY RN TAKES HMS *PARAMORE* AMONG THE ICEBERGS

Whatever traits Edmond Halley FRS may have lacked, naked courage, social confidence, persuasiveness, and intellect were not among them. One still stands in awe of how this academic scientist, whose seagoing adventures to St Helena had been purely as a passenger, persuaded the Admiralty to give him captain's rank and command of one of King William III's armed ships. HMS *Paramore* was only a "pink", a three-masted cockleshell under 60 feet long and under 80 tons burthen, but she was still a king's ship, and Halley her blue-coated commander.

Halley's motive behind *Paramore*'s voyages was to build up an exact picture of the terrestrial magnetic field in the north and south Atlantic, which could, hopefully, be dovetailed into data that sea captains reported from the China Seas, the Pacific, and elsewhere. Yet when he set sail from Deptford on 20 October 1698, he was a novice as a practical naval commander. His wardroom officers, slowly ascending the promotion ladder, were not wholly happy with this civilian who now commanded them. Halley, however, brought *Paramore* home after her initial English Channel survey, and initiated a court martial for Lieutenant Edward Harrison, who had objected to being upstaged in command by a landsman scientist. While the Admiralty saw Harrison's point of view, Halley was given a replacement officer and returned to sea with a more respectful crew.

Paramore's voyages included the English Channel, the mid and north Atlantic, the Caribbean, and then down into the deep south Atlantic. In the wild seas of 52° 24' south, well to the east of the Falkland Islands, where wind and water currents from the Atlantic collided with those of the Antarctic Ocean, HMS *Paramore* encountered ice "mountains", or vast icebergs, as high as Beachy Head on the south coast of Sussex, alongside which *Paramore* was like a toy boat.[3] Yet as Halley criss-crossed the Atlantic from the Americas to Africa, he built up a large body of data about the earth's magnetic field. The logic behind the *Paramore* voyage was also utilitarian, for if these hundreds of geomagnetic fixes for the local twists in the earth's magnetic field could be plotted exactly on a sea chart, they might help a sailor to fix his position when the stormy skies denied him a glimpse of the sun, moon, or stars.

Back in London, Halley made cartographic history in 1701 when he published his Atlantic magnetic chart. On one level, it was a failure, for it was soon discovered that as the earth's magnetic field changes, so the coordinates that he had carefully and so daringly accumulated after 1699 could not be relied upon in 1710. Yet his magnetic chart contained one priceless innovation, for he had the idea of connecting all places sharing the same magnetic coordinates with a flowing isogonic contour line. This made Halley the first cartographer to use such contour lines, which later map-makers would use to show the heights and depths of mountains and other features on maps – hence every modern user of a map is indebted to Captain Halley.

Professor Halley and the Great Aurora Borealis of 1716

In 1691 Edmond Halley had applied for the vacant Savilian Professorship of Astronomy at Oxford. In spite of Royal Society support, he did not get it, although this may well have been due to the turbulent politics of the day. The Catholic convert King James II had been forced to abdicate in 1689, to be succeeded by the Dutchman William of Orange, who, to complicate matters and loyalties, was married to James's staunchly Protestant daughter, Mary. The high politics of church and state weighed heavily in the academic appointments of the Stuart age. Halley, while an undisputed Protestant, fell foul of the rival "Juring" and "Non-Juring" Protestant factions, who fought each other over oaths of allegiance to the deposed King James, especially in the High Church world of Oxford.

Then in 1704, Oxford's companion Savilian Chair of Geometry fell vacant. By this time, Halley's reputation was unstoppable, especially as he had so successfully served the new Queen Anne in a diplomatic mission to the Adriatic and the Holy Roman Emperor's Court in Vienna over 1702–03. Even so, John Flamsteed could not help but snipe, claiming that Edmond "swears and drinks brandy like a sea captain".[4] Changing his blue, brass-buttoned captain's coat and his diplomat's finery for the scarlet gown of an Oxford doctor, Halley proceeded to stun his critics with a display of classical Greek erudition. He delivered a series of professorial lectures on the classical Greek geometer Apollonius of Perga, even reconstructing Apollonius's missing Book VIII from surviving sections, through Greek textual and geometrical analysis. Then in 1705, he published his *Synopsis of the Astronomy of Comets*, to which we will return in Chapter 3.

On the evening of 6 March 1716 (16 March in our present calendar) people came to tell Halley, currently in London, that the sky appeared to be ablaze. He watched the phenomenon until 3 a.m., and made a careful record of its characteristics, including its great northern crown, or corona of light, with ray patterns that reminded him of the insignia of the Knights of the Garter. In spite of his travels, and many thousands of hours spent watching the sky in both hemispheres, the fifty-nine-year-old professor had never before seen an aurora. Checking the records, he found that no significant aurorae had been seen in Europe since the one recorded by Father Pierre Gassendi in 1621, although some lesser ones had been seen recently in Paris and Berlin, perhaps when it had been cloudy in England. This absence of aurorae spanning ninety years or so corresponds with what we now style the "Maunder Minimum", when very few of the newly discovered sunspots were reported by astronomers. Victorian astronomers, after 1859, would draw the crucial connections between solar storms, sunspots, aurorae, and electromagnetic disturbances (see Chapter 16).

Hungry for data, Halley began to collect travellers' reports about the great 1716 aurora, which was observed from the north Atlantic, Russia, and even southern Europe. Yet it always appeared in the north, irrespective of location. Could it be connected with the earth's magnetic field? Even more puzzling, what could be the cause of this transient, dancing, flameless light? He recalled that, some years previously, probably around 1710, he had witnessed a strange flameless light glowing feebly in the evacuated glass chamber of an air pump in the Ashmolean Laboratory (now the Museum of the History of Science), Oxford. Halley's friend the Revd Professor John Whiteside had just succeeded in firing a small quantity of gunpowder inside the pump's vacuum chamber (presumably in the dark) and noticed that long after the small explosion had ceased, the smoke inside the chamber continued to emit a feeble glow. Could the aurora be caused by some similar strange and as yet undiscovered chemical luminescence high up in, or beyond, the earth's atmosphere? In 1716, nothing was known about the physics of ionization or flameless luminescence.

Halley studies the nebulae and ponders cosmological vastness

Similar to the strange glow of the aurora borealis was the question of the source of the feeble light emitted by the Orion, the Andromeda, and four other nebulae known by 1700. At St Helena around 1677, Halley had discovered one of the known half dozen, the nebula on the constellation Centaurus, followed by yet another in Hercules in 1714. In a pair of papers published by the Royal Society in 1715 and 1716, announcing his discovery of two new nebulae, he began to ask profound cosmological questions. What *were* these misty patches of light (now called galaxies) seen among the "fixed stars"? What made them glow, and how big and how distant must they be? Because they displayed quite large angular diameters in contrast with the pinpoint stars, they must, on purely geometrical grounds, be immensely large. Deep space, therefore, appeared to contain an incalculable number of individual stars, along with these vast glowing masses of light. Yet the small aperture, simple lens refracting telescopes of Halley's day revealed no detail in these nebulae. After him, nebula studies would have to await the development of powerful reflecting and achromatic refracting telescopes, photography, and spectroscopy, as we shall discover in forthcoming chapters.

Edmond Halley, in so many respects, set the deep-space cosmological ball rolling, especially in a foundational paper of 1720 in which he addresses the problem of the gravitational stability of the universe. If, in the wake of Newtonian gravitation, *all* objects exert a gravitational attraction upon each other, why does the universe not implode into a single centre? His suggested solution was that the universe must be infinite in all directions, so that all the stars locked each other into a stable gravitational matrix that went on to infinity.

Halley also pondered another profound question, or "paradox", taken up a century later by Heinrich Olbers, and which had even been touched upon in the 1570s by the English Copernican Leonard Digges: if we are surrounded by an infinity of light-emitting stars, why does the sky go dark at night? Halley's solution was symptomatic of his geometrical mode of thinking. If a star were a sphere, it must have some darkness around it to be visible as an individual object. This would also apply to the countless millions of individual stars only just visible in the best telescopes of 1720. Therefore, if millions of stars needed some surrounding darkness to be visible at all, there must be at least as much darkness as there were stars. Ergo, the night sky is dark. We now know this ingenious geometrical fudge is wrong. Halley and all his contemporaries believed space to be perfectly transparent to infinity, but twentieth-century astronomers discovered that much of space is occupied by dense masses of non-light-emitting gas and dust, which obscure and absorb the light of any star glow behind them, explaining the blackness of the night sky. The distant stars and galaxies are there, but screened from us by dark dust.

In addition to the above, Halley's stellar researches led him into two other fields, both of which play a fundamental role in our understanding of cosmology, and both of them owed much to his passion for collecting and analysing historical data from earlier astronomers. The first of these was his study of changes in the brightness or stability of certain stars. Like all European astronomers, he was aware of those "novae", or new stars (now known to have been exploding supernovae) that suddenly flared up then disappeared in 1572 and 1604: the stars recorded by Tycho Brahe and Johannes Kepler respectively. In addition to these, Halley began to search out the presence of stars whose luminosity or colour in his day were different from what earlier, including classical and medieval, astronomers had recorded: objects now known as variable stars.

In addition to stars that varied in light output, Halley suggested in 1718 that some stars had also changed their geometrical position over time. Comparing the catalogues of Ptolemy, Tycho Brahe, and others, Halley concluded that the bright stars Aldebaran, Arcturus, and Sirius in particular now occupied slightly different positions from those recorded in the past. He surmised that they might be "in all probability the nearest to Earth",[5] and any independent motion in space of either of these stars, or the sun and solar system, was creating a line of sight displacement. Halley had discovered that what had from time immemorial been believed to be the "fixed" stars were not actually fixed – given enough time – but possessed what would come to be styled "proper motions" of their own. These motions really can be caused by proximity to the earth – as Halley speculated – as the solar system itself moves through space, or by the stars' own independent motions in space. Once alerted to the phenomenon in 1718, the astronomers of Great Britain, Europe, and then America would catalogue many hundreds more of these proper motions. By the nineteenth century, these motions would assist astronomers in selecting candidate stars for stellar parallax displacement analysis, and would show the stellar universe is a geometrically and physically diverse place, anything but fixed. For this, we must thank Edmond Halley.

Chapter 3

Could a Comet Have Caused Noah's Flood?

There are many parallels between astronomy and medicine, one of the most obvious being a tradition of collecting case histories. Doctors record disease patterns and patient profiles, while the astronomer monitors the nature of the heavens, and several notable discoveries – such as Edmond Halley's discovery of the proper motions of the stars – were made by comparing ancient and modern data. Living as he did when the physical sciences were progressing rapidly, Edmond Halley's fact-gathering, geometrical, deeply physical turn of mind found him ideally placed to see connections and produce profound insights. We saw in the previous chapter how his work on the aurora borealis, meteorology, geomagnetism, and several branches of astronomy enabled him to pioneer whole new fields of research. He even applied science to history itself, when in 1691 he used a combination of facts extracted from Roman history and the latest data about the moon's gravitational action upon the tides to calculate the spot where Julius Caesar might have landed on the Kent coast on 25 August 55 BC. One area in particular where Halley joined the use of historical and modern data to address a contemporary scientific problem was his work on comets.

Changing views about comets, 1580–1720

From classical times to the Renaissance, learned views about the nature of comets had remained relatively static. Aristotle, the universal philosopher of classical Greece, gave what became the accepted explanation in his *Meteorologica*, around 350 BC. It stated comets were not astronomical, but rather meteorological phenomena (from the Greek *meteōros*, "in the air"). He opined that when foul fumes rose up from stagnant earthy places and reached a given altitude in the upper air, the heat of the sun caused them to combust, so as to resemble a transient fiery star. By their nature, therefore, comets were relatively *local* objects, and it is hardly surprising that astrologers regarded them as omens. Crucially, comets were believed to be much closer to the earth than the moon, the distance of which had been known with relative accuracy since antiquity, thanks to the geometry of right-angled triangles. What first changed our knowledge of comets, however, was not a new theory so much as an early example of that phenomenon that would lie at the heart of progress in astronomy and all the physical sciences over the forthcoming four centuries: improved technology.

When the "Great Dane" Tycho Brahe saw the comets of 1577 and 1585 in particular, he began to measure their positions against the fixed stars, with a set of angle-measuring instruments whose graduations were (especially by 1585) much more accurate than those of his admired near-contemporary Nicholas Copernicus. Tycho tried to measure the distances of the comets by means of parallax angles, expecting to find they were much closer than the moon. Instead, even his great quadrants and sextants, with their precision hair's breadth graduations, were too imprecise to enable him to triangulate the comets' distances, leading to the conclusion that they must be much more remote than the quarter-million miles of our satellite. (I discuss Tycho's work in more detail in *Stargazers*, Chapters 5 and 6.) In short, comets were in planetary space, and not in our atmosphere. So what could they be, how generated, and how did they move with relation to the sun and planets?

During the seventeenth century, comets became objects of intense interest to astronomers. If they were in planetary space, moving in strangely curved orbits, they must be crashing through the nine concentric crystalline spheres that antiquity told us carried the sun and planets around the earth. So did the crystalline spheres not exist? If not, what was the precise invisible force that carried the planets through the empty void around the earth or the sun? One question was popping up after another. Why did no two comets look the same? Some remained dim and misty. Others, such as that of 1618, developed a great, sweeping scimitar-like tail; while that of 1677, meticulously observed by Robert Hooke, was long, thin, and arrow-straight. All appeared to materialize unexpectedly from nowhere, and after some weeks would just drift away.

By the 1660s, two physical aspects of comets were occupying the attention of Europe's astronomers: what orbital geometry governed their motions, and how did they burn and emit light? Comets were known to display a distinct relationship with the sun, as they grew bigger and brighter when closest to it. But were comets ejected out of the rotating sun, and shot into space into curved orbits? Did they come from the depths of space and were drawn into the sun? Or were they drawn in by a solar force of attraction and passed round the sun, only to be flung back into space, like a stone from a sling? In his *Cometographia* (1668), Johannes Hevelius hypothesized that a comet's head might not be spherical so much as disc-shaped, which could explain why, when the disc was flying edge-on, the comet appeared to accelerate, whereas when the disc turned face-on to the orbit, it appeared to retard, to produce a braking action.[1] Whichever option one chose, a lot of progress had been made since 1580, as telescopes, combined with increasingly accurate measuring instruments and a rapidly advancing analytical mathematical geometry, had disposed of the classical flaming atmospheric object explanation for ever.

Dr Robert Hooke takes comets into the chemical laboratory in 1677

In his *Cometa* of 1677, Robert Hooke FRS, Professor of Geometry at Gresham College, London, and one of the foremost advocates of the new instrument- and experiment-based approach to science, announced his groundbreaking researches into why comets emit light. Having made detailed high-powered telescopic studies of the bright comets of 1664–65 and 1677, he showed that comets possessed common structural features. There was always the bright head or nucleus, pointing directly to the sun. From it sprang a stem or "medulla" in the midst of the tail, which originated in the comet's head. From the head also came a complex series of streamers, the different parts of which formed the comet's tail.

Examining the nuclei of several comets at high magnification for that period (using a telescope that modern scholars calculate magnified around 173 times), Hooke noticed that they were equally bright all round, not just the half facing the sun. Ergo, the nucleus must be self-luminous, and not just lit up by sunlight, like the moon. Yet if this was the case, what was the source of the comet's light? Hooke argued that from all appearances the comet was *not* on fire in the assumed sense, but was being somehow eroded away by some other, albeit light-generating process: one that he suspected was chemical in its action. Yet how could this operate? By 1677, most European scientists accepted that planetary space was probably empty and airless, although there might be some sort of ether that conveyed light and solar heat. On the other hand, how could a corrosive light-generating process take place in vacuo? To Hooke's ever-inquisitive, experimentally oriented intellect, this suggested new possibilities in physics, chemistry, and combustion theory, enabling him to design a laboratory experiment.[2]

He first took a wax ball, which he encrusted with fresh iron filings. Then, suspending it on a long silver wire, he gradually immersed the ball into a 2-foot-tall laboratory glass cylinder filled with sulphuric acid. The acid attacked the metal on the wax ball, releasing great streamers of bubbles – of what we now know (but which was still unknown in 1677) to be hydrogen gas. Examining the reaction carefully, Hooke was struck by the flow-pattern of the streamers, curving gracefully around the front of the ball, then bending around the back to form a central stem and streamer patterns: astonishingly like the structure of a comet. (Hooke was very accurate in his description of this process. I have checked it myself and reconstructed his experiment. It really does work.)

Were the increasing tail and brightness of a comet, therefore, the product of some chemical action taking place between the approaching comet and some corrosive agency associated with the sun? Hooke used the alchemical term *menstruum*, or solvent, for this agency. Yet if a comet was being somehow eroded away by approaching the sun, it was only reasonable to speculate about what the comet might be made of. This point would be brought up by Halley and the Unitarian astronomer William Whiston in their ideas about comets as possible causal agents in shaping the earth's surface, and even in bringing about the biblical Flood of Noah.

COMETS TAMED AT LAST: 1680–1705

What was the shape of a comet's orbit? By 1680, most European astronomers agreed that, irrespective of their chemical composition and dissolution, comets moved in orbits whose curvature was proportionate to an angled slice through a cone, or "conic section". But were they open-ended curved orbits such as the parabola or hyperbola, or closed, like the circle and the ellipse?

Between 1680 and 1682 two conspicuous comets blazed across the heavens, and that of 1682 was studied especially closely by John Flamsteed at Greenwich, by Halley in London, and by other astronomers on the Continent. Their main concern was to determine its precise position with relation to the fixed stars, to try to reconstruct the geometry of its orbit.

After 1687, however, Europe's understanding of cometary orbits was to change for ever, for in his *Principia Mathematica*, Book III, of that year (the volume for which Halley acted as publisher), Isaac Newton would lay down the essential orbit-forming laws for comets, based on the application of gravitation theory under the inverse square law of attraction to existing cometary observations. In particular, in Book III Newton provided an extensive geometrical analysis of the orbit of the 1682 comet. By 1695, assiduous observer that he was, Halley was both computing orbital curves and collecting observational data for some twenty well-observed comets. He came to the conclusion that comets orbited the sun. Some, after swinging around the sun, might fly off into space if their orbits were open curves, and be gone for ever. Others, if their orbits were closed, such as being ellipsoidal, would come back in future. Unlike the planets, however, comets moved in "very Eccentrick" or elongated orbits.[3]

Halley's *A Synopsis of the Astronomy of Comets* (1705) drew all of the recent gravitation theory-based knowledge of comets together. His scientific historical instincts were very much to the fore, as he presented a list of well-recorded comets observed between 1337 and 1698. He also associated the bright comet of 1682 (Halley's

Comet) with previous cometary apparitions of 1607, 1531, and 1456, suggesting that they were all the same body, and that it would return again in 1758. However, he suggested that some other historic comets he listed might also be periodic, which we now know are not. His *Synopsis* was only a small book of twenty-four pages, but it represented a milestone in our modern understanding of astronomy.

Noah's Flood, the ancient earth, comets, and the saltiness of the sea

Edmond Halley's love of data collection and analysis led him to see many connections across the physical sciences and history. But one of the most remarkable was his work linking biblical chronology to cometography, oceanography, climatology, and gravitational physics.

It is clear that several scientific men, most notably Robert Hooke, were discussing, from the 1660s onwards, not only the possible physical causes of Noah's Flood, but also the age of the earth itself. Much of this was taking place before the Royal Society – which contained many Anglican clergymen and even a few bishops – and no one was prosecuted for heresy. It is all too easy today, in our "post-classical" world, to forget that seventeenth-century Greek-language-trained scholars – be they astronomers, bishops, or both – would have been quite familiar with pagan stories about great floods, such as that in Plato's *Timaeus* creation narrative and elsewhere, and how they might relate to the biblical Flood.

Could there have been more than just *one* flood that had shaped the earth, and could these floods have extended across vast aeons of time, long before the traditionally assumed Genesis date of *c.* 4004 BC? Nor was it just classical Greek literature that spoke of great chaos, for European – usually Jesuit – scholars were finding roughly parallel stories in newly translated Chinese and other Oriental literature. Robert Hooke – a friend of bishops and the recipient of a Lambeth Palace Doctorate from the Archbishop of Canterbury – cites many of them in his "Earthquake Discourses" presented to the Royal Society from the 1660s onwards.[4]

Hooke had suggested that the earth might be vastly old, and had gone through numerous geophysical changes, in the long ages separating the divine creation *ex nihilo* from God's planting of the Garden of Eden. These probably occurred, he said, because the earth's poles were gradually rolling around with relation to the plane of our orbit, causing the oceans to wash over the continents and constantly refigure the surface of the planet.

By 1691, Edmond Halley was developing these ideas. He would even come to suggest establishing the age of the globe itself, if the increase in the saltiness of the oceans over a given number of years could also be measured. Although he did not specify dates or weights, suppose we took, for example, a gallon of sea water in 1700, boiled it dry, and carefully weighed the resulting salt. Then we did it again in 1900 and noted the increase over time. We would have had only to calculate backwards, thereafter, to establish when the young, un-eroded earth was covered with pure, sweet water: geophysics's first natural chronometer dating technique.

Believing that comets were substantial, gravitationally responsive bodies, Halley wondered whether the impact resulting from the earth colliding with comets, or the gravitational tides generated by near misses with them, had exerted powerful moulding influences upon the surface of our planet. Could not the very "without form and void" chaos out of which, so Genesis tells us, God framed the earth for the Garden of Eden, have been the disturbed debris left over from such a cometary episode? And could a subsequent, post-Eden cometary encounter have been the immediate (albeit God-directed) agency behind Noah's Flood?

These ideas would be explored further when William Whiston suggested in 1696 that cometary tails might well contain water vapour, and the Deluge possibly be occasioned by a comet cutting across the earth's orbit. Such an encounter could not only result in a prodigious terrestrial downpour, but also cause subterranean waters to burst forth, to produce tidal waves a mile high. The new knowledge of comets captivated the imaginations of late-seventeenth and early-eighteenth-century astronomers, who saw no incongruity in the idea of the Divine Will working through natural or secondary causes. Comets had come a long way since the days when they were but fiery stars blazing menacingly above London or Nuremberg.

Halley next started to calculate how much water would be needed to wholly inundate the earth in the Great Flood. In a lecture delivered to the Royal Society in 1694, and published in 1724, he pointed out that the Genesis narrative does not provide any account of where the Deluge waters came from, or eventually went to. Beginning with the growing body of meteorological data of his day, Halley took the rainfall for the wettest part of England – about 40 inches per year – and scaled it up. But if one applied this rainfall for forty days and nights across every square mile of the earth's surface, this would result in a global inundation of only 22 fathoms, or about 132 feet deep: a quantity that was insufficient to flood the earth as it was known by 1700. So had there been a long period of geological change between the creation and the moral teachings of Genesis, which begin with Adam and Eve? Had rolling poles, cometary encounters, earthquakes, drenching vapour tails, and pre-Noachian floods moulded the pre-Adamite earth in a way that God had not seen fit to inform the Mosaic writers of Genesis, but which the divine gift of ingenuity had permitted later generations to discover for themselves?[5]

EDMOND HALLEY: THE ASTRONOMER ROYAL AND THE LONGITUDE, 1720–42

Until the late-Victorian era in the nineteenth century, men who held some kind of public office – a military command, an ecclesiastical benefice – did not retire. Such offices were deemed equivalent to pieces of freehold property and valid for life. If an incumbent became too infirm to fulfil his duties, he simply employed a deputy to act in his stead, while continuing to draw the bulk of the income accruing to the office. Only by being aware of this circumstance can one understand how, when the Revd John Flamsteed, the first Astronomer Royal and Director of the Greenwich Royal Observatory, died in office on 31 December 1719, he was seventy-three years old. It also explains why the government of King George I in 1720 was willing to appoint Edmond Halley, now in his sixty-fourth year, to succeed Flamsteed. Halley was as energetic, persuasive, and determined as ever. He had treated Flamsteed shabbily, by publishing the Astronomer Royal's forty years of "uncorrected" observations without his consent in 1712, and he was on Newton's side in the great Sir Isaac's own appalling treatment of Flamsteed. But Professor Halley had his eyes firmly on the Greenwich job: a circumstance that so enraged Mrs Margaret Flamsteed, her deceased husband's assistant, amanuensis, and publisher of his "corrected" observations in 1729, that she sold all the Observatory instruments to prevent Halley from getting his hands on them. It was all above board, for when John Flamsteed had been appointed Astronomer Royal in 1675, he moved into a brand new, completely empty Royal Observatory building. Having been granted an erratically paid salary of £100 per annum, it was clearly up to him to find his own instruments, either purchased with his own money or personally gifted by his friend Sir Jonas Moore FRS, and even to cover the cost of publishing his official observations. Hence, in 1720, his widow was entirely within her rights to sell off all the astronomical hardware, especially as Flamsteed said in 1710 that, over the years, he had been obliged to spend £2,000 of his own money to keep the Royal Observatory operational, being the "son of a rich merchant" of Derby. It was not generous when the son of another "rich merchant", Halley, published Flamsteed's incomplete observations and said unflattering things about him in the "Preface".

Halley may have wanted the Astronomer Royal's still badly paid, £100-per-year job, but he drove a tough deal when it came to instruments, forcing the government to provide £500 to re-equip the Observatory: a circumstance for which British astronomy has been grateful ever since. But why did Oxford Professor Halley want the job in the first place? He was permitted to retain his Oxford chair and its £150 yearly salary and elegant house in Queen's Lane, Oxford, enabling him to move between Greenwich and Oxford as required, and appoint a deputy to do his more routine teaching. But, I suspect, he had another motive.

Following the tragic loss of Sir Cloudesley Shovell's victorious homecoming fleet off the Isles of Scilly due to an error of longitude in 1707, Parliament passed the Longitude Act in 1714, offering a stupendous prize of £20,000 to anyone who could provide a reliable method of finding the longitude at sea to within 34 miles, or half a degree, on the earth's surface. Nothing in the Act debarred a government officer – such as the Astronomer Royal – from winning the prize. It had been well known for 200 years that two possible methods were available for finding the longitude, which is a ship's east or west angular separation from a fixed point on the earth's surface, such as the Greenwich Observatory. One method was for the ship to carry a precision clock, or chronometer, set to the local astronomical time of the home country. If this clock said – for example – that it was noon in Greenwich, but the local time by the sun on board ship was 9 a.m., it meant that the ship must be 45° of longitude west of Greenwich. This is based on the fact that the sun appears to move through 15° of sky per hour due to the earth's rotation, and 3 × 15° equals 45°. Thus we "lose" time when we travel west, and appear to gain it moving east; and why New York is five hours behind British (Greenwich) time, and Tokyo nine hours ahead. If the clock were accurate to a few seconds, one could even measure the longitude differences between areas as close to each other as Anglesey in west Wales and Dublin in eastern Ireland. Yet no one would be able to produce a chronometer of such accuracy until John Harrison finally secured the last instalments of the prize in 1773 (he had already received around £14,315 in earlier instalments).

The other method, and the one that supplied the very rationale behind the 1675 foundation of the Greenwich Observatory, was to use the hourly motion of the moon against the stars as a natural clock: the "lunars" method. In addition to its daily rotation across the sky, with the stars and planets, the moon has a large, quite independent motion of its own: one that produces the whole 28-day cycle of phases with relation to the sun and earth. As the moon is only a quarter-million miles away from the earth, and the constellations are at infinity, it must mean that the moon's position among the stars, at any moment in time, will appear slightly different from different points on the earth's surface. For example, if the moon were close to the bright star Aldebaran in Taurus as observed from Greenwich, then at the same time a sailor off southern Ireland would see it in a slightly different place, and a sailor approaching Boston Bay, Massachusetts, in yet another. If the tiny moon-star angles could be accurately measured, each sailor could calculate his exact position, and take due precautions when approaching Ireland and Boston, regarding known rocks or other navigational hazards. Knowing one's longitude would greatly reduce the guesswork in contemporary navigation, guesswork that had led to the deaths of Sir Cloudesley and hundreds of his fellow sailors.

3.1 The Royal Observatory, Greenwich, seen from the west. Although the long telescopes and the general layout of the building are of Halley's time, the characters in the picture appear to be in Victorian dress. (E. Dunkin, *The Midnight Sky* (1891), p. 152. Author's collection.)

Before this elegant astronomical fact could be turned to practical use, however, five things were necessary. Firstly, land-based observatory instruments had to be capable of measuring celestial angles to within a few arc seconds. Secondly, the moon's motion had to be established with critical, predictable accuracy for several years ahead (which became possible with Newton's gravitational lunar tables). Thirdly, the tables computed for two or three years ahead had to be printed so that every ship could sail with a reliable lunar almanac on board. Fourthly, accurate shipboard angle-measuring instruments, such as the octant and sextant, had to be mass-produced relatively cheaply. And fifthly, navigating officers needed to be trained how to make the necessary observations, use the tables, do the complex calculations, and establish the ship's precise position. All of this meant not just technology, but high-precision, progressive, and commercially viable technology. Halley used his £500 government grant to commission a set of instruments that set new standards of angle-measuring accuracy and acted as prototypes for future generations of instruments. They included an 8-foot-radius iron and brass quadrant, a transit telescope for delineating the exact southern and northern meridians, and critically accurate pendulum clocks.

3.2 Mural meridian quadrant. The 8-foot-radius brass quarter-circle is attached to an exact meridian, or north-south, oriented wall. The telescopic sighting arm and precision micrometer are then used to measure the exact altitude (Declination) of an astronomical body. The resulting angles can be used either for time-finding or for the compilation of precise astronomical maps and tables. (Drawing by Allan Chapman.)

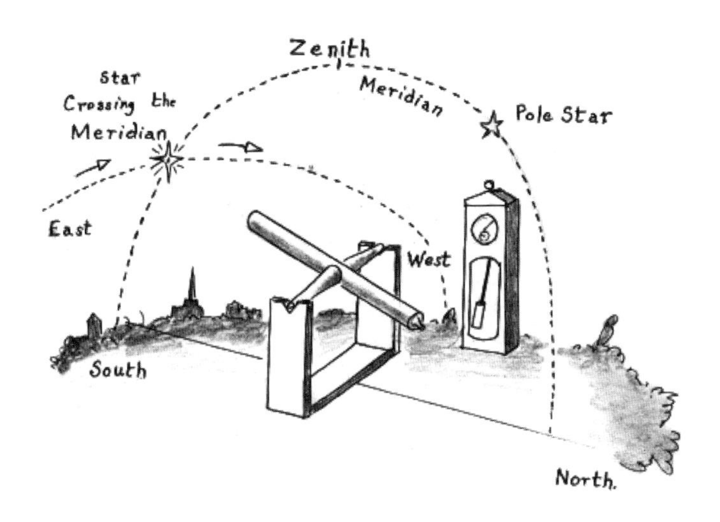

3.3 Transit telescope. The transit consists of a precision refracting telescope mounted upon a strong and inflexible axis, in the manner of a cannon. Its east-west pointing trunnions sit in carefully-engineered "V" bearings, so that the telescope describes a 180° arc across the sky. This arc, which runs through the northern pole and zenith, defines the southern and northern meridians. Transit instruments were used in conjunction with mural and other fixed quadrants to establish the exact time, to regulate the Observatory clocks, and to cross-check the meridian alignment of the mural quadrant. (Drawing by Allan Chapman.)

All were designed and built by Halley's friend, the physicist and inventor George Graham FRS Halley then went on to use these instruments regularly to observe the moon and stars down to 1739. Halley and Graham, however, did not realize that even these superlative instruments were not sufficiently accurate to measure the tiny angles necessary. That would have to await later generations of even more accurate instruments, designed and built by John Bird, in the mid-eighteenth century.

Religion and politics, a merry life and a sudden death

The first sixty years of Edmond Halley's life saw constant upheaval in Great Britain. Born when the Cromwellian "Saints" still ruled

England, Edmond was four when King Charles II restored the monarchy. Then there were upheavals around Charles's Roman Catholic brother James's inheritance of the throne, which he abdicated in 1688. Protestantism was secured by William and Mary's reign, and by that of Queen Anne in 1702, but when Queen Anne died in 1714, outliving all of her children, the throne once again was vacant.

Would there be a French-backed Roman Catholic invasion via Scotland or Ireland, spearheaded by James Francis Edward Stuart, the "Old Pretender"? Then what sort of a monarch would Protestant George, Elector of Hanover, be? Only with the Georgian succession after 1715 would there be lasting stability. And Roman Catholicism apart, British Protestantism was itself a divided house, with High, Low, and Latitudinarian Anglicans, High Tories and Whigs, to say nothing of the Baptists, Presbyterians, Quakers, Congregationalists, and other non-Catholic Trinitarians whose legal (but not full political) rights had been assured under the Toleration Act of 1689. Before one can begin to make sense of Edmond Halley's world, therefore, one must appreciate the extent to which religion, politics, and identity coloured all aspects of both public and private life.

There was such unbelief in varying forms, depending on what one doubted, such as Socinianism, Arianism (doubting the Holy Trinity), Deism (God no longer active in the world), and on to full-blown atheism. Religion was often seen as relating to one's political allegiances, such as the old recusant Catholics, supposedly loyal only to the pope (making them potential traitors in popular opinion), whereas High Church Anglicans were loyal king's men, and dissenters "Whiggish" or Parliamentary in sympathy and possibly susceptible to dangerous foreign ideas. Terms such as "sceptic", "atheist", "Papist", and "Socinian" were standard abuse words of the day.

Biblical interpretation was also a contentious topic, for while the mainstream Christian churches were never "fundamentalist" in the modern sense (for a critical understanding of Scripture extended back to the early Church Fathers), the interpretation of certain biblical passages was often much disputed. For example, how one read the Genesis creation story, or Noah's Flood, in the light of modern evidences could often make fur and feathers fly. Perhaps this is why some of Halley's contemporaries accused him of being a "sceptic", especially in the wake of his geophysical ancient earth and cosmological infinity publications. Yet one looks in vain through Halley's voluminous writings for any sceptical remarks, and as Sir Alan Cook made clear in his excellent Halley biography, the explicitly religious statements he did commit to paper were more to do with *thanking* God than *denying* him. Halley gave thanks to the Almighty, for example, for HMS *Paramore*'s safe deliverance after almost being overwhelmed by a mountainous storm wave in the south Atlantic on 26 February 1700; "but it pleased God She was righted again".[6]

Halley's temperament also played a part in the abuse that some people hurled at him. In an age of high seriousness, when learned men were expected to behave with stern gravitas, Halley's natural light-heartedness and easy manner probably riled many contemporaries, as did his decades-long rude health and vigour when headaches, agues, fevers, the stone, and other maladies are displayed prominently in the writings of his contemporaries, such as the eternally unwell Flamsteed and the moody Newton. In short, a man who implied that Genesis might not supply us with an exhaustive account of the creation and the Flood, and was also healthy, successful, invariably got what he wanted, and preferred jokes to agonizing was at risk of attracting criticism. Consequently he acquired a reputation for being unorthodox in his beliefs.

Though he was sixty-nine years old by the time that George Graham completed his new set of Royal Observatory instruments, Halley's iron constitution and relentless drive meant that he was able to use them to good effect for over fourteen years. We have little detail about his married life, but Edmond and Mary Halley were together for an incredible – by seventeenth-century standards – fifty-four years, and a son and two daughters survived into adulthood. Mary died in 1736. Sometime before his death, Halley suffered a mild stroke, but continued to observe the heavens and enjoy convivial dinners following Royal Society meetings. His lack of teeth, however, obliged him to eat fish rather than tough meat.

Death came suddenly. Always fond of drink and good company, it is said that, against his doctor's orders, he sank a large glass of strong wine – and promptly expired. That was on 14 January 1742, at the start of his eighty-sixth year. He died at the Royal Observatory, still in office as Astronomer Royal. He was buried in the same vault as Mary, in Lee Churchyard, a short walk from the Royal Observatory.

I hope that one day I might meet him in a place above the starry sky to share a joke and a merry laugh!

Chapter 4

"Let there be more *light." How Telescope Technology Became the Arbiter in Cosmological Research*

Since the time of Tycho Brahe in the 1580s, many new instruments have contributed to our understanding of the universe. These have included precision angle-measuring scales, engineering technology, photography, spectroscopy, and the twenty-first-century satellite and Mars Rover technologies. Underpinning them all has been the telescope, be it the simple spectacle lens instruments of Harriot and Galileo, the Hubble Space Telescope, or the deep-space invisible energy-detecting radio telescopes of today. No other single scientific device has done more to reveal the nature of the universe and to teach us that progressive technologies – each one a designed improvement upon its predecessor – can open up yet more wonders, ad infinitum. This progressive dimension was to be incorporated into non-astronomical technologies, such as the microscope, which continues to reveal yet more marvels within the realms of the very small, nowadays with the aid of lasers and electrons, as I show in my *Physicians, Plagues and Progress.*[1] But how did these early telescope technologies develop?

Long telescopes on tall poles

For the first thirty-odd years of the telescope's existence after 1608, the basic technology remained the same, as small, spectacle-sized lenses were mounted in tubes. Harriot, Galileo, and others soon realized that if the instrument had a long-focus lens, or object glass, in its far end, and a short-focus lens directly in front of the observer's eye, then one could obtain a higher magnification. The power of the telescope was (and still is) computed by simply dividing the focal length of the eye lens into that of the object glass.

Take a 30-inch-focus object glass and divide it by a 1-inch-focus eye lens, and the resulting instrument will magnify thirty times. Should the eye lens focus at a half-inch, then the power will be 60 times, and so on.

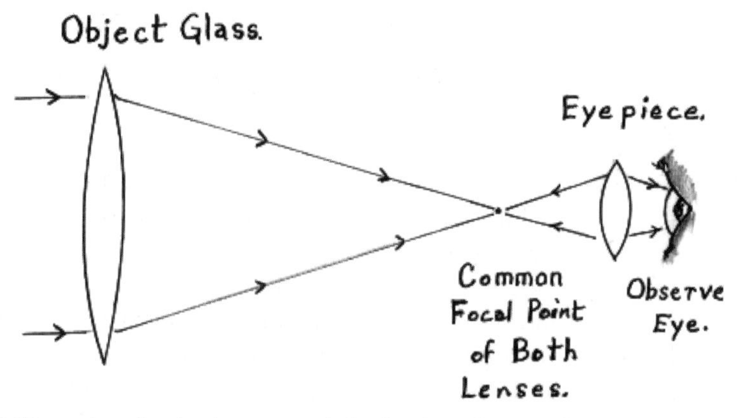

4.1 The optics of a simple astronomical refracting telescope. A convex object glass brings light to a point focus. A smaller convex eyepiece is adjusted so that its focus coincides with the focus of the object glass in a straight line. This will give a magnified image of a distant object. (Drawing by Allan Chapman.)

Early telescopes failed to produce good images beyond thirty or fifty times, however (Harriot's best telescope was ×50) because of the lenses themselves. A simple object glass, of 1½ inches in diameter, simply cannot gather enough light to focus to a single point source so as to facilitate a magnifying power much beyond 30 to 50 times. These early instruments were limited by the physical constraints bearing upon glassmaking technology in *c.* 1630, for clear, bubble- and streak-free pieces of glass beyond 2 inches at most were not viable. The optical breakthrough came via an unexpected source: the burgeoning luxury goods industry. Growing prosperity, innovation, and cross-fertilizing technological spin-offs impacted upon astronomical research in the 1640s in a way that has occurred on many other occasions in the history of Western civilization. In this case, the breakthrough came from the development of large, fashionable mirrors, backed in those days by a layer of metallic mercury, sealed in to give high reflectivity.

The Dutch astronomer Christiaan Huygens tells how he and his brother Constantijn visited a glass house in Holland to inspect large blank sheets of mirror plate glass. In a plate of glass measuring, perhaps, 2 by 3 feet, the Huygens brothers were able to find regions of pure, clear, unstriated glass several inches across. When these were carefully cut out with a diamond, they made good blanks for lenses. Small lenses for spectacles were traditionally made by grinding the small glass blanks with an abrasive such as fine sand and jeweller's rouge into a concave or convex iron plate so as to impart a curve onto the glass. But bigger lenses needed heavier, semi-automatic tools to ensure the correct sustained pressure. The Huygens brothers, Johannes Hevelius in Poland, and others in France, Italy, and England developed treadle-powered optical lathes, making an abrasive tool impart exact curves, with strong yet gently controlled pressure, on to glasses of 3, 4, or more inches across.

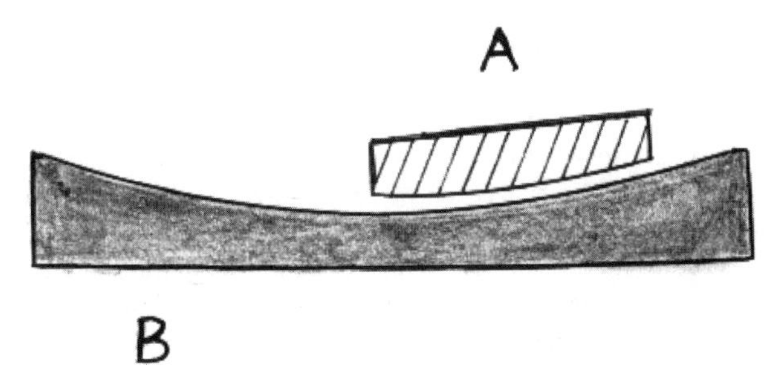

4.2 Figuring a convex lens. A transparent glass disc "A" is gently ground down under hand-pressure into a concave iron plate "B". A coarse abrasive, such as sand, is used to produce the correct curvature, followed by increasingly mild abrasives, such as finely ground chalk or jeweller's rouge. These would be used to polish the glass to the brilliance required for a good lens. (Drawing by Allan Chapman.)

Especial precision was needed when imparting very *slight* curves to the glass, for such slight curves could throw the focal point back to 12, 25, 50, or more feet, to facilitate magnifications of ×150 or ×200, resulting in a large "prime focus" image of, say, Jupiter or the moon.

4.3 A long refracting telescope. To obtain very high magnifications and large "prime focus" images of lunar and planetary surfaces with the single-element lenses of the day, seventeenth-century astronomers and opticians made refracting telescope object glasses of very long focal length. The "prime focus" image was further magnified by the eyepiece lens. Robert Hooke used such a telescope, of 36 feet focal length, to draw lunar craters at about ×173 magnification in October 1664. Elevation would be achieved by an assistant hoisting the telescope up and down a pole, while a small stand at the rear steadied it and allowed precision adjustments to be made. Such instruments could only really be used on a clear, windless night. (Drawing by Allan Chapman.)

The Royal Society (to which Christiaan Huygens was elected a Fellow in 1663) owns three lenses of 122, 170, and 210 feet focal length, with diameters between 7.75 and 9.10 inches, made on such a machine. The glass of these dinner-plate size lenses is remarkably clear, and they are truly exquisite specimens of early big telescope technology, as I know from examining them. How far these prodigiously long-focus Huygens lenses were used in practice is not known, but all the great discoveries for almost a century after the 1640s were made with instruments of 30 to 80 feet focal length, the long tubes being suspended from tall masts, and adjusted with an elaborate arrangement of ropes, pulleys, and levers. Observations of Saturn's rings and the belts of Jupiter, Robert Hooke's detailed lunar surveys, early Orion, and other nebula studies and more were all done with such long telescopes.

All done with mirrors: the early reflecting telescope

Because of the sheer unwieldiness of these long telescopes, which could only really be used in a dead calm, ingenuity led opticians and astronomers to explore telescopes in which a concave mirror might be used as an alternative to a glass lens. A mirror only needed one optical surface, the reflecting surface, whereas a lens needed two. And a mirror need not have a pure interior, as a lens had to have, for light does not pass *through* a mirror, but is entirely reflected from its surface. While shining mercury-backed glass might have been adequate for fashionable mirror surfaces, such as those in the Hall of Mirrors at Versailles, it could not be used for telescopes, for it was technically impossible to *surface*-silver a piece of glass before the 1850s. If the light were permitted to pass through the glass and then be reflected back through it again (as in a domestic mirror), all sorts of distortions and aberrations occurred that were too tiny to matter when trying on clothes but would ruin a magnified image of Jupiter.

At least four opticians were experimenting with using a concave mirror to produce a focused image of a distant object by 1670. These included Robert Hooke and Isaac Newton in England, Father Laurent Cassegrain in France, and Professor James Gregory in St Andrews, Scotland. Each used a secondary mirror to reflect the light from the main mirror to the eyepiece lens and into the observer's eye. Cassegrain used a convex curve secondary mirror and Gregory a concave one, whereas Newton employed a plain, flat secondary. Both Gregory and Cassegrain used their second mirror to send the light back down the tube, then through a circular hole in the large, or primary, mirror, and into the eye lens. The Newtonian telescope, however, used a plain, flat mirror set at 45° to direct the light out of a hole in the side of the tube.

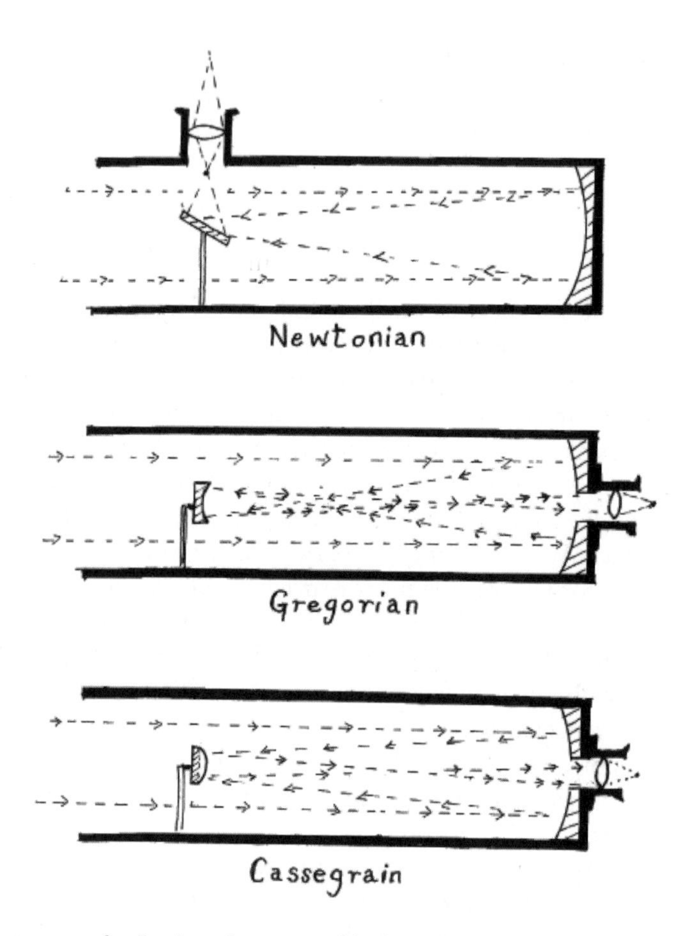

4.4 Three types of reflecting telescope used in the eighteenth century.

Newtonian. A concave primary mirror reflects the light back up the telescope tube. Here, it is intercepted by a small flat mirror set at 45°. The light focused by the second mirror is sent out through a hole in the tube, to be magnified by the eyepiece lens.

Gregorian (such as the telescopes made by James Short). The primary mirror sends the incoming light back up the tube into a concave secondary mirror. This sends the light back down the tube, and out through a hole in the primary mirror, and into the eyepiece. Gregorians give a "right way up" image, and can be used for terrestrial viewing.

Cassegrain. Similar to the Gregorian, but in this instance the secondary mirror has a convex hyperbolic curve and a slightly different optical geometry.

(Drawing by Allan Chapman.)

Hooke tried various configurations. They would all come into astronomical research in various ways, perhaps the most enduring being the Newtonian reflector, made and beloved by generations of both professional and amateur astronomers.

But what was the advantage of a mirror (reflecting) telescope over a lens (refracting) telescope? The secret lay in the reflector's potential to harvest far more light from the night sky, for as there was, and still is, a physical limitation on the viable aperture of lens diameters, it is – at least in theory – possible to make a mirror of any diameter one chooses. The problem in the early days of reflecting telescopes was the material out of which mirrors could be made, for as one could not surface-silver glass until the nineteenth century, optical-quality mirrors had to be made out of a heavy metal alloy of tin, copper, and a few additives, cast in a mould, then laboriously ground and polished to the correct optical curvature. The alloy came to be known as speculum, or mirror metal, and the resulting telescopes as speculum reflectors.

In the early days, however, metallurgical knowledge was inadequate, so these speculum mirrors tarnished rapidly and lost reflectivity, especially in urban environments where the sulphur fumes and smoke from coal fires soon turned mirrors dull yellow or grey. So while the optical configurations for the Newtonian, Gregorian, and Cassegrain reflecting telescopes were all correctly in place by 1680, it would not be until the next century that both the metallurgical and manufacturing technologies rendered them truly practicable.

John Hadley and his Newtonian reflecting telescope

Isaac Newton's original mirror telescope was constructed in 1668, and in 1671 he made another. This 1671 instrument is now one of the Royal Society's greatest possessions, but it is quite tiny. Its now heavily tarnished speculum mirror is 2 inches in diameter, in a tube only 6.25 inches long. Even so, when new it was reported to provide clear, sharp images of both terrestrial and celestial objects of around 35 times magnification. And there things stood until 1719–20, when John Hadley developed Newton's design.

John Hadley was born in London, the son of a well-to-do Hertfordshire landowning family. He was an accomplished mathematician with a full grasp of Newton's gravitation theory, and possessed a strong practical turn. At the 12 January 1721 meeting, he displayed a Newtonian reflecting telescope before the Royal Society that was a considerably upscaled and improved version of Newton's original design.[2] Assisted by his brothers George and Henry, Hadley first melted and cast a copper–tin alloy of speculum metal 6 inches in diameter, then figured it into a concave optical curve by means of abrasives, finishing it by imparting a bright luminosity to the mirror.

Hadley's ingenuity and manual dexterity had enabled him to achieve two significant improvements. Firstly, he devised a way of imparting not a spherical but a *parabolic* (another conic section) curve to the optical, reflecting surface, because a parabola allowed a higher definition and greater image clarity. Secondly, he developed a method for giving the parabolic curve a very high-reflecting polish. Both innovations played a vital role in making the reflecting telescope scientifically viable.

Hadley's 6-inch-diameter mirror had a focal length of 62 inches, and was placed at the bottom of a hexagonal wooden tube and mounted on a convenient, easily adjusted wooden stand. The Royal Society tested the telescope optically and mechanically, comparing its images of the planets with those seen through a 123-foot-focal-length refractor set up on a tall mast at Wanstead in rural Essex, 7 miles from London. It was concluded that the telescopes produced images of similar quality, though Hadley's reflector images were perhaps less bright. On the other hand, Hadley's telescope had huge advantages. Its short, 5-foot tube and easily adjustable stand made it quick and easy to use, in contrast with all the ropes and fittings of the 123-foot refractor. Hadley's reflector could also perform when it was windy, and gave much better images in twilight.

4.5 John Hadley's reflecting telescope, 1721–22, with a 6-inch-aperture speculum metal mirror of 62 inches focal length in a wooden tube. (Drawing by Allan Chapman, based on Hadley's account to the Royal Society, *Philosophical Transactions* 32 (1723), and in Robert Smith, *A Compleat System of Opticks* (1738).)

John Hadley had taken Newton's ingenious device and transformed it into an instrument that would have a profound effect upon astronomy and cosmology, extending from William Herschel's 7-foot Newtonians after 1770 to the home-built telescope of choice for countless amateur astronomers from the 1850s into the late twentieth century.

A Golden Guinea an Inch: James Short Turns the Reflecting Telescope into Big Business

As we shall see shortly, astronomy and science were becoming fashionable by the 1730s, and various opticians, in London, Edinburgh, and elsewhere were manufacturing versions of James Gregory's reflecting telescope design of 1663. The Gregorian reflector was very compact and convenient to use. Unlike the purely astronomical Newtonian (where one looked into the tube from the side and got an upside-down, right-to-left image, which was immaterial when looking at celestial objects), the Gregorian was pointed straight at the object, just like a lens telescope, and produced a right-way-up image.

Whether you wanted to study the moon, view ships out at sea, admire your country estate on a balmy June evening, or follow the winning horse at Newmarket races, the Gregorian was for you: big ones for the moon, pocket-size ones for the races. (I have used a beautiful *c.* 1750 pocket-size Gregorian in the Museum of the History of Science, Oxford, its mirror still bright, to read the dust-jacket titles of books in Blackwell's shop window 50 yards away.)

The undisputed "king" of the Gregorian reflector was, like its inventor, a Scotsman, the Edinburgh-born James Short. He brought together three admirable traits: scholarship, science, and a shrewd business acumen. The orphaned James distinguished himself as a fine classical scholar, first at Heriot's, then at the Royal High School, and went on to Edinburgh University, studying Divinity and preparing for the ministry. But the lectures of Edinburgh's great mathematical professor Colin Maclaurin, plus his delight in using, repairing, and making scientific instruments led him in another direction. By experimenting with speculum metal alloys, and then developing a technique to figure them into beautiful, highly reflective parabolic curves, he mastered the art of manufacturing Gregorians in Edinburgh and sold them to a well-heeled market.

By the time that he was twenty-six in 1736, James Short had already amassed a fortune of £3,000 in Bank of Scotland stock, from the sale of, among other things, about 180 Gregorian telescopes of his own construction. Then two years later he strategically moved to premises in Surrey Street, London, after appointing an Edinburgh agent to look after his Scottish investments and banking a hefty £500 profit. In London his brilliance was quickly recognized both academically and scientifically. He was elected Fellow of the Royal Society, and moved equally among the leading scientific men of London, while St Andrews University bestowed an MA degree upon him, and the Royal Swedish Academy of Sciences then elected him a Foreign Member.

Short's telescopes were excellent as optical instruments, and also things of beauty that any gentleman would be proud to display on his sideboard. The tube, in polished brass, might be 2, 3, or more inches in diameter, depending upon the size of the primary mirror, and its length might be 12, 24, or 36 inches. All would be mounted on a turned-brass pillar stand and held up by three curved legs: the famous pillar and claw design.

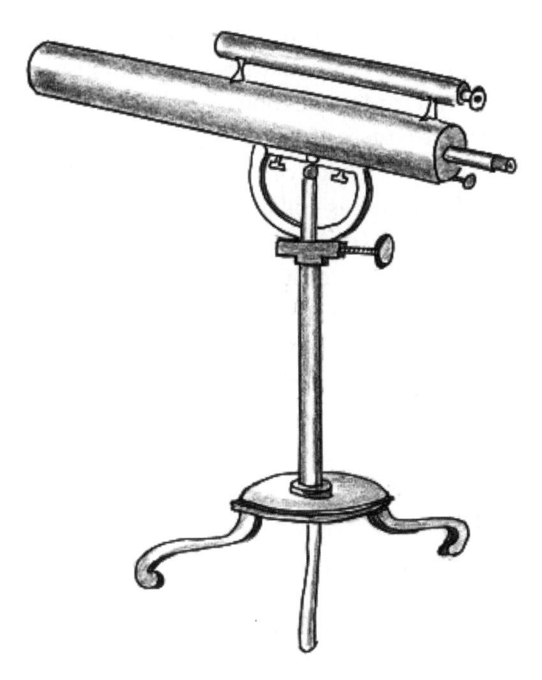

4.6 Gregorian reflector on a "pillar and claw" or three-legged folding stand. This type of telescope was popularized by James Gregory. The picture shows an instrument of 2 feet focal length, c. 1750. Such telescopes were mass-produced to a fine state of finish. (Drawing by Allan Chapman.)

Short found the time to do all of this workshop work while being an active academic Fellow of the Royal Society, and also handling his burgeoning finances, as he appears to have employed out-workers, who made the brass tubes, stands, and fittings. Out-workers, in the large "jobbing" London instrument trade, might even have cast and rough-ground his mirrors, so that all he needed to do was give them their final perfect polish and adjust the whole instrument ready for sale and use. Such out-working had long been standard practice in the watch- and clock-making trades, and a watch with London on its face could well have had its gearwheels cut in the Lancashire village of Prescott, near Liverpool, which had a long-established trade in watch parts, which it exported to London.

In addition to beautiful brass-mounted Gregorians for gentlemen's houses, which were said to sell for a guinea (or £1 and 1 shilling) an inch of tube length, James Short undertook commissions for very large telescopes. These included a great instrument with a colossal 18-inch-diameter mirror for the king of Spain, for which he charged £1,200, and another large one for Uppsala University, Sweden. His great telescope for the Duke of Marlborough's Blenheim Palace Observatory, Oxfordshire, is still preserved in Oxford's Museum of the History of Science. Its 18-inch-diameter parabolic mirror still retains a good lustre, being mounted in a 12-foot wooden tube. It cost the duke 600 guineas in 1742.

James Short never married, though the gossips of the day associated his name with several ladies, including a Miss Elliot. He died in London in June 1768, in his fifty-eighth year, worth £15,000 to £20,000.[3]

John Dollond "perfects" the refracting telescope *c.* 1760

Optical theory, especially in the wake of Newton's magisterial *Opticks* (1704), stated that whenever light was bent, or refracted, such as when passing through a glass lens, the spectral colours that combined to form white light were unavoidably split or dispersed. This resulted in red or blue colour fringes – *chromatic aberration* – around bright objects, such as the moon, as the colours could not all be brought to a single focal point. One way of minimizing this aberration was to use an object lens of very long focal length, and hence, the very long pole-mounted refracting telescopes of the day. Reflecting telescopes, by contrast, did *not* disperse the light into aberrated colours, because the light was surface-reflected from the mirror and did not penetrate the body of the mirror, as it did with a lens. This is why, by 1740, opticians were paying so much attention to perfecting the mirror telescope.

One country gentleman scientist who had also trained as a barrister, Chester Moor Hall, applied the same lateral thinking to the lens problem. Though appreciating that the lens splitting the light into spectral colours was an unavoidable natural phenomenon, Moor Hall conjectured in 1729 that if one had a telescope object lens composed of *two* discs of glass, one might be so adjusted as to cancel out the errors of the other. Being a capable optical mathematician, he computed the necessary matching curves for a pair of lenses mounted one behind the other. One lens would be made from light crown glass, and the other from heavy flint glass, the different chemical compositions of the two glasses giving them different refractive indices, or light-bending powers.

Here we enter the world of London's jobbing optician community, for, not possessing the skill to grind and polish each lens himself, Chester Moor Hall contracted them out. Being cautious not to let his secret out, however, he contracted the making of each component to a separate craftsman, one going to Edward Scarlett in Soho, and the other to James Mann senior, near St Paul's Cathedral, their premises being about a quarter of a mile apart. But to Moor Hall's ill-luck, both Scarlett and Mann subcontracted, and – even more bad luck – to the same man: George Bass, who quickly rumbled Moor Hall's secret. Unable to undertake the commercial development of this colour-distortion-free achromatic lens himself, Bass seems to have hawked it around to several manufacturing opticians in London, and the secret gradually leaked out within the trade. What is truly amazing, however, is that Moor Hall – a trained barrister – did nothing about it, apparently preferring a quiet life to lawsuits.

The man who developed and advanced Chester Moor Hall's brilliant invention was a London optician of Huguenot descent, John Dollond, partly in conjunction with his son Peter. A skilled optical experimentalist, Dollond began around 1755 to undertake an experimental investigation into the bending of light by Venice crown glass, water-filled glass prisms, and heavy flint glass. He was also spurred on by a letter from Professor Samuel Klingenstierna of

Uppsala, in which Klingenstierna pointed out some optical errors in Newton's conclusions about light refraction and dispersion.

Guided by his experimental findings in tracing the colours and angles of light passing through his glass and water-filled prisms, John Dollond designed and constructed an achromatic, or colour-distortion-free, lens around 1758. It had several crucial advantages over the traditional long-focus refracting telescopes. Firstly, by bringing all the incoming light to an exact and defined focal point, one of Dollond's achromatic glasses did not squander part of the light on false colour fringes and blurs. Secondly, by 1758, matched crown and flint glass object glasses of 3, 4, or more inches in diameter could harvest much more light from the sky, bring all the rays to a common focal point, and produce a bigger field of view, combined with a brighter, sharper, undistorted image. All of a sudden, it was possible to have an astronomical telescope with a 4-inch-diameter object glass of a mere 5 feet focal length that would produce images in every way superior to those of a traditional instrument of 120 feet focal length suspended from a pole. Such telescopes, owing to their greatly enhanced optical efficiency, could tolerate much higher magnifications without image quality breaking down, as well as increasing what later generations would call optical resolving power, the ability to see very delicate detail. And a standard Dollond telescope, with a 5-foot tube and folding tripod mount, was not only capable of being used on windy nights, but was also portable. Achromatic telescopes with even larger lenses could be made to be mounted upon permanent mounts in observatories.

John, and particularly his business-minded son Peter, obtained a patent for the achromatic lens. While eighteenth-century patent law was less precise than today, and could sometimes be evaded by the ingenious, the Dollond family came to make a large fortune from their telescopes, to establish a great and enduring British optical dynasty.

"Every gentleman must have one!" Benjamin Martin, lecturer and entrepreneur, makes scientific instruments fashionable

One ingenious gentleman who not only did his best to wriggle around the Dollond achromatic lens patent but even manufactured telescopes signed "Doland" was Benjamin Martin. Born in Surrey in 1705, the son of a modest farmer, Martin embodied so many of those traits I admire. As far as we can tell, he was wholly self-educated, good with his hands, and a lucid, gifted, and quite prolific writer, who turned commercial scientific lecturing into a serious business. A showman and an entrepreneur to his fingertips, he combined lecturing to paying audiences with publishing, and with the manufacture and sale of scientific instruments: hence the "Doland" telescopes. He also took premises in Crane Court, off Fleet Street, London, only a few yards from where the Royal Society held its meetings, though the Society rejected his candidacy for the Fellowship.

Given his lecturing in Bath, London, and elsewhere, his enormous publication output – he was a pioneer of the publishing format of building an encyclopaedia in weekly instalments – and stock of scientific instruments advertised for sale, it is clear that Martin used out-workers to manufacture his instruments. He marketed orreries, microscopes, telescopes, early static-electrical machines, globes, and other pieces. He also recognized the sales strategy of both charging his well-off audiences to hear his lectures, in the same way as an actor or musician would, and of demonstrating with elegant instruments. An audience that had been so captivated then wanted to buy the book of the lectures, and possibly purchase fully working duplicates of the lecture demonstration instruments. It was a clever format.

In many respects, Martin was in the business of selling scientific culture. By 1750, there would have been quite a market, especially among people with abundant new money made in City trading, East India Company dividends, early industrial ventures, provincial banking, and land speculation. In their new silks and satins, such people wanted their portraits painted, edifying-looking books with which to line the libraries of their country and metropolitan residences, and beautiful devices of brass, glass, and polished mahogany to demonstrate their grasp of science. These devices would impress dinner-party guests, whether it was a fine new Gregorian, Dollond (or "Doland") telescope to display the moons of Jupiter on a starry evening, or, if it were cloudy, a microscope with which to raise gasps of horror when exhibiting a prepared slide of a dead flea. How better could one signal one's grasp of the arcane wonders of the Newtonian universe than by demonstrating one of Mr Martin's "Grand Orrery" planetary machines in the drawing-room to guests over coffee, after one had first attended his lecture, and then bought his books? Joseph Wright's beautiful paintings of "An Experiment on a Bird in an Air Pump" and "A Philosopher Lecturing on the Orrery" from the 1760s capture perfectly this new world of scientific instruments, wonders, and sophistication.

While he was the most famous and dynamic of the popular science lecturers and writers, Martin was not alone. By the time of his death – sadly as a bankrupt – in 1782, telescopes, astronomical books, and a polite knowledge of contemporary astronomy had joined objets d'art, fine paintings, and well-laid-out houses and gardens as the hallmarks of a cultivated gentleman. Be your telescope by James Short, Peter Dollond, or Benjamin Martin, every true gentleman *must* now have one!

Chapter 5

The Rector and the Organist: Gravity, Star Clusters, and the Origins of the Milky Way

The people of 1760, when the young King George III ascended the British throne, were aware of living in a very different universe from that understood by their ancestors in the days of "Good Queen Bess", who had died in 1603. Instead of the universe being constituted of the nine spheres of classical astronomy, with the earth in the centre, the early Georgians realized they were on a spinning planet rotating around the sun, and the sun itself was but a star, perhaps like any one of the millions in the sky that even a pocket telescope could reveal on a clear night. But what was the Milky Way, with its countless myriads of stars, and what was the fluffy glowing stuff of which the Orion Nebula was made? Why did some stars congregate into dense clusters, as in the Pleiades and Hyades, while other stars seemed evenly planted from each other, almost with the regularity of flowers in a well-tended garden? Most fascinating of all, what was this strange, invisible, yet mathematically calculable force of universal gravitation that seemed to permeate the cosmos, causing motions as diverse as the regular rotation of the planets around the sun and the falling of a stone to earth? The new universe seemed so infinitely vast, full of empty space between the glowing stars, and governed by an exact law, that in its perfection, it could only have emanated from the mind of an all-powerful supreme beneficent Deity.

Thomas Wright of Durham and Eighteenth-Century Speculative Cosmologies

In many ways it had been medieval Europe, between the thirteenth and fifteenth centuries, that had first wrestled with the theological and philosophical implications of cosmological vastness, infinity, and the nature of time, especially with thinkers such as Thomas Bradwardine, Nicole de Oresme, and Nicholas of Cusa. By the latter part of the seventeenth century the cosmology of Descartes, and in particular that of Newton, opened up all manner of speculations about infinity, especially in the wake of the newly revealed telescopic deep space cosmos. In 1692, for example, the scientific theologian Revd Dr Richard Bentley delivered the first series of lectures, endowed under the will of the Honourable Robert Boyle (who died in 1691), aiming to demonstrate the compatibility of science with the Christian faith. Bentley suggested that a universe that was perfectly and infinitely spread throughout space would be gravitationally stable, as each gravitational field would be counterbalanced by another equal field: an idea with which, as we saw in Chapter 2, Edmond Halley was familiar. Bentley discussed the matter with Newton, and considered different theoretical permutations of homogeneity and un-homogeneity, but concluded that the beneficent hand of the Creator maintained gravitational stability. By the eighteenth century, the size and shape of the starry heavens (or galaxy), whether it extended in one plane more than in another, and whether the stars might vary in size – as in brightness – and hence in their gravitational power, all came up for discussion in a series of essentially speculative cosmologies, based upon Newtonian theory and very limited telescopic data.

From at least as early as 1734, the Durham clockmaker, astronomer, and philosopher Thomas Wright had been wrestling with the subject, and in 1750 published his *An Original Theory or New Hypothesis of the Universe.* His cosmology was deeply theological, in that he identified cosmological stability with Divine Providence. Wright speculated that the universe possessed a gravitational centre, which he connected with the Godhead, and over the years he developed theories about the structure of the Milky Way, and why the stars appeared dense in the surrounding Milky Way band but less so above or below that band. Wright drew upon recent astronomical discoveries, proposing that Halley's proper motions of stars against one another, announced in 1718, might suggest that the universe, including relatively near and distant stars, rotated around the gravitational centre. Were there even spherical shells, or zones, of stars surrounding the gravitational centre, of which our Milky Way was but one? He almost touches upon what twentieth-century cosmologists would style "multiverses", or perhaps an infinity of universes known only to God, through which souls might pass as part of a drama of condemnation and redemption.

Wright is significant astronomically because of the attention he paid to the structure and possible extent of the Milky Way star system. In the mid-eighteenth century, two Germans also speculated upon the Milky Way. Johann Heinrich Lambert, in *Cosmologische Briefe* ("Cosmological Letters"), 1761, considered the universe's possible lenticular structure. To envisage this, think of two saucers held edges on: wider in the middle and tapering off to the long, extended edges. Then the great metaphysical philosopher Immanuel Kant, inspired by what he had read about Thomas Wright in a Hamburg newspaper, speculated in 1755 about the possibility of a dynamic universe shaped like a flat disc, whose stellar inhabitants were in a constant state of decay and regeneration as things exploded and reconstituted. One can see traces of these various systems in the much more mathematically cogent nebula cosmology of Pierre-Simon Laplace of 1796, which we shall encounter in Chapter 9. In all of the above speculative cosmologies, however, the philosophical imagination clearly raced ahead of the observed evidence, which, even in 1760, was decidedly thin. Before serious *scientific* progress could be made in cosmology, a truly vast corpus of observed and cross-checked primary data would first have to be harvested from the night sky, which would, in itself, constitute astronomy's huge debt to William Herschel after *c.* 1780.

The Revd John Michell: the Pleiades Cluster, "dark stars", and gravitational "black holes" in 1783

The Revd John Michell FRS belonged to the extensive and venerable company who did much to further astronomy and several other sciences between the sixteenth and early twentieth centuries: the Church of England parson–scientist. These gentlemen, highly educated, devout, humane, and intellectually curious, did much to advance botany, natural history, fossil geology, geomagnetism, chemistry, technology, archaeology, experimental physics, and even medicine. John Michell made major contributions to several sciences in addition to straight astronomy, especially geomagnetism and gravitation physics. He was a Fellow of Queens' College, Cambridge, and the university's Woodwardian Professor of Geology until 1764 – when at the age of thirty-nine he married and relinquished his bachelor chair – and gravitation physics was central to his entire vision of nature. He did pioneering research into the physical causes of earthquakes, devised a torsion balance to measure gravitational attraction in the laboratory (see Chapter 8), and like so many other Natural Philosophers before him, pondered the nature of cosmological vastness, infinity, and the gravitational stability of the universe. Michell was not just a cosmological theoretician, for he constructed a large reflecting telescope for his own use: an instrument later acquired by Herschel.

One particular class of object that fascinated the astronomers of the eighteenth century and would become a lifelong study for Herschel was the nature of double, triple, and whole clusters of stars. Given a presumed uniform distribution of stars at the creation, why had so many stars moved into close proximity to each other? Especially fascinating was the Pleiades cluster in Taurus. With the naked eye, one saw the Seven Sisters stars in a close group. Then, as Galileo, Hooke, and subsequent astronomers had discovered, each successive improvement in telescopic power revealed countless numbers more. Gravity, under the inverse square law, was the obvious candidate for the star-drawing force, with the effect that the addition of each new star to the cluster augmented its combined gravitational power, to draw in yet further stars.

In November 1783, Michell read a paper before the Royal Society, published the year after,[1] in which he asked what would happen to the light from a large, gravitationally powerful mass in space: a mass 500 times bigger or denser than our own sun. Would its pull of gravity be so powerful that even its own light was unable to escape into space? The accepted model for the transmission of light in 1783 was not Robert Hooke's wave of 1665 but Newton's particle stream of 1704. If the light consisted of countless millions of tiny, discrete particles, streaming away from their source, would not each particle be affected by gravity, and hence be prevented from escaping its source? Michell spoke of these purely theoretical entities as "dark stars"; they possess distinct parallels to what modern cosmologists have identified as "black holes", although, as we shall see in Chapter 25, a true "black hole" is fundamentally different from just a dense cluster of stars.

After holding various academic appointments in Cambridge, Michell moved to Yorkshire with his second wife in 1767, his first wife having tragically died only a few months after their marriage. Here he became rector of St Michael's Church, Thornhill, near Leeds, where he spent the rest of his life and conducted some of his most far-reaching researches. He died in April 1793, in his sixty-ninth year. He was undoubtedly one of Cambridge University's true yet largely unacknowledged scientific giants of the post-Newtonian age.

Charles Messier: comet hunter and nebula cataloguer of the *Ancien Régime* in Paris

Throughout much of the eighteenth century, comets were of particular interest to astronomers, as works such as Hooke's *Cometa* (1678), Newton's *Principia* (1687), Book III, Whiston's *New Theory of the Earth* (1696), and Halley's *Synopsis* (1705) had posed all manner of questions about them. No one now believed that they were atmospheric, but their size, composition, and, most of all, their orbits still left many questions unanswered. Their unpredictable and erratic nature made them challenging objects to study, and this, no doubt, was a motive behind such comet hunters as Charles Messier and later Caroline Herschel.

The son of a noble court functionary in eastern France, the fourteen-year-old Charles Messier was first drawn to the study of astronomy by the brilliant six-tailed comet of 1744. Possessing the right intellectual skills and social contacts, he began to work for the illustrious hydrographer Joseph-Nicolas Delisle in Paris as a draughtsman. A born observer, Messier started to search for comets, and King Louis XV called him his "ferret of comets". The assiduous Messier, who would become a French *académicien* and a Fellow of the Royal Society, would ferret some thirteen comets out of the Parisian skies and publish accounts of them.

Charles Messier's enduring fame, however, came not so much from the comets he discovered as from his list of celestial nuisances that a comet hunter might well mistake for comets. As most comets first appeared as small smudges of light, it was all too easy, in the early stages of a cometary apparition, to be fooled by one or more of that large number of smudges called nebulae, those glowing cloudy patches to which Halley had drawn attention in 1716. So Messier, and his younger friend Pierre Méchain, began to make a list of these tiresome objects. By 1781 the list had grown to 103 "Messier objects", which turned out to have far more cosmological significance than the comets for which they might be mistaken. For many of these nebulae are now known to be *galaxies*, star systems, vastly remote from the Milky Way. They would become the research focus of the life's work of an astronomer, eight years Messier's junior, who would survive him by five years: Wilhelm Friedrich Herschel of Hanover.

THE ENTERPRISING OBOIST: HERSCHEL COMES TO ENGLAND

The career of William Frederick Herschel, as Wilhelm Friedrich would become in England, constitutes a wonderful counter-example for those historians who see English history in terms of rigid class barriers and people knowing their places. Herschel was one of a veritable army of highly talented men and women, both British- and European-born, whose skills, ingenuity, and charm enabled them to ascend the social ladder. Their ranks included writers, engineers, industrialists, entrepreneurs, painters, architects, musicians, and scientists. In the case of Herschel, music and his gifts as an impresario gave him his initial modest fame. After 1781, his brilliance as a scientist would escalate him to royal favour and world renown, and lay the basis of a new approach to cosmology. This approach was based not on metaphysics or theology, but upon painstaking empirical research made with great telescopes of his own devising – with the odd speculative leap thrown in. As is clear from reading Herschel's writings and considering his career and friendships, he possessed charisma and an ability to captivate people: essential attributes for a successful performing musician and concert promoter, who, before he made his first significant telescopic observation, had already risen well above the ranks of what were sometimes referred to as "fiddle-scrapers".

William Herschel was born in November 1738, probably in the barracks in Hanover, where his father Isaac was military bandmaster. He and his brothers Jacob, Alexander, and Dietrich were trained as musicians and were accomplished on a variety of instruments, while little sister Caroline was simply taught to keep house. Father Isaac, however, was a deep-thinking and wide-reading man, whose interests extended beyond performing music to philosophy, languages, mathematics, science, and general culture. William and his brothers, but sadly not Caroline, were brought up to share his intellectual passions.

5.1 Sir William Frederick Herschel. (Print in author's collection, c. 1788.)

By 1738, Hanover enjoyed close ties with England, including sharing monarchs, as King George II was already the second Hanoverian king to ascend the British throne. Following the death of Protestant Queen Anne in 1714, the nation feared a Roman Catholic invasion, and distantly related George of the solidly Lutheran and potentially

Anglican House of Hanover was invited to become king: the first British King George, whose son, George II, would succeed him in 1727. Their descendants still wear the British crown.

Scientific history, being just as entwined with politics and theology as any other aspect of our cultural past, brought the Herschel brothers to England in April 1756.[2] The contemporary war with France caused fears of a French invasion of Kent, and the Hanoverian regiment was brought over to help strengthen the local defences. No invasion came, and Herschel, stationed in Maidstone, set about mastering English, to complement his native German, and possibly, French. Yet William's approach to learning English was not to pick up such basic language as he might need to have a chat in the pub or read a newspaper, but rather to read tough books. Before long, he tells us, he "was enabled to read Locke on the Human Understanding": this 1691 classic was one of the foundational texts of the British empirical tradition of science and philosophy.[3] Although the regiment moved back to Germany without firing a shot in anger, William and his brothers saw that England was a land of opportunity for enterprising and gifted people. Leaving the regiment under questionable circumstances, after more campaigns in Europe – musicians not being officially classed as combatants – William returned to England in 1759.

Georgian England was not only prosperous – it also had the biggest and most upwardly mobile middle class in world history, and such people, if they were sufficiently interesting, mixed with aristocrats and gentry in a way that would have been shocking in much more socially stratified Continental Europe. People of all social groups wanted music, entertainment, and pleasure. Entertainment venues – including lectures by men like Benjamin Martin – were commercial box-office concerns, not courtly or princely venues; and if a prosperous shopkeeper could afford a few shillings for an admission ticket to Vauxhall Gardens, he would find himself on equal terms with an earl.

The Hanoverian oboe player was also a skilful violinist and a fine keyboard player, which gave him versatility in the business of earning his living. William Herschel prospered, and at the same time read widely and in several languages, while his musicianship, conversation, and obvious social skills began to win him invitations. In addition to London, he began to enjoy success in the provinces, and became a concert director in Leeds and other towns. He was also received in the houses of the Yorkshire gentry. As a composer, performer, and concert promoter, Herschel was in demand by the age of twenty-five, with promises of salary increases if only he would stay in Yorkshire. Then in 1766, he won an open audition on the new organ of the fashionable parish church in Halifax, and was offered the post of organist. But after only a few months, he received an even better offer.

Herschel the fashionable church organist and musical impresario of Bath

The builder of the magnificent new organ in Halifax church was the Swiss-German John Snetzler, and he went on to build another splendid instrument in the new Octagon Chapel, Bath. By 1767, however, Bath, full of wealthy residents and invalids coming to take the waters for their health, had a problem. The magnificent medieval Abbey Church had now so many good Christians buried, several deep, beneath its flagstones, that in wet and warm weather it often stank. In an age when bacteria were not understood, but it was believed that noxious "effluvias" were agents of infectious disease, many fashionable folk deserted the Abbey and began to worship in private or subscription chapels elsewhere. One of these was the Octagon Chapel. The "Octagonians", in Bath and elsewhere, sat in pews around an eight-sided chapel, with the pulpit in the centre, and an organ loft and places for other musicians in a gallery. Here, the great and good of Georgian England could hear well-paid popular preachers, and listen to beautifully performed musical interludes in clear, sweet air.

As an organist, Herschel was a great success. When not playing for worship, he promoted concerts in the fashionable Assembly Rooms and Pump Room. Eighteenth-century audiences craved novelty, so Herschel, just like his contemporaries Haydn and Mozart on the Continent, constantly needed to compose new music. This would include organ voluntaries and anthems for the Octagon, and suites, symphonies, minuets, and songs for the secular venues, as people chatted and drank their tea or wine – just as in London. Back in 1762, Herschel the commercially minded twenty-four-year-old émigré had composed a humorous song, based upon the current sensation gripping London: the supposed haunting of a house in Cock Lane. "The Ghost will be heard in Cock-lane to-night"[4] is written in the fashionable rollicking style known as a "catch", and almost certainly it was written to make money. The term lingers on when we speak of a catchy tune today. Sister Caroline would later record that William enjoyed "glees and catches".[5]

William Herschel was very successful financially, and the money rolled in. Between December 1769 and December 1771, his income rose from £316 per annum to £400, and he took an elegant house at 7 New King Street, Bath, at a rent of 30 guineas (£31.50) a year.[6] Thirty guineas was, in itself, as much as many working men earned to keep a family for a whole year, while £400 a year made Herschel an undisputed gentleman, with an income the same as that of a well-beneficed clergyman, a successful provincial solicitor, or even a cathedral dignitary. In 1773, he even recorded, "I paid £15. 10. 0 for a handsome suit of clothes, it being then the fashion for gentlemen to be very genteelly dressed."[7] By this date, when he was thirty-five, it is clear that he knew his place in Bath society, and it was far from being an underling. In that same year, however, he also records spending money on a copy of James Fergusson's *Astronomy* and buying some astronomical instruments.

From organ pipes to telescopes, from acoustics to optics, and on to cosmology

Herschel had no doubt first encountered astronomy from his father back in Hanover, and probably knew the constellations, planets, and the basic principles of the science, but in the early 1770s he wanted to embark upon the business of practical astronomical observation.

One author whose works he studied in Bath was the Revd Dr Robert Smith, formerly Master of Trinity College, Cambridge, and a protégé of Newton. In his *Harmonics* (1749), Smith had analysed and experimented upon the nature of sound, including attempting to relate certain musical pitches to physical vibrations. He also discussed the design and function of different musical instruments, such as those using strings, reeds, or pipes, and how pipes or strings of given lengths always produced sounds of the same pitch. (An organ pipe of 8 feet, for example, will always produce Middle C on the scale, whereas a 4-foot pipe will sound at C one octave above Middle C.) Mathematics and music have always been intimately connected.

Herschel was not just a successful musician; he was also fascinated by acoustics and the physics of sound. In this respect, so his sister Caroline recorded, he would often "retire to bed with a bason of milk or glass of water, and Smiths 'Harmonics' and 'Optics'".[8]

In his *A Compleat System of Opticks* (1738), Robert Smith had treated light in a similar fashion to how he would treat sound in *Harmonics*, and one can see how, in the 1770s, Herschel's ever-curious mind took him from sound to light: from organ pipes to telescopes. In addition to discussing the prevailing – primarily Newtonian – theories of light and the spectrum, Smith went on to discuss the use of lenses, microscopes, and refracting and reflecting telescopes. He even discussed astronomy and cosmology, including current speculations about the Orion Nebula, of which he published a picture based on that of Christiaan Huygens.

Herschel now wanted a telescope of his own, and ordered lenses from a London optician, but what he could see only left him hungry for more. Realizing that reflecting telescopes with relatively large metal mirrors ground to a parabolic curve could give brighter and better images than modest refractors, he resolved to make a reflecting telescope for himself. No doubt his skill on musical instruments, such as the violin and the organ, were part of a wider manual dexterity that encouraged him to try his hand at making a Newtonian reflecting telescope. He was not the only aspiring reflecting telescope maker in Bath, and he saw an advertisement in a Bath newspaper where a presumably unsuccessful amateur maker wanted to sell off his mirror-making kit.[9] Herschel bought the equipment, and soon found that he could make remarkably good parabolic mirrors of 6 or so inches in diameter. Within a short time, he had built himself a good Newtonian reflecting telescope.

By the mid-1770s, he had been truly bitten by the observational astronomy bug, and wished to see the heavens with his own eyes. One can only admire his prodigious energy. Organizing and directing concerts, hiring performers, composing, playing the chapel organ, teaching private pupils – all filled his days and evenings; and increasingly, the hours of darkness came to be occupied by observing the heavens. Where did he find time to sleep?

Brothers Alexander, Dietrich, and Jacob were often in Bath as they travelled about performing music. Then in 1772, William had gone home to Hanover and brought their twenty-two-year-old sister Caroline to England. Their father had wanted to give Caroline an education similar to that of the boys, but her mother had insisted that domestic drudgery was to be her lot. Yet Caroline possessed both an intellect and a musicality in no way inferior to those of her brothers, and once in England, she blossomed. A tiny lady, standing less than 5 feet high, and with a complexion sadly marred by smallpox, she was found to have a splendid soprano voice, and William began to coach her as a performer. But when observational astronomy began to claim more and more of his spare time, she revealed her talent for astronomy as well, as we shall see in the next chapter. Then in March 1781, from their new address, 19 New King Street, Bath, William made a discovery that would not only transform the Herschel family, but also Western astronomy.

Bath, 13 March 1781: William Herschel discovers a "comet"

Caroline Herschel was deeply attached to her brother William, who had rescued her from a life of domestic drudgery in Hanover, and opened up a whole new world of opportunity before her, as we shall see in Chapters 6 and 9. After her arrival in England, she became, in many ways, the chronicler of the Herschel enterprise. She tells us that in 1777, William joined the very genteel Philosophical Society in Bath, and we have record of research papers that he delivered to it and the Bath worthies who became his friends, such as the physician Sir William Watson FRS By the time of his fortieth birthday on 15 November 1778, it is clear that William Herschel had come a long way since playing the oboe in the Hanoverian Foot Guards band.

5.2 19 New King Street, Bath, where William Herschel discovered Uranus, 13 March 1781. From a photograph by John Poole. (R. S. Ball, *Great Astronomers* (1895), p. 204. Author's collection.)

Caroline was not present, however, on the night of Tuesday 13 March 1781 – they had only just moved to 19 New King Street, and she was winding up at their previous house in Rivers Street – when William noted a "curious either nebulous star or perhaps a comet"[10] in the field of his homemade 7-foot-focus Newtonian reflecting telescope. It was in the constellation of Gemini. He quickly reported his discovery not only to Dr Watson and other friends in Bath, but also to the Revd Dr Nevil Maskelyne, Astronomer Royal, at Greenwich, and to the Revd Dr Thomas Hornsby of Oxford University's brand-new and magnificently equipped Radcliffe Observatory, who confirmed Herschel's discovery.[11]

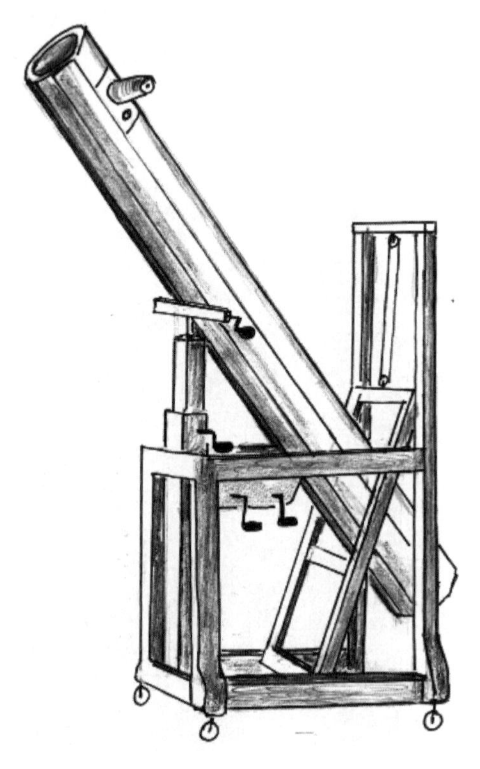

5.3 A Herschel 7-foot-focus reflecting telescope with a 6½-inch-diameter speculum mirror. William Herschel discovered Georgium Sidus (Uranus) with such a telescope in 1781. It shares several design features with Hadley's telescope of 1721–22 (4.5 above). The mirror was mounted in an octagonal mahogany tube within a wooden stand. Pulleys and cords were used to raise and lower the tube in the vertical, brass castor wheels being employed to move the whole instrument through large east-west angles. A handle actuating a fine screw was used to track an object under observation across the field over a small angle. Such telescopes were much in demand after 1781, and Herschel began to manufacture them for sale, at around £100 each. Many almost identical instruments survive in museum collections. William made a similar wide-field instrument of 5 feet focal length for his sister Caroline's cometary researches. (Drawing by Allan Chapman.)

Gemini is a winter constellation, and cannot be seen in summer because the sun occupies this region of the zodiac as the days get longer. While sightings of the "comet" would be made over the ensuing weeks, it would not be until September 1781 that – returning darkness and clear skies permitting – the celestial twins would reappear in the pre-dawn sky. Hornsby in Oxford secured fresh observations and accurate position measurements over 25–26 September, still preserved in the Radcliffe Observatory manuscript "Observing Registers", while others followed from elsewhere.[12] The comet's position near to the star H Geminorum had hardly changed since spring, while it still looked the same round misty object that it had in March. Comets, on the contrary, moved quite rapidly, grew brighter, and sprouted tails. So what had Herschel discovered?

As 1781 moved into 1782, and the "comet" stayed the same and merely crept along the zodiac, it came to be realized that the Bath organist had done something that no other human being in history ever had: he had discovered a hitherto unknown planet rotating around the sun. Mercury, Venus, Mars, Jupiter, and Saturn had been known since the dawn of recorded history, being bright naked-eye objects, whereas one needed a telescope to see the newcomer. And while modern archival research has led scholars to realize that previous telescopic astronomers had seen the "comet", they had simply recorded it a normal star, and it was not until Herschel's time that it was recognized as a new *planet*. Patriotic as ever, as the news of his discovery began to reverberate around Europe and the new USA, Herschel christened his new planet Georgium Sidus ("George's Star"), after King George III, the sovereign who ruled both Great Britain and Hanover. In the nineteenth century, however, it would be renamed Uranus, after Ouranos, the Greek sky god.

Chapter 6

William and Caroline Herschel Fathom the "Construction of the Heavens" from an English Country Garden

The discovery of a new planet propelled Herschel to celebrity status, and everybody wanted to meet him. The Astronomer Royal, the Revd Dr Maskelyne, invited Herschel to bring his home-built telescope to Greenwich, where they observed together, Maskelyne declaring the superiority of Herschel's telescope to those made by London's best craftsmen. William was also invited to address the Royal Society, which elected him a Fellow on 7 December 1781, while he formed an enduring friendship with Sir Joseph Banks, President of the Royal Society and hereditary Lincolnshire landowner and baronet. And not surprisingly, King George also wanted to meet him.

Historians have often portrayed George III, as the mad king who lost his American colonies. Yet the forty-four-year-old George was far from mad in 1782 when he invited William and Caroline to Windsor Castle. He loved science and had an impressive collection of scientific instruments, and a private Kew Observatory at nearby Richmond. He was also a discerning art collector, accomplished musician, and farmer: a man of intellect and culture, though dementia would cloud the end of his eighty-two-year lifespan.

After the royal meeting, the Herschels gave up music and Bath, and moved to a house in rural Datchet, near Windsor Castle, settling permanently at Observatory House, Slough, in 1786. William was offered the title of "King's Astronomer" (not to be confused with Maskelyne's "Astronomer Royal"), with a pension, not a salary, of £200 a year. Caroline would receive £50 annually as her brother's assistant. A pension was an acknowledgment of past achievement rather than a stipend for continuing work. Being so close to Windsor Castle also meant that William and Caroline could visit the king and royal family to show them interesting objects in the sky and discuss astronomy.

6.1 Caroline Lucretia Herschel, as an old lady. Sadly, no pictures of the younger Caroline are known to have been made, perhaps because her complexion was severely damaged by smallpox. (R. S. Ball, *Great Astronomers* (1895), p. 207. Author's collection.)

William always asserted that his discovery of Georgium Sidus (Uranus) was not due to chance or luck. As he told Dr Charles Hutton in a written account of his own life, he had already begun his systematic sweeps of the night sky, conducted to discover new objects, sometime before he saw Uranus. He would have found the new planet sooner or later; it just so happened that by 13 March 1781 he was systematically sweeping that sector of the sky currently occupied by the new planet.[1] His decision to sweep the heavens was part of his wider project of trying to make sense of how the stellar system was constructed. Yet before he could do this, he needed telescopes of unprecedented power.

William Herschel's telescope technology

The reflecting telescope with which Herschel discovered Uranus was of modest dimensions, even for 1781. It had a metal parabolic figured mirror of almost 6½ inches in diameter and of 7 feet focal length, mounted in an octagonal wooden tube set on a wooden stand made portable by small castor wheels. It was very similar in dimensions to that of John Hadley in 1721. But what made Herschel's telescope so special, and won the praise of the Astronomer Royal and the Royal Society, was the superlative quality of its optics. Herschel wrote and published detailed descriptions of the mechanical parts of his instruments, but he cannily never divulged the secret of how he made his mirrors: largely because it rapidly became a valuable trade secret. Over the next thirty years, it would earn him many thousands of pounds in client commissions.

One must note William's new financial circumstances as King George III's "Astronomer", for while his and Caroline's pensions brought in a joint £250 a year, this was a couple of hundred or more pounds a year less than he had been making from music in Bath. On the other hand, his now celebrity status enabled him to develop a new string to his bow, as aristocrats, rich clergy, country gentlemen, British and European universities, and even Italian and Spanish royalty all wanted Herschel telescopes. These instruments became cultural status symbols in their own right. Large telescopes for the prince of Canino in Italy and the king of Spain earned him £2,310 and £3,150 apiece, while the Museum of the History of Science, Oxford, has a lovely 7-foot mahogany tube reflector with a 6-inch mirror. From his surviving list of prices – including several dozen privately commissioned telescopes – Herschel generally charged over £105 (100 guineas) for a 7-foot telescope. He was a wealthy optical impresario after 1782.[2] But he would not have made every part of every commercial telescope with his own hands. Instead, as was standard in the scientific instrument-making trade, he would have employed skilled out-workers – carpenters, brass and metal fitters – to make all the standard parts. I suspect that his own personal input, similar to that of James Short, would have been largely confined to putting the final, perfect figure and polish on the main mirror, making the critical optical adjustments, and general quality control.

But how would an un-apprenticed "outsider" such as Herschel learn how to make high-quality mirrors in the 1770s? Books, such as Smith's *Opticks*, could teach one a good deal, including how a good telescope mirror had to be made from a speculum alloy of tin and copper, with a few trace substances such as arsenic. One could then either build a modest metal-smelting and casting furnace at home, as Herschel did, or pay a local metal founder to cast the mirrors for you. Such local commercial furnaces were common in the eighteenth century, as iron, bronze, tin, and other "smiths" used them to manufacture metal fittings for a miscellany of useful artefacts, from window catches to anvils, for the local trades. Caroline's accounts tell us they were melting and casting speculum mirrors in Bath, for on one occasion the mould broke, and everyone had to jump to avoid the stream of red-hot metal pouring out onto the flagstone kitchen floor. The 6- or 8-inch metal mirror blank would next, like the lenses described in Chapter 4, be abraded against a convex iron tool, using a mixture of wet abrasives such as fine sand and chalk. When the optical figure was right – after many patient hours of slow grinding – jeweller's rouge would be employed to bring the mirror's concave parabolic figure to a beautiful, brilliant polish. This was not the end: Herschel, no doubt from sheer experience and practice, perfected the skill of making a mirror that was simply exquisite in a way that no book could teach. It was perfected, no doubt, between music lessons and concerts, on long cloudy nights in Bath.

Observing with a Herschel telescope

So how did Herschel go about using one of his telescopes to sweep the heavens? All of his and Caroline's telescopes were mounted in the "altazimuth" plane. Unlike with telescopes mounted on a polar or equatorial axis, one could not track an object in the sky by only applying one single guiding motion; instead, an altazimuth-mounted telescope required two motions to be applied when following a celestial object across the sky. One motion guided the instrument from the east to the west, while another controlled the up and down movements. Large angular movements, up and down, would be performed easily by hand for a 7-foot telescope, but Herschel's great 20- and 40-foot-focal length telescopes required winches, ropes, and machinery, operated by an assistant.

This might look like an unnecessary complication, but for his detailed sweeps of given zones of the sky, Herschel developed an observing technique that overcame it, and yet retained the simpler engineering of the altazimuth mount. Everything in the sky rises in the east and sets in the west, passing through a great arc, most obviously seen in the sun's daily journey. At morning and evening, at rising and setting, the arc is at its most acutely angled with regard to the horizon, but to the south, in and on either side of the meridian, the arc appears relatively flat. Therefore, if most observing were done facing the south, or a little to the east or west of the meridian, it became quite easy to track an object through a small angle of the sky. By using a guiding screw – such as was fitted to all Herschel's telescopes – it was possible to keep an object in the telescope's field of view with little more than one gentle motion of this screw, aided occasionally by a vertical adjustment.

Herschel's sky sweeps were very methodical and carefully planned in advance. Already having a complete familiarity with the summer and winter constellations of the night sky, he would position his telescope facing south. Before even looking into the eyepiece, he would have divided the sky into a series of long, narrow, east-to-west gauges, rather like long ribbons. Each one would fit into an exact place vis-à-vis its fellow gauges, and correspond to precise coordinates on the star charts. The east-west longitudinal or "right ascension" dimension of the gauge would correspond to a given number of degrees of arc on the star maps, and the narrow breadth would correspond to a few degrees of declination, or up and down. To envisage how these gauges all fitted together, imagine a 12-inch-diameter globe of the sky (or earth), perfectly wrapped, from equator to poles, with a series of edge-to-edge bands of ribbon, each a half-inch wide.

Sweeping the sky by this method required relatively little manipulative effort by the observer. Herschel would simply align his south-facing telescope to correspond with a prearranged gauge or zone selected for a given evening's observing, then apply his eye to the eyepiece, and look carefully. The earth's rotation would do the rest, as every star, planet, cluster, and nebula within the selected zone made its stately passage through the telescope's field of view. All that Herschel needed, in addition to an intimate knowledge of the established star patterns, was leisure, patience, and a good overcoat. He was already observing by this method in Bath, and this is what he meant when he told his friend Dr Hutton that his sweeps would have revealed Uranus eventually. It was methodical, systematic observation that added a new planet to our knowledge of the solar system, rather than good luck.

Stars, the Milky Way, and the "Construction of the Heavens" after 1784

As William used his increasingly mighty instruments to methodically sweep the sky, zone by zone, whenever the sky was clear, he was not especially concerned with planets and comets (although sometimes he did observe them), but with deep, stellar space. How were the stars distributed in space? Why were the stars dense in some regions and sparse in others? What were these fluffy patches of light, the nebulae (or what we now call galaxies)? Why did stars sometimes form clusters of varying degrees of density, such as the Pleiades (a fairly scattered open cluster) and the intensely compacted Hercules cluster, which was listed as number 13 in Charles Messier's catalogue of comet hunters' nuisances? Why were some stars so close together as to appear double? Was one star in the pair closer to us than the other – due to a line-of-sight effect – or were they a true gravitational pair, rotating around each other? Most baffling of all, what were those objects that revealed distinct pale glowing bluish-white discs, yet which must, owing to the absence of any geometrical parallax, be in the deepest regions of space: the planetary nebulae? How vast must these bodies be, subtending such large angular diameters yet being so remote?

As we have seen, Herschel was not the first astronomer to be interested in these various objects. But where he was a major innovator was in his methodical technique of collection – by sweeping each gauge of sky visible from Bath or Slough. Once a gauge had been swept and harvested, its population would be counted, quantified, classified, and arranged. Herschel's observational technique had close parallels to those used by the field botanist. For just as Sir Joseph Banks, P.R.S., who had sailed to Polynesia with Captain Cook in 1768, had studied the fauna of Tahiti by collecting and classifying different species of plants and animals to build up a natural history of the Pacific island chains, so Herschel did likewise with the strange population of deep space. Before the rational understanding of any natural phenomena can commence, a great body of facts must be collected, classified, and made sense of: stars and nebulae no less than flowers and butterflies. So, one might say, William and Caroline Herschel worked from an English country garden to study the exotic fauna of the celestial garden.

By 1784, this sweeping technique was already revealing some fascinating results, as each zone of the northern heavens had been meticulously gauged and measured at least once. On 17 June 1784 Herschel read a paper of monumental significance before the Royal Society entitled "Account of Some Observations Tending to Investigate the Construction of the Heavens".[3] In it, he attempted to suggest a structure – or "Construction" – for the galaxy of stars of which we are a part. Unlike the cosmologies of Wright, Lambert, Kant, and others, however, his cosmology was founded less upon a priori philosophical principles and more upon systematic, telescopic observation, although inevitably, like the cutting-edge cosmologies of today, it was not free from speculation. While early gauges may have been made with smaller telescopes, he tells us that by 1784 he had perfected a great Newtonian reflector of 20 feet focal length with a polished speculum mirror of 18.70 inches in diameter. It was set up in the garden at Datchet in "a meridional situation" and used in the manner outlined above.

In his great 1784 paper, Herschel proposed that the starry heavens stretched across space in a relatively flat striatum, or plane, which appears to us as the Milky Way. The sun and our solar system are probably at or close to the centre. Therefore, when we look horizontally, as it were, our eyes are carried into the densest mass of stars, receding to a glowing infinity all around us. When we look up or down, or out of the plane of the striatum, however, we see fewer stars. So far, Herschel's idea has parallels to the cosmology of Thomas Wright, but it is now borne out primarily by observation. Having no concept of dark, light-absorbing dust and gas clouds in space in 1784, Herschel then suggested that those dark bands within the Milky Way that appeared star-less resulted from the striatum having a break or split in it. (The recognition that space was not pure and transparent, but contained light-less matter that obscured those stars that were behind it, would have to wait more than another century.) Having addressed the physical structure of the stellar universe in his great paper of 1784, Herschel's subsequent cosmological papers dealt primarily with the strange population of objects that occupied it in addition to straightforward stars.

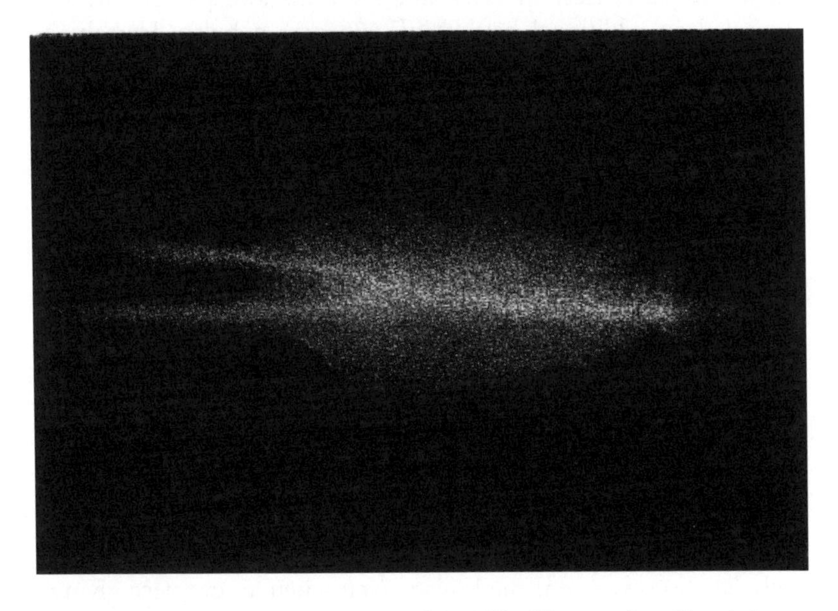

6.2 William Herschel's "Construction" of the Milky Way, or galaxy. Herschel placed our solar system at the centre. The split he depicts on the left-hand side is now known to be an area covered by light-obscuring matter. (*Philosophical Transactions* 75 (1785); redrawn in J. P. Nichol, *Views of the Architecture of the Heavens* (1843), p. 19. Author's collection.)

"OH HERSCHEL! OH HERSCHEL! WHERE DO YOU FLY? / TO SWEEP THE COBWEBS OUT OF THE SKY" [4]

Although this amusing nursery rhyme variant was written in the early nineteenth century by the Revd Professor Adam Sedgwick about his Cambridge friend Sir John Herschel, it applied with even more force to John's father, William; both father and son were concerned with "sweeping the cobwebs out of the sky" – the cobwebs being the nebulae. We shall meet Sir John again in Chapter 10, but what about the cobwebs?

William Herschel had been given a copy of Messier's new catalogue of 103 nebulous objects by his friend Alexander Aubert soon after its publication in 1781 (in the *Connaissance du Temps* for 1784), yet by the time of the "Construction" papers, presented to the Royal Society in 1784 and 1785, his sweeps had already revealed an additional harvest of 466 new nebulae. Herschel's great 20-foot telescope with its superb 18.70-inch mirror had shown that many objects in Messier's catalogue were, in fact, dense star clusters. As the decades rolled on, and the sky came to be swept and re-swept with relentless thoroughness, Herschel would eventually clock up a list of well over 3,000 new nebulae and star clusters, which would form a bedrock of observed empirical data that would serve cosmological astronomers right through the Victorian age.

By the mid-to-late 1780s, Herschel had come to realize that, like Messier's objects, the nebulous cobwebs in the sky appeared to be of several distinct types. Some of them were composed of dense masses of individual stars that, in a less powerful telescope, emitted a combined generalized glow – as do the combined millions of light bulbs of London, Paris, or New York on a space satellite night-time image of the earth. Other nebulae, however, remained misty even when an eyepiece giving a thousand times magnification was used on the 20-foot. Other non-compacted nebulae were composed of glowing filaments of hairy stuff, luminous "chevelures" as they were sometimes styled, or "true nebulosity". Being at the cutting edge of both science and language, one sees Herschel struggling to find the right words to describe what he was seeing. But if there really was "true nebulosity" or "shining fluid" in the depths of space, what could it be made of, what was the source of its pale light, and what physical laws could it obey? Herschel's nebulae opened up a whole series of questions in contemporary physics and chemistry. In 1785, even a chemically specific gas was a new and by no means universally acknowledged concept in the wake of Antoine Lavoisier's recent chemistry researches. No one had any idea how the handful of then known gases, such as "oxygene", nitrogen, hydrogen, and carbon dioxide, might behave in the infinite vacuum of deep space, or how they might emit light.

Could the nebulae be composed of individual reactive atoms, as terrestrial gases were assumed to be, or did they consist of something so insubstantial as the glow of the aurora borealis, or the electrical flashes of light discharged from a Leiden jar? Atoms, if physical entities, should be susceptible to gravitational attraction; what conceivable laws might lie behind electric glows in deep space? Newton's laws were demonstrably "universal" within the solar system, but did they still operate in the Orion Nebula or Hercules star cluster? Herschel's big telescope cosmology was throwing up problem after problem for eighteenth-century science, and so speculations came to be built upon observations. The deeper Herschel probed into space, the more bizarre it appeared to be.

"Shining fluids", glowing rings of light, star clusters, and gravity: the Herschelian universe

Over the two or three decades following his Construction paper of 1784, Herschel described a universe that was as strange and unfamiliar as the fauna of any newly discovered South Sea island. Not unlike natural fauna in island chains, his sweeps had indicated that nebulae and clusters did not generally occur in isolation, but in strands or clumps of varying density, sometimes extending through several degrees of arc of sky – or billions of miles of space. In fact, he likened them to rock strata in the earth's crust. This then begged the question of whether the nebulae were physically associated with the individual bright stars that often appeared in the same field of view, or whether they were vastly more remote, and only appeared adjacent to the stars due to a line-of-sight effect. The bright nebulae in Orion (Messier 42) was a case in point, as the clouds of shining fluid nebulosity lay in the same field as numerous individual stars.

One of the puzzles that haunted William Herschel, as it would his son John, was whether insubstantial "shining fluid" could actually exist. If the physical creation were composed of matter, and by extension, of billiard-ball-like atoms, then shining fluid was a logical contradiction. A likelier explanation was that the nebulae were dense star clusters and bands of packed stars, just like the densest regions of the Milky Way, yet were so remote that only their overall combined glow was visible. If true, one was faced with the prospect of an even greater cosmological vastness than that of the self-contained one big galaxy predicated in Herschel's 1784 Royal Society paper. Were the nebulae and dense clusters part of our big Milky Way galaxy, or were they separate from it?

A further puzzle implicit within Herschel's big telescope cosmology was gravitational action in the depths of the stellar (as opposed to planetary) universe. It was reasonable to assume that gravity did operate in stellar space, yet it would not be until 1830 that Félix Savary in France and Sir John Herschel in England demonstrated this to be so, from the behaviour of double star systems. By 1830, several of the close-pair double stars discovered by William Herschel thirty or forty years before showed signs of measurable angular rotation around each other. Knowing the period that had elapsed between two measurements – made, say, in 1800 and 1825 – it was possible to calculate not only their precise gravitational relationship, but also the mass of each star.

So by 1830, gravity was demonstrated as a truly universal force, applying to the remotest star systems just as it did to the moon and planets. This realization would give a powerful analytic tool to scientific cosmologists. Whether gravity applied to the nebulae would remain a moot point until the late 1840s. Although both William and John Herschel classified several thousand nebulae and clusters they jointly discovered in the northern and southern hemispheres between 1780 and 1838, their classification was based only on purely visual criteria. Some Herschelian nebulae, for example, were diffuse, others concentrated, while others were round, elliptical, or elongated. But it would be Lord Rosse, with his 72-inch-mirror telescope of 52 feet focal length brought into use in 1845, who discovered that a dozen or so nebulae resembled "whirlpools" or spirals, which suggested a gravitational rotation, as we shall see in Chapter 14. The most baffling were the so-called "planetary nebulae", suggesting from their perfectly spherical shape and large angular diameters that they must be as big, if not bigger, than our own solar system, yet vastly remote. (We now know that they are glowing shells of gas, expanding outwards from an anciently exploded supernova star.)

One such object especially intrigued Herschel: that pale ring of light with a dim star at its centre in the constellation of Taurus, known to modern astronomers as NGC (New General Catalogue) 1514. What could that ring be made of and how could it glow, especially as its central star was so dim and clearly incapable of illuminating it? We now know that it too was a supernova remnant, the dim central star being the exhausted cinder of its former self.

Virtually all of the above discoveries were made with William Herschel's 18.70-inch, 20-foot-focus telescope. King George III, however, had granted Herschel £2,000 with which to build a truly giant telescope with a 48-inch-diameter mirror of 40 feet focal length, which came into use in his Slough garden in 1789. In spite of its theoretically massive power, though, the great instrument failed to yield the momentous results expected from it, primarily because, by the engineering standards of 1789, it was simply too cumbersome to use effectively. It was too far ahead of its time, and it would take good half-century until Victorian heavy engineering technology made such vast structures viable, as we shall see in Chapter 14.

6.3 William Herschel's 40-foot-focus reflecting telescope with its 48-inch-diameter mirror, 1788–89. (J. N. Lockyer, *Stargazing* (1878), p. 294. Author's collection.)

Observatory House, 1784: an account by a visiting French savant

Barthélemy Faujas de Saint-Fond from France visited England and Scotland in 1784. He was hosted by Sir Joseph Banks, and met everybody who was anybody in science. After dining in London with Sir Joseph on a fair early autumn evening, he set out at 7 p.m. to undertake a three-hour carriage drive, along the excellent 20 miles of road to Windsor, to visit the Herschels.

6.4 Observatory House, Slough, from the garden. (R. S. Ball, *Great Astronomers* (1895), p. 212. Author's collection.)

What makes his account of his *Journey* so fascinating is the incidental detail he includes. His journey was made on a Sunday evening when the roads were safe, and free from highwaymen. This was *not* because the highwaymen were a pious bunch and singing hymns in church, but that the Sunday evening road was far too busy with traffic to enable them to pick on isolated carriages: Londoners of "all ranks, who, having passed the day in the country, return at night to London, to be ready to resume their usual occupations on Monday morning". The sheer prosperity, courtesy, and freedom of England deeply impressed Faujas, as it did many other Continental visitors, for "it certainly showed a degree of wealth and extent of population of which one has no notion in France".

Faujas arrived at Datchet (he mentions nearby Slough, although Herschel did not move to Slough until 1786) at 10 p.m. Upon being shown into the house by a servant, he was surprised to find "a young lady seated at a table" in a room "adorned with maps, instruments of astronomy… and a large harpsichord". This was Miss Caroline Herschel. She had by her a copy of Flamsteed's great Latin star atlas, *Historia Coelestis Britannica* (1725), a pendulum clock, and a device by which her brother outside at the telescope could signal astronomical information to her, using a dial controlled by strings. Faujas was then introduced to William, who showed him the "ever-memorable telescope with which the eighth planet Georgium Sidus [Uranus] was discovered".[5] Faujas felt deeply honoured, not only by his warm reception by the now world-famous Herschel, but by being allowed to spend two full hours observing the heavens with this iconic telescope.

Then, after being shown other instruments, Faujas was taken to Herschel's very new 20-foot reflector with its 18.70-inch mirror. A "domestic" worked the basic tracking apparatus, and Faujas was shown some double stars, along with a bright star "in the foot of the Goat" (Capricornus), the light of which, when projected from the eyepiece onto a piece of paper, made it possible discern the lines of small letters written upon it. (As the stars in the "foot" of Capricornus are quite dim, one suspects that Herschel showed him the bright variable star Deneb Algedi, in the Goat's tail.) The French savant was also amazed to see Jupiter "much larger than the full moon" and the belts beautifully displayed. Herschel was clearly a very gracious host and patient teacher, who showed and explained everything to Faujas, for Herschel's "boundless complaisance was never wearied by my ignorance, and the importunity of my questions" right through to dawn. Faujas then left for London at 8 a.m. to meet Sir Joseph Banks, who would show him Kew Gardens. He supplies us with a delightful glimpse of the Herschel astronomical establishment at the outset of the astronomer's fame.

Sir William Herschel, Knight Guelph

By long custom, Dr William Herschel (with his honorary doctorate from the University of Glasgow, 1791) has been referred to as *Sir* William. Modern archival research has shown that this may not be a correct title, for the knighthood conferred on Herschel by the

Prince Regent in March 1816 was not an English but a Hanoverian title. Herschel was a civil (i.e. non-military) Knight of the Guelphic Order, third class. This would have given him a status in Great Britain's Hanoverian territories, but it would not have conferred upon him the legal title of "Sir" in England.[6] Irrespective of his formal title during the last six years of his life, from the ages of seventy-eight to eighty-four, his standing as the first great deep-space observer and cosmologist remains wholly unblemished. William and Caroline would become the astronomers of the Romantic age, to which we will return in Chapter 9. But first, we must look at the eighteenth-century attempts to establish the precise dimensions and gravitational characteristics of the solar system, along with the true shape of the earth. These would give rise to a series of explorations and acts of extraordinary courage that would make Newtonian gravitation truly global.

A Herschel telescope postscript

In the summer of what I think was 1970, I had the pleasure of handling what I understand was the last Herschel telescope in private hands. It was owned by Alan Sanderson of the Liverpool Astronomical Society, a retired banker, amateur astronomer, microscopist, and model railway enthusiast, to whom I had obtained an introduction from Alan Whittaker of the Salford and Manchester Astronomical Societies, who was himself also a retired bank manager. I have clear memories of my arrival at Mr Sanderson's large house at Hunt's Cross, Liverpool, where he lived alone as a widower. The house was really a museum, full of telescopes, microscopes, a large Bassett-Lowke model railway layout, and piles of books everywhere: on shelves, on the treads of the wide staircase, and even on top of the grand piano. The "star" exhibit was a still-working 7-foot Herschel telescope. Although I was visiting on a summer's afternoon, with no stellar objects visible, I recall looking through it at some distant trees and obtaining quite a decent image. Alan Sanderson told me that he had bought it in the 1920s or 30s, following a dinner party at a house in Liverpool. When handed his umbrella on leaving, he noticed that the "umbrella stand" in his host's hall was really the mahogany stand of a Herschel telescope. He asked his host whether he had any more parts from it. Receiving an answer in the affirmative, he asked to see them. He said he paid his host £5 for their "umbrella stand" and its ancillary bits: two weeks' wages for an office clerk in 1930. I understand that following his death some years later, the telescope was acquired by the National Maritime Museum, Greenwich.

Chapter 7

Measuring the Heavens and the Earth in Eighteenth-Century Europe Part 1: In Pursuit of Venus: Astronomy's First Great International Adventure

We saw in earlier chapters how Edmond Halley was one of the first scientists to undertake long-haul and sometimes dangerous expeditions to far-flung parts of the earth, in pursuit of new factual data. It had been Halley on St Helena in 1677 who first realized that the transits of Mercury and Venus across the disc of the sun could be used to calculate the earth–sun distance, or "Astronomical Unit". This measured distance, through the application of Kepler's Three Laws of Planetary Motion (1609–18) and Sir Isaac Newton's Law of Universal Gravitation (1687), could then be used to calculate the distances of all the planets, to give firm dimensions to the solar system.

It had also been Halley who realized that the slower-moving and larger Venus, when seen in transit across the solar disc, would be a better planet for determining the "solar parallax" than small, swift Mercury, although Venus transits were very rare, and there would not be another until 1761. (Venus transits occur in pairs, eight years apart, once every 112 or so years. The first observed transit had been seen by Jeremiah Horrocks in Lancashire in 1639.[1]) Halley had given wide publicity to the forthcoming transit of 1761 long before his death in 1742, while Joseph-Nicolas Delisle in France, an admirer of Halley and Newton, would take up the cause in the eighteenth century. What was the "solar parallax", and how could one use it to measure the distance of the sun?

IN PURSUIT OF THE SOLAR PARALLAX

Everything hinged upon geometry and the critically accurate measurement of angles – both of the track of Venus across the disc of the sun and of each individual astronomer's observing location upon earth. For the 1761 and 1769 transits, these observing stations would be as far-flung as South Africa, Tahiti, Siberia, Harvard, southern Canada, South America, Lapland, India, and all across Europe. The theory was simple, and no more in its basics than the upscaling of the techniques by which a land surveyor or military engineer measured terrestrial distances for making a map or directed a cannon to hit a distant target. Let us begin with a terrestrial measurement. If a surveyor or a gunner laid off a straight line on the earth's surface – say a line of 100 yards in length – and from each end in turn he viewed an object, let us say 500 yards away, then his two positions would produce an angular displacement. When reckoned against a fixed standard, such as astronomical or magnetic north, he could easily employ the two angles to construct a set of triangles from which the object's distance could be calculated with great accuracy.

To measure the solar distance one simply scaled up this technique. Let us say one had an astronomer in Greenwich, England, at the latitude 51° north, and another at the Cape of Good Hope, at the tip of South Africa, at 33° south. If one knew the exact diameter and circumference of the earth (and both were known with considerable accuracy by 1760), one could easily compute the length of a notional straight line passing through the earth and connecting the two – in this case, just over 5,400 miles apart. One could calculate similar lines between, say, Tobolsk in Siberia and the Cape, or India and South America. All were easy to compute when one had established the exact latitude and longitude of the observing stations, which was an easy task for an accomplished astronomer.

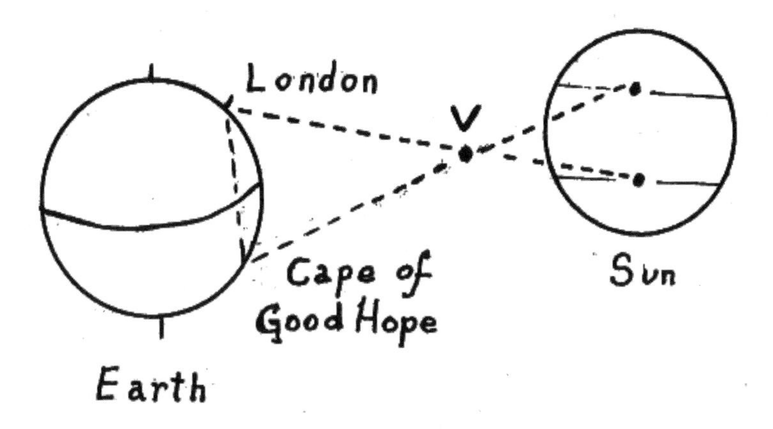

7.1 Using Venus to measure the solar distance. Astronomers in London and South Africa respectively would see Venus (marked with a "V") occupying slightly different positions on the surface of the sun. Having calculated the length of a notional straight line passing through the earth and connecting the two locations, and measuring these positions with great accuracy, they could use them to triangulate the distance of Venus. And once that was known, Kepler's and Newton's Laws could be used to work out the Venus–sun, and hence earth–sun, distance. (Drawing by Allan Chapman.)

So just think of these lines running between places on the earth as ends of a long surveyor's baseline with which to measure the solar system.

Now what an astronomer at each end of such a baseline had to do was threefold. The first task was to observe, as exactly as local conditions would permit, the very second when the dark, round, silhouette of Venus first bit into the solar disc. These were the "first" and "second" contacts. Then its passage over seven hours or so should be recorded, as it crossed the sun, then similarly when it left the disc, at the "third" and "fourth" contacts. Secondly, the exact solar latitude position was recorded, between the sun's equator and pole, at which Venus transited the sun. Thirdly, an exact timing would be made, with an accurate pendulum clock, of how long Venus took to transit.

All of these observations would have geometrical features unique to each position, be it Paris, India, or Cape Town. When all the observations had been made, and the astronomers had all returned to Europe, their results were published, to be passed around internationally. From all this data, gathered from north, south, east, and west, skilled mathematicians could calculate the distance of Venus from the earth. Once the earth–Venus "yardstick" had been established, then Kepler's and Newton's Laws could be used to compute the Venus–sun distance, and from that, the distance of every planet in the solar system. This was done by reckoning each planet's already-known precise orbital characteristics, and computing the proportions using Newtonian gravitational criteria.

The Venus transits of the 1760s also tell us something about the nations of Europe in previous centuries. The seventeenth and eighteenth centuries were times of spasmodic wars between the nations of Europe, and especially between Britain and France, but these wars rarely had anything to do with wider cultural values. Whether you were English, French, Polish, Italian, or German, you all shared a common base culture. (To a very large extent, we still do.) This was rooted in the political ideals and practices of ancient Greece and Rome, and the ethics of the Judeo-Christian religion. A Catholic might sometimes shoot a Protestant in a commercial or dynastic war, and vice versa, yet nonetheless Europeans shared the wider bond of Christendom. Europeans also shared the same literature, philosophy, fashions, music, art, and science. In the same way that a well-paid Dutch Calvinist artist would gladly paint a French Catholic aristocrat, so the astronomers, scientists, and doctors of Europe enjoyed open and often warm friendships that bridged nationhood and religious denomination. The thirteen colonies on the eastern seaboard of North America, then the United States after 1776, similarly shared this European culture.

None of this meant that Europe's and America's ladies and gentlemen of learning and culture were not patriotic, and loyal to their monarchs and Christian denominations. It just meant that they acknowledged common bonds, and could cooperate on matters of shared cultural interest, such as astronomy. Such cultural ease and confidence is sometimes hard for us twenty-first-century people to appreciate, with our recent recollections of modern wars of hate and brutal ideologies. This explains how the often war-torn Christendom of eighteenth-century Europe could allow such successful cooperation in the organizing, observing, and analysis of the Venus transits of the 1760s, and how a Spanish Jesuit and an English Anglican astronomer could work together. The fortunes of war at sea, however, often made getting to selected observing stations difficult, especially as in 1761 Europe was locked into the Seven Years War.

Venus in transit, June 1761

In early May 1760 in France Joseph-Nicolas Delisle, inspired by Halley's earlier publications and his own chart for the Mercury transit of 1753, had published his mappemonde, a world map showing the best global locations from which to observe the forthcoming 1761 transit.[2] The sheer international scale of the transit observations Delisle envisaged is evident from his surviving mailing list of persons and institutions to whom he sent copies of his mappemonde: some 200 in all. In total, expeditions from most European nations, along with colonial America, set out in pursuit of Venus across four continents in 1761. Yet this is where the polite deliberations of the savants and the meticulous work of the observatory hit the realities of wars and storms. While both the Royal Society and the French Académie had issued their expeditions with Safe Conduct Licences, to be shown to potential belligerents on either side, they counted for little on the high seas and distant shores, where prize money could be made by daring sea captains seizing enemy vessels.

Take, for example, Charles Mason and Jeremiah Dixon, who were contracted by the Royal Society to sail to Bencoolen, India, to observe the transit. Their ship, HMS *Seahorse*, had scarcely left Plymouth before encountering the French vessel *Le Grand* in the English Channel, and a ding-dong sea fight ensued. It ended in a draw, with many dead on both sides, and the battered *Seahorse* limping back into port. Mason and Dixon had been scared out of their wits and refused to ever go to sea again. Only when the Royal Society threatened legal action did they agree to undertake the shorter voyage to the Cape of Good Hope. Mercifully for them, the delay had made it impossible to catch the Indian Ocean winds – essential in the days of sail – necessary to get them to Bencoolen by June 1761, so South Africa was a compromise. Fortunately, Mason and Dixon succeeded in travelling without mishap on their second voyage, and a successful transit observation was made, followed by a safe return home. Having got their sea legs, Mason and Dixon put to sea again: Dixon to Hammerfest and Mason more locally to Ireland in 1769. They were to win geographical immortality in North America for their survey of the geodetic line between the states in the 1760s.

On the opposite side of the English Channel, the French Roman Catholic priest and Augustinian canon astronomer Alexandre-Gui Pingré, assisted by Denis Thuillier, suffered at the hands of the English. Both had enjoyed a safe voyage out to their selected observing station, the island of Rodrigues in the Indian Ocean, and secured a good observation. Their troubles began when an English privateer (or an officially licensed wartime pirate) Captain Robert Fletcher hove into view. Paying scant attention to the Royal Society and British Government official *passeport* that Pingré produced, Captain Fletcher treated the Frenchmen most "uncivilly", looting their possessions, then burning their ship. Stranded on the island, they had to await the arrival of a French vessel to begin their voyage home to Europe, where they once again were attacked on the high seas by a British vessel. Both men finally arrived safe in Paris in the spring of 1762.

Whereas Mason and Dixon, and Pingré and Thuillier travelled to observing stations deep down in the southern hemisphere, Jean-Baptiste Chappe d'Auteroche journeyed overland to the far north. His destination was Tobolsk in Russian Siberia, and although it was no further north than Scotland's Shetland Islands, its importance as an observing site came from its easterly longitude and the short duration of Venus's complete passage across the solar disc. Instead of taking a direct route, via Holland, Germany, and Poland into Russia – or taking a ship directly to St Petersburg – the moving theatre of the Seven Years War obliged Chappe to describe a great arc down through south central and Eastern Europe, making a great geographical sweep through Vienna, Warsaw, Krakow, St

Petersburg, and Moscow. As he had to be at Tobolsk, at 58° north, before 6 June 1761, this inevitably predicated a winter journey. Chappe had many adventures on the way, most notably a night-time incident when travelling by sledge in snowbound Russia.

Waking in the night, he found himself alone in the makeshift camp, his guides and sledge-men having disappeared, and with hungry wolves howling nearby. Arming himself with a pistol, he began to trace the footprints of his vanished guides. The incident had a happy ending, for the men had simply been invited to a party in a nearby village and were enjoying themselves. They carried on next day to Tobolsk. Finally, as they approached Tobolsk from the west, they were faced with crossing the frozen River Tobol, in which, as the season was well into spring, the ice was beginning to crack and melt. Either they would have to wait until the river had melted sufficiently to cross by boat – and no doubt be so delayed as to miss the transit – or take a gamble. Chappe had the sledges driven at full speed across the frozen river, the ice cracking beneath them, and bits of it sinking under the water. They reached Tobolsk in one piece, just before the ice broke completely.

The governor and archbishop of Tobolsk treated Chappe splendidly, even providing buildings and a protected observatory compound. But the local folk regarded him with suspicion, believing that his celestial observations were being used to invoke spirits and that he was responsible for severe local flooding. In the few precious days between his arrival and the Venus transit on 6 June, however, he was able to accurately measure his latitude as well as establish his longitude east of Paris, by using the motion of Jupiter's moons, whose longitude-finding uses will be described shortly. In this isolated northerly part of Russia, he made geomagnetic, meteorological, and other observations, including swinging a special pendulum to discover the strength of the local gravitational field (see the next chapter). Finally, he enjoyed a perfect sighting of the transit, followed by a safe return to Paris.

The most ill-starred of all the 1761 long-distance travellers was another Frenchman, rejoicing in the glorious name Guillaume Joseph Hyacinthe Jean-Baptiste Le Gentil. He left France in March 1760 to give himself plenty of time to get to Pondicherry, India, for

June 1761. But the hazards of war intervened. The ship carrying him had to dodge several British men-o'-war on their outward voyage, although they found some respite when calling into the French Île de France islands of Mauritius in the Indian Ocean. Between the Île and India, however, they learned from other vessels that Pondicherry was about to fall to the British, and were obliged to return to Mauritius. At anchor off the Île de France, Le Gentil's ship was destroyed by a hurricane. Hearing that French reinforcements were on their way to retake Pondicherry, Le Gentil obtained a passage on *La Sylphide*, only to find the retaking of Pondicherry had failed and the Union Jack flew over the fort, causing *La Sylphide* to put back to sea. On 6 June 1761, Venus transit day, Le Gentil enjoyed a brilliant, clear, sunny sky: the only snag was that he was obliged to stand helpless on the quarter deck of *La Sylphide* as she gently pitched and rolled in a calm sea. The slightest movement of the vessel made it impossible to use a telescope, let alone set up a pendulum clock with which to obtain a precision timing.

Le Gentil, Pingré, and other astronomers all reported using newly observed and computed sets of lunar tables – giving the moon's position against the stars – as a way of attempting to determine the longitude of the vessels in which they were sailing. This lunars method of longitude-finding, which we encountered in Chapter 3, went back in theory to Petrus Apianus in 1524, though it would not be until the mid-eighteenth century that astronomical technology and precision angle-measurement had reached a sufficiently high level of perfection as to make it practicable at sea. We will return to navigation and accurate latitude and longitude-finding later in this chapter.

In total, the Venus transit of 6 June 1761 was measured and reported by some 120 observers worldwide, from France, Britain, Germany, Spain, Russia, and colonial America. I am not aware of any women reporting an observation from 1761, but eight years later Countess Macclesfield, wife of the President of the Royal Society, would report their joint observations of the 1769 transit from their Shirburn Castle Observatory in Oxfordshire.

Let us conclude this section with more about the adventures of Le Gentil. Having failed, through sheer hard luck and the chances of war, to make a proper scientific observation of the 1761 Venus transit, he was determined to catch that of 1769. Rather than return home to France and risk further travel problems, he decided to spend the intervening eight years in the Far East, which he put to good scientific effect, collecting a large body of data about natural history, oceanography, ethnology, and other subjects. All these eighteenth-century astronomers were *gentlemen of science*, equally at home making an astronomical observation, dissecting some exotic creature, learning a hitherto unknown native language, studying ocean currents and meteorology, or collecting plants and butterflies. They were also gentlemen of wider culture, being able, depending on their talents, to play the harpsichord, judge the merits of a painting, or recite Homer in the Greek from memory.

Venus transits the sun in 1769

When all the observations from 1761 had arrived safely back in Europe, and been printed and circulated among the scientific community so that the business of calculation and collation of results could begin, the overall response was one of disappointment. Instead of a reduced body of observations producing a precise figure for the solar parallax with only a tiny margin of error, it soon became apparent that the diversity of results was too wide to permit the computation of a definitive parallax. The 1761 transit observations gave solar parallax values that ranged between 8.28 and 10.6 arc seconds: all tiny angles, yet resulting in solar distances that varied to the extent of several millions of miles: an unacceptably large margin of error by the high accuracy standards of the 1760s.

The main culprit, it was generally agreed, was the unexpected appearance of what came to be styled the "black drop". When the black disc of Venus either entered or left the brilliant solar edge, instead of two nice sharp black surfaces crisply separating from each other to enable the astronomer to make an exact timing of the contacts, a peculiar black blob formed between Venus and the darkness beyond the sun's limb. Timings could vary considerably depending on the size and duration of the drop for a given location. There was no "black drop" in reality, but no one, in 1761 and 1769, realized that Venus had a very dense (now known to be highly toxic) atmosphere, and that sunlight passing through it was refracted, thus producing a black blurring effect.

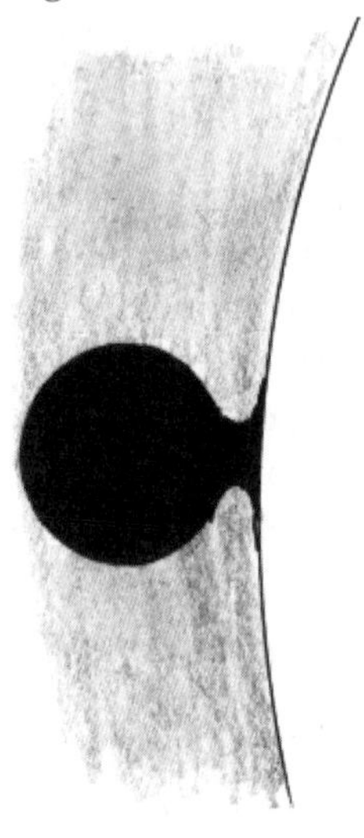

7.2 The "black drop". As the limb, or edge, of Venus approached each contact with the sun's edge, a black filament would appear, making it hard to measure the four contacts to the exact second. This was one of the first indications that Venus possessed a dense atmosphere, which we now know refracts and distorts the light around the image of the planet. (Drawing by Allan Chapman.)

Some 150 astronomers reported their observations for the Venus transit of 3 June 1769, and while the majority were in or around Europe, there was, nonetheless, a series of mainly French, British, Russian, and Spanish expeditions covering the earth's surface from which the transit would be visible. As the 1769 transit was only properly visible in the northern hemisphere, however, all the observing stations were north of the equator. The wide spread of observing stations was to increase as much as possible the length of the survey baselines, especially in an east–west direction. Not all observing stations would see the entire transit: some would see the "ingress" (or start, as Venus entered the solar disc) and others the "egress" (or exit from the solar disc). Observers down in the south, such as India, would see Venus at a different latitude on the solar disc from those in St Petersburg. Of supreme importance was accuracy of angle measurement and timing, irrespective of one's station. When all the observations were collated back in Europe, an accurate total parallax value could hopefully be computed. At least that was the theory, and being forewarned about the "black drop" and building upon the collective experience gained in 1761, astronomers had high hopes for 3 June 1769. They enjoyed an advantage not known in 1761: peacetime. By 1769, France and England were at peace, and while no one could avoid risks occasioned by storm, disease, or natural hazard, at least no one was worried about naval or military attack. Both the scientists and the political and military personnel were all on good terms.

Alexandre-Gui Pingré had recommended that for the 1769 transit, an expedition should be dispatched to the South Seas – some 170–180 degrees across the world from Europe, giving a baseline several thousand miles long east to west, and made even longer when the observations of the Russian expeditions to Central Asia were weighed in. So who were the intrepid men who set out to pursue Venus in distant places in 1769?

In contrast to his cold 1761 Siberian location, Chappe d'Auteroche was dispatched to sunny California in 1769. All went well, his expedition sailing across the Atlantic and Caribbean, crossing Mexico overland, then going by sea to California. Excellent results were obtained – then fever broke out, killing Chappe and most of his companions. Only the expedition's engineer got back home to France, carrying their observations with him.

Then the curiously named Hungarian Jesuit priest Father Maximilian Hell, at the request of the king of Denmark, was dispatched to Vardø, on the Barents Sea, in the chilly North Cape Arctic Circle of Norway, where, thanks to the far-northerly midnight sun, he was able to observe the entire approximately 8-hour transit from start to finish. Father Hell obtained good results, and made it safely home. The Englishmen William Wales and Joseph Dymond received £200 apiece to take an expedition to Hudson Bay, Canada, while Jeremiah Dixon and William Bayly went to observe in Hammerfest, also on Norway's polar North Cape, not far from Vardø. Father Pingré, who had such a hard time in Rodrigues in 1761, now led a French team to Cape Francis on the Caribbean island of St Domingo (Haiti) and secured good results.

Perhaps the most famous of the 1769 expeditions was that led by Lieutenant (later Captain) James Cook, RN, with the astronomer Charles Green and the adventurous naturalist and Lincolnshire baronet Sir Joseph Banks P.R.S. and Banks's Swedish physician friend Dr Daniel Solander FRS Sailing from Plymouth on 25 August 1768, Cook's HMS *Endeavour* enjoyed a safe and speedy passage around Cape Horn and up to the island of Tahiti: an island in an archipelago only discovered a few years before, and christened the "Society Islands", after the Royal Society. Tahiti, which the romantics of the age subsequently came to think of as a tropical island paradise, with its friendly natives and spectacular land- and seascapes, was so far around the globe from Europe that it was said that if you sailed any further, you would start coming back. To make sure that stray clouds did not spoil the contacts as the two sides of the disc of Venus entered and exited the solar limb, Charles Green, the official astronomer, Captain Cook, Sir Joseph Banks, and Dr Solander made independent observations. They used separate sets of instruments on different island locations. While they were fortunate in seeing the transit, the British and other expeditions were still bedevilled by the appearance of the "black drop". Green recorded a 48-second "drop" and Cook a 32-second one,[3] while most of the other astronomers recorded various periods when it caused the loss of vital seconds.

Despite this, the combined results for the 1769 transit gave a much narrower spread of error than that of 1761, covering a range between 8.43 and 8.80 seconds of arc. This narrow error spread, falling within tenths or hundredths of a single arc second, emphasizes how accurate angle-measuring had become by the 1760s, even for astronomers working in makeshift and temporary observatories in Tahiti, Siberia, or the Caribbean. From the various mathematical reductions made of the 1769 transit data, eighteenth- and nineteenth-century calculators came up with earth–sun distances of around 92,796,950 to 93,000,000 miles.[4] These are very close to the modern value, obtained by other methods, where the error-spread is between 92,960,000 and 93,000,000 miles. What made these figures possible was a constantly advancing precision instrument technology. But before looking at the instruments needed to observe both sets of Venus transits, let us return to Le Gentil and his post-1761 adventures.

LE GENTIL AND THE 1769 TRANSIT

Choosing, as we saw above, to spend the intervening eight years in the East, Le Gentil did original work on many subjects, including Brahmin astronomy, much of which he reported back to Paris. For the 1769 transit, he had considered observing from Manila in the Philippines, but at the behest of Parisian friends he returned to his original site of Pondicherry, India, as the city and region had been returned to France in 1763 as part of the Treaty of Paris post-war settlement. Le Gentil arrived at Pondicherry in March 1768, giving him well over a year to make his preparations. The new French Governor, Jean Law de Lauriston, of Scottish descent, was hospitable, even obtaining from the British commercial community in Madras a fine 3-foot telescope with one of the new high-definition "achromatic" object glasses, the type developed, as we have seen, by John Dollond (see Chapter 4) just before Le Gentil left Europe. Governor Jean Law also had an observatory constructed for Le Gentil on the old Fort, even doing some observing with him, and treating him as an honoured guest. Le Gentil commented upon the pure, still, transparency of the air in Pondicherry, which was perfect for astronomical observation, and 3 June 1769 was preceded by a long run of beautifully clear days and nights.

At an observing station so far to the south-east as Pondicherry, however, it would only be possible to observe the end or egress of the transit: this is why so many North American, Canadian, and high-latitude Scandinavian stations had to be selected for 1769, where they stood a good chance of seeing most or all of it. In the months before the transit, Le Gentil had enjoyed ample opportunities to use Jupiter's satellites to establish the exact latitude and longitude of his Pondicherry observatory. All transit observers needed to establish geographical coordinates with precision so that the great triangles through the earth and through and between their stations could be used to fix their triangulation baselines. Because of the geographical position in Pondicherry, the sun rose on the morning of 3 June with Venus already on the sun's disc. And at the crucial time of egress about 7 a.m., a single cloud drifted across the sun. By the time it had gone, the transit was all over. After nine years away from France, Le Gentil had missed *two* Venus transits.

It took Le Gentil almost another two years to get home to France, disembarking at Cadiz in Spain, and travelling overland to Paris. As no word about him seems to have reached France since shortly after the first transit, he found himself presumed dead – for many Europeans simply sickened and died in the East Indies – and his relatives busy carving up his estates and property. But the scientific community showed him great kindness, with César-François Cassini de Thury (or Cassini III) and the Académie des Sciences finding apartments for him in the Paris Observatory. He then got married, and had a daughter who gave him much delight. He died in 1792, aged sixty-seven. By this date, the French Revolution was in its third year, though mercifully the fanatics of the Reign of Terror had not yet seized control of the country.

Practical observation, Venus, and the longitude

The years between 1761 and 1769 witnessed significant changes in telescope technology. Most of the observers of the 1761 transit used instruments of a fairly conservative type by that date. They were either long-focus, simple refractors – Le Gentil took such a telescope of 15-foot focus with him to Pondicherry – or Gregorian reflectors. These reflectors were mirror telescopes of the type popularized by figures such as James Short, whom we encountered in Chapter 4. By 1769, however, there were some twenty-six Dollond-type achromatic refractors recorded as having been used by observers. Although these new, large-object-glass achromatic refractors were very expensive, cost was less of a problem with official, nationally sponsored observations.[5] While Dollond tried to use his patent for manufacturing telescopes to control the market, his or other imitators' instruments went on to open, public sale, so that French, Russian, or Scandinavian scientists could have access to these superior-definition telescopes.

The solar disc, with Venus upon it, would have been observed directly, the sun's brilliant glare being diminished by means of dense glass filters. Such a mode of observing is strongly disapproved of today, insofar as the focused and concentrated solar light (and heat) can easily shatter a filter, with potentially serious consequences for the observer's eyes. But no such mishaps were recorded in either transit.

Irrespective of what type of telescope was being employed, however, the astronomer would meticulously time and measure the exact passage of Venus across the solar disc, in accordance with the technique described above. Of course, this was the high hope, though as we have seen the blurring of the "black drop" often made it impossible to make exact timings of these contacts.

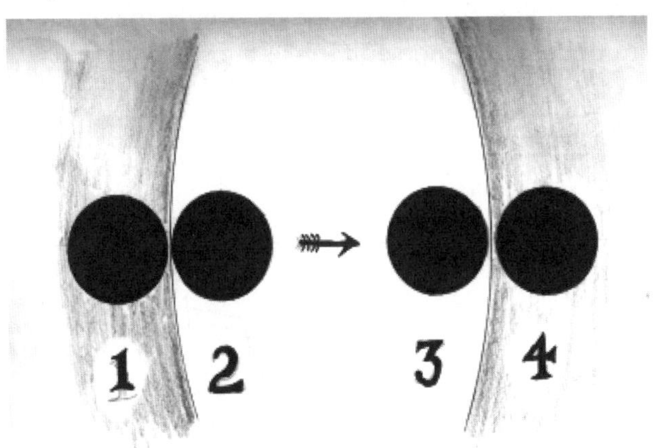

7.3 Observing the four contacts of Venus on the solar disc. Two are at entry, or "ingress", and two at exit, or "egress". These contacts had to be timed and measured with the greatest accuracy. (Drawing by Allan Chapman.)

The observer might also use the two hairlines (or "wires") of a filar micrometer, controlled by a precision screw and mounted into the telescope eyepiece, to measure the position of the transit path with reference to the sun's own north and south poles, to establish the solar "latitude" of the transit. The length of these transit lines, and their latitudes, would all be different, depending upon whether the astronomer were in Hammerfest, Tahiti, Mexico, Pondicherry, or with the Russian expeditions to Kola and Ponoi. But all of this data, and their diverse baselines and coordinates, would be essential in the calculation of a correct solar parallax.

In addition to telescopes and micrometers, two other instruments were essential for every set of observers, be they in the Paris Observatory or in a tent in Tahiti. One was a precision brass quadrant, with its own micrometer adjustments and telescopic sight; the other was a fine clock. The quadrant was used to take the altitudes of bodies in the sky, hopefully to within a few arc seconds. While this instrument would not have been used to observe the actual transit, it was essential to help establish the exact latitude and longitude of the observing station. Hopefully, in the days or weeks prior to the transit, the astronomer would take daily altitudes of the sun. Firstly, this would enable him to lay down an exact meridian, or north–south line, for his location. Secondly, the quadrant would allow him to use this meridian to take the solar altitude at the very moment of noon, so that he could establish his precise latitude on the earth's surface.

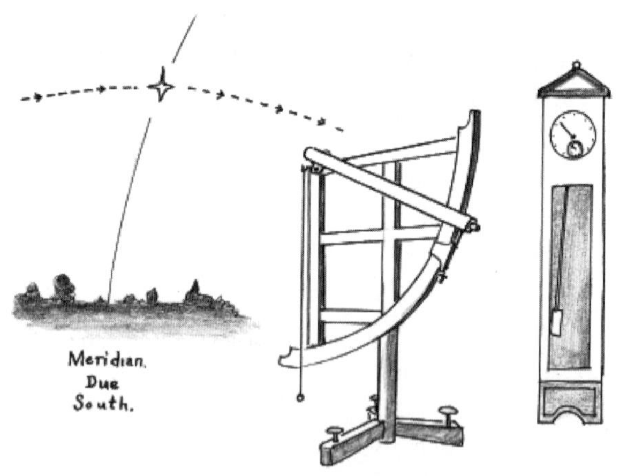

7.4 Portable 2-foot-radius brass quadrant and regulator clock. Precision quadrants of this type were carried by most transit of Venus expeditions. They were used in exactly the same manner as the large mural quadrant shown in Chapter 3, their scales and micrometers being employed to establish the exact meridian for the transit observers. This in turn allowed them to regulate their accurate pendulum clocks, local-time the transit, and establish their longitudes with reference to Greenwich or Paris. (Drawing by Allan Chapman.)

The pendulum clock, upon arriving from London, Paris, or St Petersburg, would be set up and its long pendulum minutely adjusted to make its swings delineate exact seconds for that particular location. Once set up, the clock would be used to time the precise duration of the transit, as Venus moved from one side of the solar disc to the other. We will look at the "geodesical" curiosity of why a clock pendulum should behave differently in Hammerfest, Greenwich, Pondicherry, or Cape Town in the next section.

How did an astronomer in some far-flung location determine his exact longitude, east or west of Greenwich, say, or Paris? Without a knowledge of the longitude of the place of observation, the observation itself could not be correlated with everybody else's. The favourite method was to employ the telescope to observe the motions of Jupiter's moons, first discovered by Galileo in 1610 and named the "Medicean Stars" after his Florentine patron Cosimo de Medici, the Grand Duke of Tuscany. Galileo had recognized that the regular orbital periods of these satellites could be used as natural clocks, and long before 1760 their motions had been precisely tabulated and understood with relation to Newtonian gravitational criteria.

Each of these "Jovian satellites" passed behind the body of Jupiter with every rotation, so that with a good telescope it was possible for an astronomer to note the very second when any one of the four bright moons actually appeared to "touch" the body of the planet. By 1760, published tables of these motions for Greenwich, Paris, and elsewhere had long been easily available, enabling them to be used to make accurate timings. Yet depending on whether you were east of Greenwich in Pondicherry or west of it in California, you would see a measurable time discrepancy, where the satellite contact was ahead of or behind the time given in the table. If you could measure the exact discrepancy for an individual location, then you could easily establish your position with reference to Greenwich, Paris, or St Petersburg, thereby providing the longitude of your observing station, to complement your quadrant-derived latitude. In this way all the transit astronomers could fix their exact observation points on the earth's surface, to allow the survey "baselines" to be established.

Because this "Jovian satellite" method required a rock-solid place of observation and a good pendulum clock, it could not, alas, be used at sea. Finding one's longitude at sea had presented serious problems for navigators since the fifteenth century, although it was during the 1760s that two possible techniques suggested by Petrus Apianus in 1524 were at last brought to a working perfection, as we have seen. It had taken a good 240 years for the astronomical, mathematical, and precision technological requirements in both methods to be fulfilled. The lunars longitude method required the angle between the moon and certain stars to be measured with great accuracy, by the shipboard navigator. By the 1760s the lunar tables of Tobias Mayer, Abbé Nicolas-Louis De La Caille, James Bradley and others, computed from observations made with state-of-the-art observatory instruments, were achieving the necessary levels of accuracy. Using a newly developed on-board octant or sextant to measure the moon–star angles from the quarterdeck, followed by a few hours of patient calculation, one could obtain a decent value for one's longitude at sea. This was especially so after 1767, when the Astronomer Royal, the Revd Dr Nevil Maskelyne, issued his first *Nautical Almanac*.

The second shipboard longitude-finding method was to use an accurate watch or chronometer to carry Greenwich, say, or Paris, time reliably across the globe in spite of storm and stress. By comparing his local time, taken by the sun, with the "home" time shown on the chronometer, the navigator could turn this discrepancy into a precise angle east or west of home. It was the Yorkshireman John Harrison who perfected the first seagoing chronometer in a series of timepieces after 1735, and while Captain Cook found Tahiti and then western Australia on his transit of Venus expedition of 1768–71 using the lunars method, he carried chronometers in his two further voyages during the 1770s.[6]

What is so striking is the speed with which astronomical precision had increased since the time when the young Edmond Halley voyaged to St Helena in 1677, and in the years following the Venus transits of the 1760s. Improved pendulum clocks, micrometers, achromatic telescopes, George Graham's – then John Bird's – new instruments for the Greenwich Observatory, the perfection of the lunars method of longitude-finding at sea, and the very first reliable – while still extremely expensive and scarce – marine chronometers after Harrison, all were part of this advancement of learning. As precision measurement was always the foundation upon which astronomical progress was made, and all of these technological devices spread through the normal commercial channels across the world, the savants of Europe were well placed to ascertain the dimensions and details of the earth and the heavens.

Chapter 8

Measuring the Heavens and the Earth in Eighteenth-Century Europe Part 2: Pendulums, Planets, and Gravity: Creating the Science of Geodesy

From classical Greek times onwards, an aspiration to exact measurement and quantification had always lain at the heart of astronomy, and once a reliable value for the earth–sun distance had been established by the latter part of the eighteenth century, the proportions of the solar system could fall into place. The elucidation of these proportions, in the wake of Kepler's Laws and Newtonian gravitation, helped to facilitate a number of physical investigations. Such investigations included studies of the precise elliptical nature of lunar, planetary, and some cometary orbits, and after 1800, of deep-space binary star system orbits. These investigations also helped to provide a context for understanding that mysterious gap in the solar system between Mars and Jupiter: a gap that first Father Giuseppe Piazzi then Karl Harding and others after 1800 came to populate with a growing swarm of minor planets, or "asteroids". Were these remnants, perhaps, of a planet once orbiting the sun between Mars and Jupiter, which had somehow broken up into fragments in the remote past? The foundation and development of geophysics would also come to play a central role in the wider enterprise of solar system measurement and quantification.

The curious behaviour of M. Richer's clock: Cayenne, Brazil, 1672

In 1672, the French astronomer Jean Richer, in conjunction with colleagues back in Paris, spent ten months attempting to triangulate the distance of Mars, at one of the Red Planet's biennial "close" approaches to earth. Richer was working from the river estuary island of Cayenne, off the coast of Brazil just north of the equator. The logic behind their thinking was the same as that behind the Venus observers ninety years later: if the earth–Mars distance could be measured accurately, the solar parallax could be derived, and from there the other planetary distances by means of Kepler's Laws. What puzzled Richer, however, was that when he set up his fine, carefully regulated Paris Observatory pendulum clocks in Cayenne, they lost 2 minutes and 29.5 seconds each day, and he had to readjust the timepieces to make them beat exact seconds in Cayenne. Then back in Paris, the pendulum bobs had to be readjusted to their original positions to make the clocks keep Paris time. As we saw in Chapter 2, when Edmond Halley was at St Helena between 1677 and 1679, he also found that his pendulum clock ran slower on top of the mountain than it did at sea level.

Consequently, it came to be realized that while the earth might be round, it was not a perfect sphere. If the pull of gravity became weaker with distance from the earth's "point mass" centre, then Cayenne, at 4° 45' north, must be further away from the centre of the earth than Paris, at 48° 50' north. Factoring in these and other geophysical observations, Sir Isaac Newton would conclude in *Principia* (1687) that "the Earth is higher under the equator than the poles, and that by an excess of about 17 miles".[1] (The accepted modern figure is about 26 miles.)

Richer's chance discovery at Cayenne would lead to innumerable pendulums being swung across the globe, from South Africa to China to the Americas, and into the north polar circle over the next couple of centuries, as astronomers and pioneer geophysicists conducted countless observations and experiments across the surface of the planet, and even down the deep coal mines of Industrial Revolution Britain, such as at Harton Pit, South Shields, in north-east England in 1854. At the bottom of Harton, a pendulum could be swung 1,260 feet closer to the earth's centre than at the surface. This information could also be used to calculate the earth's density.[2]

Even at Cayenne in 1672, however, Richer had recognized that if one were trying to use the swinging pendulum as a "gravity meter", or gravitational-pull-measuring instrument, one could get more accurate results by swinging a free pendulum rather than one attached to the going-train of a clock, where driving pressure from the clock's weights might cause minor aberrations in the swing. A "free" pendulum is a normal pendulum disconnected from a clock mechanism, where the astronomer simply pulls it to one side and lets it go, allowing it to swing until the earth's pull of gravity eventually stops it. If one knows the exact length of the pendulum and times its swings against the seconds hand of another clock or spring-driven stopwatch, then one can count the exact number of swings that it will perform in a measured minute of time. It will perform more swings at Spitzbergen at 78° north up near the north pole than it will in Borneo on the equator (Borneo being almost 26 miles further away from the earth's centre than Spitzbergen, making the pull of gravity in Borneo slightly weaker than at the pole).

GEOPHYSICS BY DEGREES AND THE SHAPE OF THE EARTH

It is not surprising that Delisle and other Frenchmen had been so enthusiastic about observing the transits of Venus in 1761 and 1769 and establishing the physical constants of our planet, for precision large-scale surveying had long been a specialism of French astronomers and physical scientists by that time.

Attempts to establish the planet's physical dimensions went back to Eratosthenes, Posidonius, and Ptolemy in antiquity, whereas in the seventeenth century Willebrord Snell, Richard Norwood, and Jeremiah Horrocks had all made computations for the length of a terrestrial degree in miles or other units. But it was in France that things took on a new level of activity and precision: first with Jean Picard in the 1660s, then with the Cassinis, the father and son directors of King Louis XIV's newly built Paris Observatory. The latter surveyed a meridian line running through the great hall of the Observatoire – the brass strip let into the floor is still there – then extending through trigonometrical points due north until reaching the English Channel at Dunkirk, and south to the Pyrenees west of Montpellier. This great meridian line was laid down by means of an elaborate set of interlocking triangles, surveyed across the French countryside by means of large sextants and quadrants equipped with telescopic sights and with "zenith sectors" (of which more will be said below). These triangles made it possible to impose stringent cross-checks upon the survey, and guarantee that the *c.* 570-mile-long meridian line did not deviate.

What the Cassinis discovered in 1701 was that of the 8 degrees of latitude over which France extended on the earth's surface, each degree appeared about 1/800th shorter as one moved south: shorter, that is, in terms of *toises* (a *toise* was 2.131 yards or 1.949 metres), the old pre-metric French measure. Now if one degree is reckoned at close to 69 British statute miles, this means that each degree, in linear terms, across the face of France, shortens by about 151 yards. What is more, in his excellently surveyed map of France, Cassini had found that his east–west longitude determinations made the country slightly narrower than had been generally thought. It is hardly surprising, therefore, that King Louis XIV – hopefully in jest – told Cassini that his survey had deprived him of more of his kingdom than had all of his enemies.

What Picard's, Cassini's, and the other survey measurements suggested was that the earth was *prolate*, or shaped somewhat like a grape, the equatorial diameter being narrower than the polar diameter.

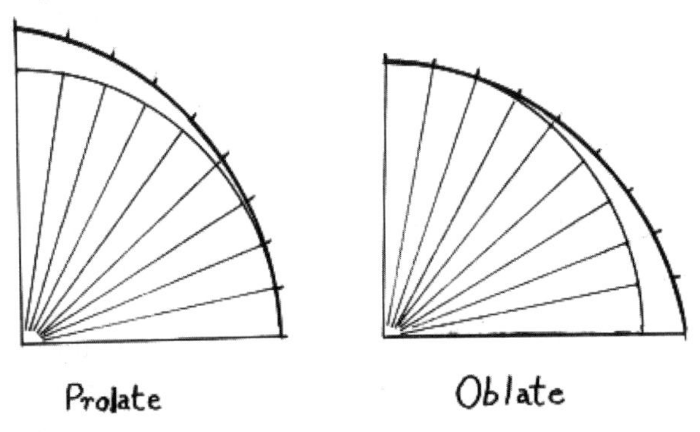

8.1 Prolate and oblate earth. The radiating lines on the drawing are 10° apart. If the earth's shape resembles a grape (prolate), a measured terrestrial degree at the pole will be longer than one at the equator. If, however (as is the case), the earth is like an orange, slightly flattened at the top and bottom (oblate), a measured terrestrial degree at the pole will be shorter than one at the equator. (Drawing by Allan Chapman.)

Yet this shape was in flat contradiction to that deriving from the swinging pendulums, which pointed to an orange-shaped or *oblate* earth: thicker through the equator than through the poles. Which geophysical shape was correct? This was taken up by French savants across the eighteenth century. They appreciated that France itself might not occupy a sufficiently large part of the earth's surface to make a decisive measurement possible, and expeditions needed to be sent further afield. This led in 1735 to a French expedition under Pierre Louis Moreau de Maupertuis being sent to Finland, to the north of the Gulf of Bothnia, in the northern Baltic Sea, over 66° north. There they would lay out a survey baseline on the frozen waters of the River Torne, utilizing the perfect flatness of the thick winter ice as a bedrock.

The method employed to ascertain one exact degree on the earth's (or ice's) surface was to set up a "zenith sector" telescope at the commencement point of the survey. By 1735 zenith sectors were well known among Europe's astronomers, and one had been used only in 1727–28 by the Revd Dr James Bradley of Oxford to discover the aberration of light.[3] A zenith sector consisted of a refracting telescope supported on a precision gimbal mount near its object glass, so that it hung down like a great pendulum. An observer lay on his back looking through the telescope directly at the zenith. A short scale of degrees and a precision screw micrometer allowed the telescope to measure the zenith distance angle of a small number of stars near to the zenith, for cross-checking and comparative purpose. This 9-foot sector, along with a 2-foot-radius portable quadrant and a precision pendulum clock, had been commissioned from George Graham, who had supplied Bradley's sector and recently re-equipped the Greenwich Royal Observatory with a set of instruments and clocks that were the talk of Europe. As he expected, Maupertuis was delighted with them, even noting how the extreme cold of Finland slightly contracted the brass scales. Maupertuis spoke with deference of both George Graham and James Bradley: all three men were Fellows of the Royal Society.

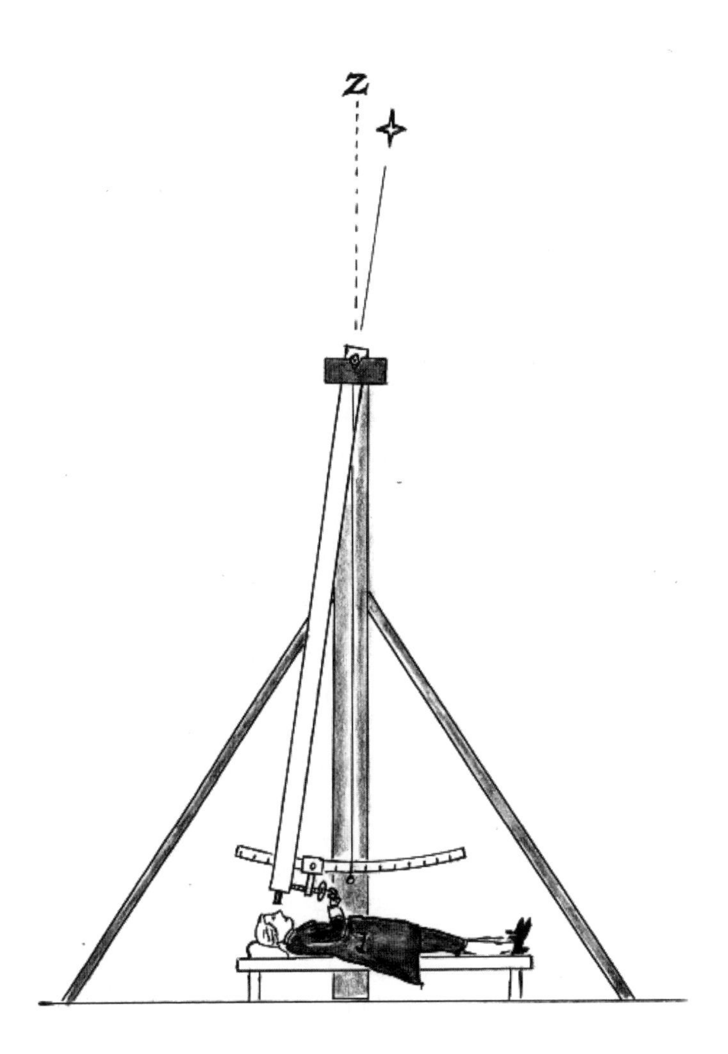

8.2 Zenith sector, as used to measure the exact shape of the earth. A telescope of around 12 feet focal length was hung like a transit telescope from its object glass end, to resemble a pendulum. When an astronomer lay on his back and looked straight up it, he saw stars as they passed across his local zenith. In this way, he could establish an exact 1° celestial angle, and use it to measure points on the earth's surface that were 1° apart. Thus French astronomers discovered that a terrestrial degree in Peru was longer, in *toises*, yards, or miles, than one in Lapland. This demonstrated that the earth must be oblate, or bulge toward the equator. (Drawing by Allan Chapman.)

From their survey base on the ice, the zenith sector was used to observe the angles of the stars of α (alpha) and γ (gamma) Draconis from the opposite ends of a surveyed baseline 55,023½ *toises* (about 66.6 miles) long, to compare a degree of arc in the heavens with a measured length upon the earth's surface (or ice). Comparing the length of the measured degrees in France and in Finland across an angle of over 20 degrees of latitude, it was immediately clear that a near Arctic polar degree was slightly larger than the ones surveyed in France by Picard and Cassini, leading Maupertuis to conclude "*la terre est un sphéroïde applati vers les Poles*" ("the earth is a spheroid flattened toward the poles").[4] This shape accorded elegantly with that deriving from the swinging of pendulums in Arctic and equatorial regions. Maupertuis further recognized that Picard's and Cassini's mainland French surveys contained a significant error. Working before Bradley's discovery of the aberration of light, they were unable to discount an error deriving from the aberration. But when this correction was applied, along with an improved value for the refraction of light through the earth's atmosphere, the oblate, or flattened at the poles, model for the earth found further substantiation, and the grape became an orange.

To establish the earth's true shape and geophysical proportions, however, Pierre Bouguer was despatched to the great flatlands of Peru to survey an equatorial degree of arc. Here, some 40-odd degrees south of France and 70 plus degrees south of Finland, and slightly south of the equator, Bouguer produced a decisive confirmation of the fact that the earth was slightly flattened at the poles and bulged somewhat at the equator. These combined French measurements, made across a good seventy years, constitute the true birth of geophysics. Everything was found to accord elegantly with Newtonian gravitational criteria, as the French – regular military hostilities notwithstanding – held Newton in no less veneration than did the English, along with Bradley, Graham, and a host of others. That warm regard was reciprocated by the British, who held the achievements of Picard, Maupertuis, Bouguer, the Cassinis, Delisle, and a host of other Frenchmen in equal respect. The Paris Académie, the Berlin and Stockholm Societies, and the Royal Society in London freely exchanged scientific results, and bestowed learned honours and titles upon one another as well.

During his surveying in Paris, Bouguer encountered a curiosity that would open up a wholly new branch of geophysics. When surveying near the towering 20,610-foot Andean mass of Mount Chimborazo, he discovered that the long, delicate silver plumb-line wire of his zenith sector did not always point to the precise astronomical zenith. Could the great mountain be attracting the relatively tiny mass of Bouguer's plumb-line lead bob, and pulling it out of true, in the same way that the earth exerts a gravitational influence upon a falling stone?

THE ASTRONOMER ROYAL, THE MOUNTAIN, AND THE VILLAGE FIDDLER

In 1775, the Revd Dr Nevil Maskelyne, Astronomer Royal and Director of the Royal Observatory, Greenwich, published a paper in the *Philosophical Transactions* that would be of fundamental importance to the new science of geophysics.[5] Fascinated by Bouguer's account of Mount Chimborazo influencing his plumb-line bob, Maskelyne, at the behest of the Royal Society, travelled to Perthshire, eastern Scotland, in an attempt to ascertain the gravitational attractive force of Mount Schiehallion, west of Dundee, a great granite mass. One reason for choosing Schiehallion was that it stood out as a single mountainous mass, rather than being part of a mountain chain, thus making its gravitational attraction more specific.

Travelling from the Thames up the North Sea – by far the most comfortable way to get to Scotland in 1774, especially if one had a collection of delicate precision instruments – Maskelyne began by triangulating a baseline upon the southern foothills of Schiehallion. From here, he observed zenith stars with that now standard geophysical instrument, the zenith sector. When satisfied with the accuracy of his observations, he moved his whole operation to the north side of Schiehallion, to an observing station whose position had been exactly surveyed with relation to the southern one.

On each side of the mountain he found that Schiehallion was exerting a gravitational tug upon his zenith sector plumb line, when compared with the known positions of zenith stars based on the accurate Greenwich star catalogues. Maskelyne next calculated that an angular discrepancy of 11.6 arc seconds in his pendulum, cross-checked north and south, "is to be attributed to the two contrary attractions of the hill".[6]

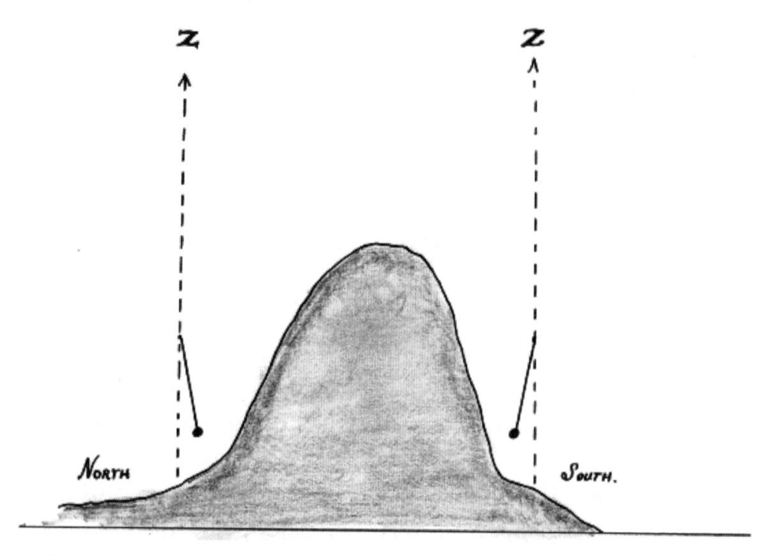

8.3 "Weighing" the Scottish mountain Schiehallion in 1774. By suspending long precision zenith sector plumb lines north and south of the mountain, the Revd Dr Nevil Maskelyne was able to measure the degree by which the mass of Schiehallion attracted the lead plumb-line bobs out of true, as compared with a zenith star. From this deflection, he computed the mass of the earth. (Drawing by Allan Chapman.)

From these figures, Maskelyne calculated that the earth – the whole terrestrial mass possessing a close parallel to that of the mountain – had a mean density of between four and five times the density of water. This figure would later be modified to 4.713 by Charles Hutton and John Playfair, based on their subsequent field survey of the mineral composition of the Schiehallion rock.

Maskelyne and his colleagues seem to have got on well with the local Scots, and before they were due to depart for London, a celebration was held. It seems to have been a very jolly affair, so much so that a fire broke out, in which the fiddler's violin went up in flames. On his return to London, Maskelyne purchased a replacement instrument, which he dispatched to Scotland as a gesture of appreciation. The replacement fiddle, it is said, was a Stradivarius.

Geophysics goes to the laboratory: Henry Cavendish and the torsion balance experiment, 1797–98

As always happens in science, once an instrument or technique has shown that some new thing can be done or discovered, human ingenuity sets about improving it. This is exactly what happened in the wake of Maskelyne's experiments at Schiehallion in 1774, published in 1775. Long pendulums beside mountains can be susceptible to all manner of interferences, from wind, weather, air pressure, and humidity. This led Henry Cavendish in the 1790s to take gravitational physics into the laboratory. Cavendish, the son of Lord Charles Cavendish FRS, was a wealthy private gentleman, and a cousin of the family of the dukes of Devonshire, one of the richest aristocratic families in Britain. But he had no care for the high life. He lived simply, dressed plainly, and was a shy, reclusive bachelor of few words, with whom even the sociable William Herschel did not find it easy to converse. He found social relationships difficult and may have been autistic. But his passion was science, and – like Newton – he possessed intense powers of concentration. By the 1790s, when he was in his sixties, Cavendish already had a string of scientific achievements behind him, mainly in gas chemistry, electrical physics, and heat, on which he had worked with his physician friend Sir Charles Blagden, both gentlemen being prominent Fellows of the Royal Society.

With ample independent wealth behind him, and all of his time his own, Cavendish addressed himself to finding a more accurate way of determining the earth's gravitational attraction than by means of pendulums and mountains, of weighing the earth under exact, controlled laboratory conditions. He devised a torsion balance to do so, an instrument that was no less a triumph of engineering brilliance than were George Graham's graduated instruments, John Harrison's marine chronometers, and James Watt's great steam engines.[7]

As Cavendish fully admitted, however, he was not the inventor of the torsion balance: that honour went to the Revd Dr John Michell FRS, the Cambridge-trained astronomical and scientific clergyman whose work on the possible gravitational physics of star clusters had helped to inspire the young Herschel, as we have seen. Michell had been the first to suggest, in 1783, the possibility of a star-mass so dense and gravitationally powerful that light itself could no longer escape from it: a Georgian "black hole". He had even built a torsion apparatus, but by his death in 1793 had not been able to obtain definitive results. Cavendish acquired Michell's torsion balance, and improved and rebuilt the instrument to make it capable of a phenomenal level of accuracy.

What is a torsion balance and how does it work? Like a conventional laboratory balance, the torsion apparatus was designed to weigh. What it weighed was not so much conventional heaviness as the gravitational attraction acting between two very large and two very small lead balls: the large balls being 383 pounds and the small just over 1.5 pounds.

The two large balls were suspended by wires from a 73-inch-long balance arm, and the two small balls from a separate one. As the two arms shared a common central pivot, about which they could rotate, it was possible to separate the four balls so that they stood at 90 degrees to each other. Then, by activating a delicate rope and pulley mechanism, the two sets of balls could be gradually drawn closer to each other: each large ball approaching a small ball. The whole purpose of the experiment was to detect a gravitational motion of a small ball toward a larger one, in accordance with Newton's Inverse Square Law of Gravity. This attraction was detected via the torsion, or twisting, of the metal wires supporting the balls, especially the small ones, where the gravitational response was naturally more conspicuous. In his detection of this torque, Cavendish displayed his genius as an experimentalist by devising a weightless pointer: a focused ray of light would be reflected from the balls, making the slightest gravitational deflection obvious, and measurable upon a scale.

Realizing the extreme sensitivity of his large torsion balance apparatus – it was sensitive even to the observer's breath and body heat – Cavendish encased the whole within a sealed chamber, which was, in turn, housed within a windowless building near his house on Clapham Common, London. When observations were being made over 1797–98, every adjustment was made by remote control using ropes, rods, and pulleys, and telescopes were specially placed to enable the observers to see inside the apparatus. The light for the illumination was provided (as was the light in the powder magazine of a warship) by oil lamps shining through external glass apertures into the apparatus. With this apparatus, Henry Cavendish was able to revise Maskelyne's values of twenty-three years before, to obtain a value for the terrestrial density of 5.48, close to the accepted modern value of 5.513. Once the terrestrial density had been obtained, it was possible to use it to calculate the density of other solar system bodies from their measured gravitational relationship with the sun.

So the quest to measure and quantify the force of gravitational attraction, which began with the French astronomers at Cayenne, and led on to that force's meticulous testing against Peruvian and Scottish mountains, finally reached its fulfilment, with astonishing precision, inside a sealed chamber on London's Clapham Common in 1798.

Chapter 9

Cosmology and the Romantic Age

In our own day and age, people generally separate the arts from the sciences as two cultures. In previous centuries, however, this cultural dichotomy did not exist. Astronomers, doctors, physicists, and engineers read poetry and played musical instruments (and many still do), and poets and painters attended lectures on astronomy and chemistry. The great English chemist Sir Humphry Davy also thought of himself as a poet, and the artist J. M. W. Turner painted heroic scenes of railway trains and ships battling against the elemental forces of nature. So how did the "Romantic Imagination" come to terms with the new cosmology?

From Daffodil Fields to Starry Fields: A Universe of Awe and Wonder

In the generality of thinking, the Romantic age of the late eighteenth and early nineteenth centuries is about wonder and the imagination. Its motifs included daffodils by Ullswater on a glorious spring morning, the lofty mountain grandeur of the Scottish Highlands or the Alps, Beethoven's *Eroica Symphony*, and even the efforts of French and Scottish doctors to attempt to revive dead bodies by the "electric force". Fossil geology suggested not only a vastly ancient earth, but also whole races of pre-Adamite creatures – styled *dinosaurs* after 1840 – populating the antediluvian tropical swamps, which by 1800 had metamorphosed into the coal measures beneath Manchester, Newcastle, and South Wales. Awe, wonder, extremes of emotional experience, terror, *Frankenstein*, ghosts and ghouls, passionate love-affairs, and opium-induced wild trances were all part of Romanticism.

However, in the predominantly literary culture of the age of Jane Austen, Mary and Percy Shelley, and William Wordsworth,

William Herschel's cosmology, along with geology and chemistry, were perhaps more mind-blowing than all the rest put together. For what Herschel and other scientists were revealing after *c.* 1784 was the very stuff of creation itself; and it is hard to out-Romanticize cosmological vastness and the questing scientific intellect, retracing the handiwork of God himself. Joseph Haydn, over 1797–98, devoted a whole oratorio to *The Creation*, and was inspired to write within it one of the most joyous choruses in the entire musical repertoire, when he set Psalm 19: "The Heavens are telling the Glory of God."

Laws of wonder: Herschel, Laplace, and the laws of gravitation

We have seen in previous chapters how William and Caroline Herschel wrestled with the bizarre yet glorious celestial fauna revealed in the cosmic garden by their telescopes between 1780 and William's death in 1822. If in the twenty-first century astronomy is advancing faster than we can comprehend, let us remember what was happening during the Romantic age. As we have seen, a new planet – Herschel's Georgium Sidus (Uranus) – had "joined" the solar system in 1781: the first such discovery in human history, as all the other planets had been known to the ancient Babylonians. Then Father Giuseppe Piazzi at the Palermo Observatory, Sicily, had made a remarkable discovery on the first night of the nineteenth century, 1 January 1801. He had observed a new planetary body, later termed an "asteroid" moving in the seemingly "blank" space between Mars and Jupiter. After Piazzi's discovery of Ceres, the hunt was on for more such minor planets, and before long Baron Franz Xaver von Zach's "celestial police", who were combing the heavens for possible planetoids (or asteroids), would discover more. What were these tiny asteroids, ranging from Ceres – no further across than France – to cosmic lumps of rock smaller than Manchester? This brought up the question of the physical stability of the solar system; for as solar system dimensions were quite well established by 1800, especially in the wake of the improved solar parallax values following the Venus transits, why was there such a big gap between Mars and Jupiter?

This sequential anomaly had been highlighted by Johann Daniel Titius and Johann Elert Bode between 1766 and 1778, and the latter's subsequently christened "Bode's Law" demonstrated that a beautiful mathematical symmetry was implicit in the known planetary distances, rooted in Newton's laws of gravitation. So what was the reason for the gap between Mars and Jupiter? Could a former planet have somehow disintegrated, and if so, could its fragments be found? This is what Piazzi, von Zach, and their fellow "celestial policemen" were trying to discover. By the end of the eighteenth century, it was becoming clear that planet earth, the solar system, and the stellar cosmos itself were not eternally unchanging. Geologists such as Robert Hooke and James Hutton had realized the earth was already vastly ancient when God created Adam and Eve in *c.* 4004 BC. Volcanoes, earthquakes, erosion, mountain-building, overlying strata, signs of changing sea levels, and curious fossils all pointed to a constantly changing earth: an antediluvian or pre-Noachian-Deluge earth.

Then, in a paper published by the Royal Society in 1801,[1] William Herschel suggested that cyclical changes in the sun might have an effect upon the earth. He did a correlation between times of abundant and scarce sunspots and changes in the price of wheat, as tabulated in Adam Smith's *Wealth of Nations* (1776). It seemed that abundant sunspots produced the best yields of wheat, abundance of crops being reflected in cheaper prices. So the sun itself became another variable in a cosmos once held to have been constant and immutable since the creation: a topic to which we shall return in Chapter 16. Of a piece with this new sense of a dynamic, changing earth and solar system was Pierre-Simon Laplace's widely read and influential *Exposition du système du monde* (1796), a model for the very origin of the solar system: the "nebula hypothesis". Earlier in the eighteenth century, Emanuel Swedenborg and Immanuel Kant had proposed metaphysically derived models for a solar system condensing out of a "nebula", but Laplace took the idea much further. As the greatest mathematician of the late eighteenth century and a pioneer of gravitational physics, he analysed the dynamics of what might occur if a nebula began to condense, rotate, and throw off filaments of matter. It might condense into an incandescent central body – the sun – and the ejected filaments into the individual planets. The fact that all the planets rotated in one relatively flat plane around the sun and all moved in the same direction suggested that they might once have been ejected from a rotating, condensing mass.

Then to cap it all was William Herschel's own observation-based deep-space stellar cosmology. When one reads Herschel's various "Construction of the Heavens" papers between the 1780s and the 1810s in the *Philosophical Transactions* of the Royal Society (see Chapter 6), one sees the emergence of a set of ideas and models for a constantly changing universe. Herschel's universe, however, was not seen to be evolving in a directional sense. It is perhaps best to think of it in demographic terms: just as individuals in whole populations of humans, animals, and plants are born, grow up, die, and disintegrate within a structurally stable demographic mass, so too did the stellar population.

In Herschel's universe, stars were most likely evenly planted like flowers in a garden by God at the creation. But that perfect symmetry was probably broken up by gravitational forces, for while all stars had a generic size, in the same way that all humans, cats, and oak trees had their generic sizes, there could be considerable variations within a group. An adult human, for example, might stand 5 feet high and weigh 7 stones, whereas another might be 6 feet 6 inches tall and weigh 17 stones. So it might be with stars. And if this size and mass variation did apply to the stars, then those more massive stars would exert a stronger gravitational pull than a smaller star, with the effect that the less massive star would be inexorably drawn to the more massive.

Then, as stars were drawn to stars, then to clumps of stars, one might end up with a star cluster, such as the ones seen in the Pleiades or Hercules constellations. The bigger the cluster became, the greater combined gravitational attraction it would exert, until it had swept a whole galactic region clean of individual stars. Should further compacting take place, the whole cluster-mass could disintegrate under its own intense gravity, blast itself apart, then recycle its starry debris back into space as a sort of dusty nebulosity. So could this process explain the existence of the nebulae? And then, just as a cloud of steam condenses into individual water droplets, could the all-pervading force of gravitational attraction cause this swirling star-debris to condense into masses that, at a given point, became self-luminous, to become discrete light sources, or newborn stars? If this process were operating across the whole galaxy or star system in different places at different times, it would result in a universe that was self-renewing yet stable across its overall mass, rather like a biological or botanical population. The universe of the Romantic age, one might say, partook in the quintessential Sturm und Drang of the age: storm, stress, grand changes, deaths, and rebirths – all because of the Newtonian Inverse Square Law of Universal Gravitation.

MYSTERIES BEYOND THE SPECTRUM: SIR WILLIAM HERSCHEL DISCOVERS THE "DARK SPECTRUM" IN 1800

Observation and ingenious experiments are wonderful things, for no one knows where they will lead. A good example of this occurred over the first few weeks of 1800, when Herschel was experimenting with different colour filters (a very dangerous process, and not to be recommended) with which to make direct observation of the solar surface at high magnification – a technique not to be recommended. Red light, he found, created a very warm sensation upon his eye. Why? He next set up an experiment in which a glass prism was used to produce a beautiful solar spectrum in a darkened room. Then, using a set of accurately calibrated glass thermometers made by the great – and recently deceased – instrument-maker Jesse Ramsden FRS, Herschel began to observe the colour temperature differences within the spectrum. Blue light was cool, whereas red light was distinctly hot; all the colour thermometers were checked against the ambient temperature of the air measured by other thermometers in the dark parts of his study, well away from the bright spectrum.[2]

Then in a simple act that was a stroke of sheer experimental genius, Herschel moved his thermometer beyond the red light into the adjacent blackness, and was struck by the fact that even in this region, the thermometer continued to rise. Only when it was moved further out did the mercury fall to that of the ambient temperature of the room. What was this invisible heat, or "black light", to which the human eye, but not the thermometer, was insensitive?

Herschel had discovered what came to be called the "calorific rays", or heat rays of the sun. We now call this "infrared" radiation, or "IR". Herschel performed this *experimentum crucis* on 11 February 1800. Then a year later, the German chemist Johann Wilhelm Ritter, working on the darkening effects of light upon silver salts, found that while blue light had the most rapid action in darkening the silver chloride, an even faster action took place in the black region beyond the blue light. This black region was christened the "chemical rays" because of its chemical (though not thermometric) action; and once again, the rays were invisible to the human eye. We now call this "ultraviolet", or "UV" radiation.

It was now beginning to look as though the sun was a much more complex physical and chemical object than anyone had previously imagined. It was also after 1807 that the English physician, Herschel admirer, and experimental physicist Thomas Young began to popularize the word "energy" (Greek *energeia*, "force", "action") as a term for these intangible natural forces that, nonetheless, exerted such prodigious power in the universe. What was this strange solar energy, and how did it cross space? Robert Hooke had proposed an "undulatory" form, or wave, in 1655, for all types of force (such as light, sound, and gravity), but his ideas had been sidelined by Sir Isaac Newton, whose model for light was that of a stream of "corpuscles" or discrete particles. Yet Young, and his doctor and physicist colleague William Wollaston, were coming to rethink light as a wave. These waves, moreover, had different amplitudes of pitch – just like the different sound pitches emitted by violin strings – which created a sense of blue or red, or invisible chemical or calorific rays, depending on their geometry and sinusoidal patterns. We will return to Young, Wollaston, and the new discoveries in light later in the chapter.

Science for Georgian ladies and gentlemen

We saw in Chapter 4 how enterprising instrument-makers such as James Short and popular lecturers and shopkeepers like Benjamin

Martin were successfully taking astronomy to paying audiences by 1750. This trend was destined to grow right through the Georgian and into the Victorian age after 1837. Science was a subject of intense cultural fascination, in the Romantic age and beyond. In addition to the growing number of popular lecturers touring the country, the late eighteenth and early nineteenth centuries saw the establishment of permanent institutions, often with elegant buildings and fine resources. These included the Royal Institution, founded in the heart of London's fashionable West End in 1799; the Surrey Institution south of the Thames, 1807; and the Royal Polytechnic Institution off Regent Street, 1838. There was also tha movement that led to the foundation of numerous subscription literary and philosophical societies, both in fashionable centres such as Bath and in burgeoning industrial cities like Manchester, well before 1800. Every town, be it an ancient cathedral city or a great manufacturing metropolis in which prosperous people were residing, wanted a "Lit. and Phil.". These institutions, behind their grand classical porticoes, were also social centres and clubs for all who could afford the subscription. They would contain a library, entertaining and social facilities, a lecture hall, often a laboratory, and sometimes – such as that in Bury St Edmunds – an observatory. Astronomy was popular, but so too were the fast-developing sciences of chemistry and geology, and careers could be made here by enterprising lecturers. That charismatic literary dandy Humphry Davy, up from a Penzance apothecary's shop counter, captivated fashionable London with his chemistry lectures at the Royal Institution after 1800, while his more sober but equally captivating and deeply devout successor, Michael Faraday, would do the same after 1821.

Science was "in fashion", and audiences packed in to hear popular lectures, a large part of these audiences being female. I own a print, dated 1809, of a chemistry lecture delivered in the classical auditorium of the Surrey Institution, where at least half the listeners are elegantly dressed ladies.

9.1 A chemistry lecture at the Surrey Institution, London, 1809. Note the large number of ladies in the audience. (Print in author's collection.)

By 1810 or so, if one lived in London, had some spare cash and were looking for edifying entertainment, one could decide between a poetry reading, a concert, seeing the legendary Edmund Kean act Shylock at Drury Lane Theatre, or attending a dazzling lecture on astronomy or chemistry. A few decades before that, one might have opted to hear the young Mr Herschel lead the band for the dancing at Vauxhall Gardens.

The Friday Night Discourses, delivered at the Royal Institution, Albemarle Street, still recapture the customs and learned theatricality of Regency science lectures, as I know from personal experience. Both lecturer and audience are in evening dress. Then – after being left to clarify his, or nowadays her, thoughts in a quiet room in the company of several bottles of spirits – the lecturer is led down the Great Gallery behind a pair of liveried footmen. On the stroke of 9 p.m., the footmen dramatically open the lecture theatre doors, and the lecturer walks in, bows low, and delivers the Discourse. There are no introductions or preliminaries. Exactly one hour later, at the stroke of 10 p.m., the invisible footmen reopen the doors, the lecturer then bows to the President, guests, and audience, turns, and exits, and is taken once more to the bottles for refreshment, while the President and special guests lead the audience into the Great Gallery. The lecturer is now brought out and introduced to visiting dignitaries by the President before mingling generally over a grand, late buffet supper. If the lecturer is of a naturally theatrical disposition, and (like myself) is abstemious, then the whole occasion is great fun, although I have been told that lecturers of a more shy disposition can find it a trifle overwhelming. This is one way in which audiences of the Romantic age could get to learn of the latest scientific discoveries, to say nothing of books, newspapers, and magazines.

A few years later, by 1830, astronomy and science would begin to migrate down the social scale, to be favourite subjects in the new Mechanics' Institutes – the "Lit. and Phil."s of the working people – as we shall see in Chapter 20.

The London physician, the Bavarian orphan, and the wonders of light

The first twenty years of the nineteenth century would place optics upon a new foundation. There had, as we have seen, been Herschel's and Ritter's discoveries in 1800 and 1801, then in 1803 the English physician Thomas Young reported a series of simple yet clinching experiments before the Royal Society, which confirmed that light was indeed a wave, and not a straight stream of corpuscles. The experiments were so simple that a child might perform them, with no risk, on a bright sunny day. Admit a narrow shaft of bright sunlight into a room, then hold a thin piece of card (or metal) edge-on into the light, and let it fall upon a white screen. You will notice that on each side of the card's shadow there are long, dark bands, containing streaks of colour. Or cut a very thin slit into a piece of card, and carefully stretch a fine human hair lengthwise down the middle of the slit. Hold up the card toward the sun, so as to admit sunlight directly through the slit and on to the white screen, and you will again notice long dark side-shadow bands, interspersed with predominantly reddish colours.

Young realized that this was caused by an interference phenomenon, as the thin card or the hair stretched across a slit split up the beam of light, scattering the waves. This was a phenomenon entirely congruent with a wave model of light, but not with a corpuscle or particle streams one. The same thing happens with sound waves, as asynchronous waves collide and cross each other, to produce an ear-jarring whine. Visually, this is like standing on a beach as the tide comes in and watching what happens when one set of waves collides with a slightly different one. This works equally well in a bathtub full of water if you disturb the surface from opposite ends at the same time.

It was from such homely experiments that after about 1807 Dr Young was able to place optical physics on a new and enduring foundation, in which each colour of the spectrum (including IR and UV light) depended upon a very precise sinusoidal wave frequency. Later generations of scientists would extend Young's spectrum, discovering long radio waves and very short X-rays, to constitute the electromagnetic spectrum.

In 1801, a Bavarian glass factory collapsed, burying, among others, a fourteen-year-old orphan apprentice boy. The rescue operation was personally led by the kindly Prince-Elector Maximilian Joseph, who dragged out the injured boy. Recognizing his obvious gifts, Prince Maximilian Joseph became the lad's patron, securing study time from his employer and paying for books and instruction. By 1818, Joseph von Fraunhofer, the now adult orphan boy, had become the Director of the Benediktbeuern Glassworks and Optical Institute, Munich, and the greatest refracting telescope maker in Europe. Through contact with Swiss and German colleagues, combined with his own experimental genius, Fraunhofer had redesigned the glassmaker's furnace, enabling him to achieve much higher melting temperatures and a homogeneous chemical composition for bulk batches of both crown and flint glass than had previously been possible. Wisely, Fraunhofer (like William Herschel with his speculum metal mirrors) kept the precise details of his industrial techniques secret. This included not only how he obtained such large-diameter glass lens blanks of such exquisite transparency, but also how he figured them into the crown and flint glass components essential for achromatic telescope object glasses. For half a century, since their invention by John Dollond around 1758, crown and flint glass achromatic telescope lenses had largely been an English preserve, but Fraunhofer's fine optics changed that. These magnificent instruments, sometimes with lenses as large as dinner plates and mounted on precision clock-driven "Fraunhofer Equatorial" mounts of his own design, were quickly dominating the observatories of Europe by 1818.

Joseph Fraunhofer was not just an enterprising optical industrialist: he was also an inspired physics researcher. Back in 1802, Dr William Wollaston in London, when experimenting with the properties of the solar spectrum, had noticed some puzzling dense black lines interspersed amid the colours. Were they colour boundaries? A decade later, Fraunhofer too came upon these lines, and studied them carefully. What Fraunhofer was after at this stage, however, was a neutral colour-free "black light" source he could use as an absolute benchmark when tracing the precise geometrical passage of rays through lenses, in a way that could not be done with spectral colours, which blended into each other without exact boundaries. Then in 1814, Fraunhofer invented the spectroscope, based upon one of his own precision surveyor's theodolites. He placed a high dispersion (or wide light scatter) glass prism of his own manufacture in a precision adjustable mount in front of the theodolite's sighting telescope. It was then possible to direct a narrow beam of sunlight down the telescope, so that each colour could be observed at the eyepiece.

The graduated degrees of the theodolite, with its delicate micrometer subdivisions scale, enabled Fraunhofer to measure the exact angular position of each colour against an absolute standard.

Whereas Wollaston had observed a handful of curious black lines in the solar spectrum in 1802, Fraunhofer's more sophisticated spectroscope enabled him to detect and measure an incredible 514 lines – with hundreds more to be discovered in the future. We shall return to the impact of the spectroscope in facilitating the new science of "astrophysics" after 1859 in Chapter 17. Fraunhofer was honoured by the University of Erlangen, which bestowed a doctorate upon him in 1822, and he was also ennobled with the Bavarian Order of Merit. He died aged thirty-nine in 1826, probably from tuberculosis, inflamed by breathing toxic fumes generated by his glassmaking experiments.

Professor Bessel and the distance of the stars

Many astronomers from the seventeenth century onwards had tried to measure the distance of at least the bright stars (which were presumed closer than the dimmer ones), but had failed. Even William Herschel, in his meticulous measurement of the positions of binary stars, had likewise failed: this was simply because the stars are so far away, and angle-measuring technology was still too imprecise before the 1830s.

The mathematical theory behind parallax measurement was itself fairly straightforward and concerned with trying to detect a regular, six-monthly displacement of a star's position. As the earth goes around the sun, in half a year its position in space will shift by about 186 million miles, or double the earth–sun distance. This gives a 186-million-mile-long geometrical baseline across the solar system against which an astronomer might detect the slight shift of a star that is closer to the earth, with respect to stars in the same field of view that are further away.

9.2

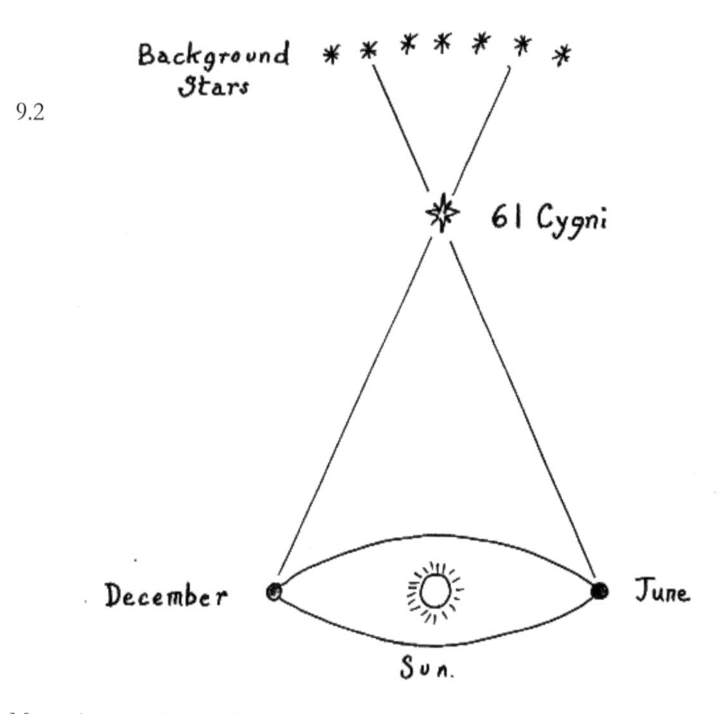

Measuring a stellar parallax. Bessel chose the star 61 Cygni, in the constellation of the Swan, because its large annual "proper motion" in space suggested that it was closer to us than other stars in the same field of view. In June and December, the earth is at opposite ends of its orbit, across a straight line some 186 million miles long. Using this like a surveyor's baseline, Bessel noted that the position of 61 Cygni changed slightly across six months with reference to the more distant stars in the field. From this tiny angle, he was able to calculate the star's distance. (Drawing by Allan Chapman.)

Yet if brightness was no guide to nearness, how could one choose a candidate star? It was here that Edmond Halley's 1718 discovery of stellar proper motions in space came in, as was shown in Chapter 2. It so happened that by 1804, Father Giuseppe Piazzi at the Palermo Observatory had discovered, from meticulous measurements, that the star 61 Cygni, in the constellation of Cygnus the Swan, had a very large annual proper motion vis-à-vis its neighbouring stars. This strongly suggested its relative closeness to the solar system.

Between 1836 and 1838, Wilhelm Friedrich Bessel of the Königsberg (now Kaliningrad) Observatory in Baltic East Prussia began to measure the position of 61 Cygni from its surrounding stars. For this research, he used a magnificent "heliometer" (Greek, "sun-measurer") telescope of 6.2 inches in diameter by Fraunhofer's Munich firm, though Fraunhofer himself had died before its completion. A heliometer was a refracting telescope on a clock-driven equatorial mount, the object glass lens of which was cut in two across the centre, and each half mounted in a precision screw-controlled frame. When the two half-circular sections were brought together to form a single glass disc, one had a single image at the eyepiece. Yet when the half-lenses were slid apart, one got a double image: one through each segment.

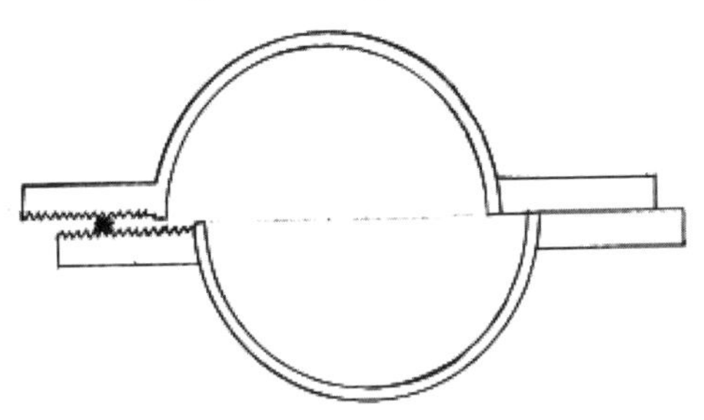

9.3 Divided lens heliometer. The telescope object glass was divided exactly into two segments capable of sliding against each other. By using precision adjusting screws the observer could bring the segments into perfect alignment, so that the two halves would act as a single lens. Sliding them apart, however, produced a split image. By this means a pair of stars that were very close together in the sky could be superimposed one upon the other, resulting in a single star image. Using the heliometer's precision micrometers, one could work out the angle between the superimposed stars to an accuracy of a fraction of an arc second. This was the basis of Bessel's measuring technique. (Drawing by Allan Chapman.)

To measure the tiny six-monthly movement of 61 Cygni, the separate star images were superimposed upon one another, and from the micrometer screw adjustments necessary to do this, Bessel could establish the displacement angles with great precision. By 1838 he was able to announce the first ever star distance. Star 61 Cygni was 10.3 light years away from the sun (an amazingly good value for the day, as we now accept its distance as 11.4 light years).

Almost simultaneously, Friedrich Georg Wilhelm von Struve, of the St Petersburg Observatory, announced a parallax for the star Vega, and Thomas Henderson in Edinburgh another for α (alpha) Centauri, from an analysis of observations that he had previously made from the Cape Observatory, South Africa.

The establishment of the first parallaxes at the end of the Romantic age heralded the fulfilment of the dream of centuries, to know the distance of the stars. A constantly advancing precise optical and engineering technology, combined with meticulous patience and profound mathematical analysis, had made this possible and taken the questing human intellect where it could not otherwise go.

CAROLINE THE COMET HUNTER

In 1788, the brother–sister domesticity of Observatory House, Slough, was upset when William Herschel decided to marry a wealthy widow, Mary Baldwin Pitt, a charming and lively lady of thirty-eight, although the fifty-year-old William was nicely off in his own right anyway. Although Caroline, who was also thirty-eight, decided to move out into nearby lodgings, she not only continued to work with her brother, and made a friend of her new warm-hearted sister-in-law, but also came to dote upon her little nephew, John, whom we will meet in Chapter 10.

In her various lodgings, such as The Crown Inn in nearby Upton, Caroline had more time to pursue her own original astronomical researches. These included a revision, corrections, and an index to John Flamsteed's monumental *Historia Coelestis Britannica* star catalogue of 1725, which she knew intimately from her years of working with William. It was published by the Royal Society in 1798: probably the first major piece of scientific work by a woman to be published by the Society. But her independent reputation was as a comet hunter. These fascinating transient bodies, as we have seen, had engaged Edmond Halley, Charles Messier, and scores of other astronomers across Europe. William Herschel had built her a special, easy-to-handle comet sweeper for her cometary researches. Between 1786 and 1797, she discovered eight comets, each of which she reported to the scientific world. The circumstances surrounding her discovery of comet number eight are outlined in a letter to Sir Joseph Banks, President of the Royal Society. In her letter to him, of 17 August 1797,[3] she says she did not trust sending her news by messenger. Instead, after only one hour's sleep, she set out on horseback in the early hours, and rode from Upton to London, a good twenty-odd miles, and then another "six or seven more to Greenwich" to the Revd Dr Maskelyne, the Astronomer Royal. Both gentlemen were good friends of William. Although protesting her ladylike modesty, Caroline clearly knew what to do when she made a discovery, and would not be gainsaid.

Following William's death in 1822, and having lived for forty years in England, Caroline decided to move back to Hanover, where she did not like the changes that had taken place since she had left as a twenty-two-year-old. Here, however, she carried out a decades-long scientific correspondence with her nephew, John, as well as with a range of European scientists, for the elderly lady was famous. In 1835, the Royal Astronomical Society elected her to an honorary membership (being a woman). In her ninety-sixth year, 1846, the king of Prussia, via Alexander von Humbolt, presented Caroline with the Gold Medal for Science in honour of her achievement, along, it is said, with a more practical gift for a ninety-six-year-old – a lovely armchair upholstered in Prussian blue. Caroline the comet hunter died on 9 January 1848, in her ninety-eighth year. She was one of the phenomena of the science and cosmology of the Romantic age. After her death, the Herschel cosmological torch would pass to her beloved nephew, Sir John Frederick William Herschel.

Chapter 10

Sir John Herschel: The Universal Philosopher of the Age

William and Mary Herschel were married at Upton parish church near Slough on 8 May 1788, and four years later, on 7 March 1792, they had their only child, a son, whom they christened John Frederick William.

John Frederick William Herschel: A Genius in the Making

John Herschel was born into a household that, had he not possessed the particular gifts of intellect and temperament that he did, could have been almost intimidating: it was also the foremost deep-space stellar observatory in Europe. His father had discovered infrared radiation only a few yards from the family drawing room when John was not yet eight, and the biggest telescope in the world, the great 40-foot reflector, stood on the lawn outside. Noisy games also had to be curtailed, as father did his observing at night and needed the day for sleep. Young John became familiar with having bishops, courtiers, and British and European scientists dropping into the family home, and he was talented enough to hold increasingly intelligent conversations with them. He also got on well with the workmen employed by his father to do the routine carpentry and metalwork for the Herschel telescopes, especially those made for commercial sale to wealthy clients willing to pay golden guineas for a "Herschel" telescope. John's fascination with tools, techniques, and making things went back almost to his infancy as, like his father, he was skilful with his hands.

Unusual as it was, Observatory House at Slough was certainly a loving home. John was cherished by his parents, though one suspects that, having what was seen as delicate health, with a tendency to pick up bad chest infections, he got a lot of mothering from Mary. His distinguished unmarried aunt Caroline positively doted on him, and this would mature into a warm personal and intellectual friendship that would end only with her death in 1848, when the then Sir John, Baronet, was fifty-six, and the father of a large family.

Though surrounded by high-powered adults, John did not lack for childhood friends, and he formed close and long-lasting friendships with Sophia and other Baldwin cousins on his mother's side. Some of his education came from private tutors, for although he was enrolled at nearby Eton College in 1800, he did not stay long. The rough and tumble of a Georgian public school worried his mother, and when she learned of her son being knocked down by a bigger boy in a fight, she promptly withdrew him from Eton. His formal education continued, more safely, in the hands of his father's friends, Mr Bull of Newbury and Dr Gretton of Hitcham, a village close to Slough. His intellectual brilliance had manifested itself early. Growing up bilingual in German and English, he easily mastered French. More important academically, though, young John took to Latin and Greek with equal facility. Dr Gretton's school was also renowned for something very unusual at that time: acquainting its pupils with natural philosophy, or science, in addition to the standard classical curriculum. John also had a private tutor for mathematics, Alexander Rogers, who instilled into him a lifetime's passion for the subject.

While John Herschel had around him people of his own age, such as relatives and school contemporaries, his first seventeen years were generally spent among people much older than him. Then, in Michaelmas Term, on 19 October 1809, he entered St John's College, Cambridge. Along with its rival and next-door neighbour Trinity, it constituted the mathematical powerhouse of the university. For the first time in his life, Herschel found himself in the company of young gentlemen of his own age on a daily basis, and quickly learned how to live within the polite rituals of an academic foundation of Anglican bachelors, including formal communal dining, attending chapel, and meeting with his college tutor. Mornings would be busy with study and tutorials, but afternoons were generally free to go on walks or ride, play informal games, or do more study. He would often see his clergyman tutor again in the evening, then do whatever he enjoyed before bed, which could include reading, attending student clubs or societies, singing, playing musical instruments, or even indulging in merry undergraduate horseplay. It was a daily ritual familiar to students in all of the colleges in Cambridge and Oxford Universities. As students have found through the generations, and still find today, college life is much more free, easy, and adult than school, for there are no parents keeping a watchful eye. It was a world of cultured gentlemen – both students and their ordained tutors – and while the ethos was quintessentially Christian, it was in no way self-consciously pious.

Although Herschel was of a hard-working and sober disposition, fired and enthused by the intellectual opportunities the university offered, he would have met and learned to get along with a real mixed bag of young English gentlemen. These would have included chaps who drank themselves silly every night, gambled, raced horses, went shooting, chased town girls, and got up to all sort of monkey tricks they hoped that their tutors would never hear of. While not in any way like that himself, young John Herschel had to learn to jog along with such fellows – as he would as a public figure in later life. Learning how to manage one's own time was always a major part of student life, for in addition to formal teaching, chapel, and dining, a young gentleman would have a lot of free time. Whether he used that time in idling around, or was a hard-reading man aiming for academic distinction, was largely up to the individual.

JOHN HERSCHEL INHERITS THE COSMOLOGICAL "FAMILY BUSINESS"

Without a shadow of a doubt, young John Herschel belonged to the latter category. His undergraduate career can only be described as brilliant, and in 1813, aged twenty-one, he graduated Senior Wrangler, obtaining not only a First Class Honours degree in mathematics, but the very top First. He also won Cambridge University's chief mathematics prize, becoming senior Smith's Prizeman. Recognizing his genius, St John's promptly elected him to a lay Fellowship of the College, a bachelor post that he continued to hold until his marriage in 1829. His youthful brilliance also saw him elected to the Fellowship of the Royal Society in 1813.

10.1 Sir John Frederick William Herschel. (R. S. Ball, *Great Astronomers* (1895), p. 253. Author's collection.)

As a Cambridge undergraduate, John made lifelong friendships, including Charles Babbage (later inventor of the first proto-computer) and George Peacock, who would join the ranks of scientific clergymen and become Dean of Ely Cathedral. Even as undergraduates, however, Herschel and his friends were demanding a broadening and toughening up of Cambridge mathematics. They founded the Analytical Society, aimed at using the Leibnizian calculus instead of the Newtonian "fluxions" method of calculation, and the "post-Newtonian" higher mathematics of Laplace and Lacroix on the Continent. We are not told whether their less able fellow students, who were already struggling to master Newton's *Principia*, appreciated their efforts to make the Mathematical Tripos yet more rigorous.

As a Fellow of St John's, Herschel would have had an elegant set of Fellow's rooms, and his own male servant. He would also be required to teach, but "pupillizing block-heads" was not an activity he enjoyed.[1] As he had no personal vocation to enter the priesthood, a common career move with many Oxbridge dons, he began to consider other career prospects, and entered Lincoln's Inn to train for the Bar, the law being a profession that many mathematically minded men found appealing. In fact, he liked it no more than "pupillizing". By instinct, he was an independently minded English gentleman. He could work with spectacular intensity and originality when following his passions – mathematical astronomy, chemistry, and his beloved optics ("my first love") – but he hated being trammelled. Fortunately, his father's earned wealth and his mother's land and money gave him economic freedom. He abandoned his half-hearted efforts to become a barrister, partly on the ground that his dusty law chambers in the Temple were not good for his bronchitic chest. It would be some thirty-five years before John Herschel was coaxed into taking a paid job – as Master of the (Royal) Mint, between 1850 and 1855 – which too had deleterious effects upon his health, and of which more will be said anon.

John Herschel would – to some degree reluctantly – find his lifetime career direction and fame in the summer of 1816. Mother and father (now Sir William, newly created Knight Guelph) were on holiday at Dawlish, in glorious south Devon. John came to join them, and Sir William, then aged seventy-eight, reminded him that there was still a lot of sky to be swept, and that he was getting no younger. Consequently, devoted son that he always was, the twenty-four-year-old John packed up his Lincoln's Inn chambers and college rooms, and came back to live at the family home, Observatory House, Slough, to take over the family business of fathoming the "length, breadth, and depth" of the universe.[2]

Optics, chemistry, photography, and a gift for friendship

In an undated note of around 1841–43, Sir John Herschel would tell his wife Margaret, "I have been *dallying* with the stars. *Light* was my first love! In an evil hour I quitted her for those brute & heavy bodies which, tumbling thro' ether, startle her from her deep recesses and drive her trembling and sensitive into our view."[3] So optics was his early passion, but astronomy broke in, obliging him to entertain new thoughts. As he was all too aware, his own father had caused the scientists of Europe to rethink the nature of light after William's discovery of "calorific" IR light at the family home in 1800, while Johann Ritter, Wollaston, Young, Fraunhofer, and others had shown that the spectrum contained a much more complex bundle of forces than Newton had realized.

With his natural gift for friendship, John had come to know Sir Humphry Davy and several other leading chemists, especially through the Royal Society. Like optics and cosmology, chemistry was changing beyond recognition from where it had stood when Sir William first began to look through a telescope. Gone were metaphysical principles such as phlogiston (a postulated fire-like constituent of combustible bodies), and in the wake of Lavoisier, who was murdered by the French Revolutionaries in 1794, chemists were now thinking in terms of unique, fundamental elements such as sodium, potassium, sulphur, oxygen, carbon dioxide, and the metals. The world, and maybe even the sun and stars, were now understood to be the products of complex combinations of these elements. Even the new galvanic battery-generated current, electricity, could be used to break up complex substances into their metallic elements and gaseous fundamentals, using electrolysis, as Davy showed in his discovery of potassium and sodium in 1807. The ancient element water was shown to be the product of two gases, hydrogen and oxygen.

Johann Ritter's discovery of ultraviolet light in the solar spectrum, mentioned in Chapter 9, derived from his prior experiments into the way that sunlight darkened silver chloride salts, emphasizing once more the close interrelationship between astronomical bodies and chemistry. Building on prior sunlight and chemistry experiments in France, Nicéphore Niépce, Louis Daguerre, and Herschel's friend William Henry Fox Talbot each independently developed technically and commercially viable processes for "light-drawing", or photography, by 1839. John Herschel would play a significant role in the early development of photo-chemistry, as we will see in Chapter 15.

Herschel's love affair with light was to be lifelong, the wonders of deep-space cosmology notwithstanding. Around 1820, while learning the practical business of deep-space observation from his father at Observatory House, he was also wrestling with optical image formation and the aberrations produced by contemporary achromatic lenses – like Fraunhofer in Germany and others across Europe, all of whom were in pursuit of the perfect telescopic and microscopic image. What John Herschel was aspiring to develop, however, was a perfect *aplanatic* lens, consisting of a crown and a flint glass component, that produced a flat field, or a field of equal clarity in all its parts. Such a lens would be a further improvement on the existing achromatic lens, and would be achieved by designing flint and crown glass components with the correct geometrical curves. He published his mathematical formulae for such telescope or microscope lenses in the *Philosophical Transactions* in 1821,[4] and *The Edinburgh Philosophical Journal* in 1822. They were intended to act as a guide for practical manufacturing opticians.

Slough, marriage, then the Cape of Good Hope

Sir William Herschel died in 1822 aged eighty-three, and by this time John had become his worthy successor. Eagle-eyed, manually dexterous, a world-class mathematician, and a skilful observer with his father's upgraded 20-foot reflecting telescope, John continued his northern-hemisphere sky-sweeps, and as a fine draughtsman made detailed drawings of nebulae and other deep-space objects. In particular, his study of the great nebula in Orion, an account of which he published in 1826,[5] was partly intended to provide a benchmark by which he, or other astronomers at a later date, might perhaps detect evolving changes in structure.

Until he was thirty-six, John the academically minded bachelor does not appear to have been especially concerned with matrimony, but then, in London, he became acquainted with the family of a Scottish Presbyterian minister, the Revd Dr Alexander Stewart. He enjoyed their genial and humorous domesticity, and fell in love with one of the daughters, Miss Margaret Brodie Stewart. On 3 March 1829, they were married at St Marylebone Church, London. The aging bachelor would be rapidly transformed into a doting family man, a loving husband, and father of twelve gifted children.

In January 1832, John's eighty-three-year-old mother Mary passed away in the family home in Slough. Devoted to his parents as he was, John's mother's passing made possible an intention that he shared with his late father: to travel south for several years to extend the Herschel sky-sweeps into the southern hemisphere. After a visit to Aunt Caroline, now living back in Hanover, and many other friends, John, Margaret, servants, and their already growing family boarded the great East Indiaman *Mt Stewart Elphinstone* in Portsmouth on 13 November 1833. Their telescopes, clocks, instruments, and household equipment had already been stowed aboard in London docks before the vessel sailed around Kent, making her last port of call at Portsmouth.

The East Indiamen were the luxury liners of the day: spacious sailing ships plying between London and India, and the Herschels took a suite of cabins, paying a fare of £500 for the voyage. They enjoyed a fast and trouble-free passage, arriving at the Cape of Good Hope on 16 January 1834. Familiar as Herschel was with the constellations of the southern sky from maps, he now began to learn about the stars by direct observation from the deck each night, as degree by degree *Mt Stewart Elphinstone* cruised south. He leased a house and estate, "Feldhausen", but soon after bought the property for £3,000. In his grounds, beneath the Table Mountain, he re-erected the famous 20-foot telescope, along with smaller transit and other telescopes. Right from the start, he discovered major differences between the northern and the southern hemisphere sky, quite apart from the constellations themselves. Most notably, they related to the distribution of nebulae and planetary nebulae: those large glowing spheres in deep space first studied in the northern hemisphere by his father.

10.2 John Herschel's 20-foot-focus reflecting telescope at the Cape of Good Hope. (Herschel, *Results of Astronomical Observations Made… at the Cape of Good Hope* (1847) and R. S. Ball, *Great* Astronomers (1895), p. 262. Author's collection.)

Especially fascinating were the two "Magellanic Clouds", which seemed to be separate from the Milky Way, and for which there was no equivalent in the northern skies. John counted as many as 919 nebulae in the larger Magellanic Cloud, and 226 in the small one. In the south, moreover, there were large numbers of double and triple gravitationally connected star systems, demonstrating once again that Newton's laws also acted in the depths of the stellar universe. The apparent asymmetry between the structures of the northern and the southern deep skies puzzled and intrigued Herschel, and he embarked on the primary fact-gathering that he realized must precede any cosmological speculations.

Nor did the Herschels lack for friends at the Cape. First among them was perhaps Thomas Maclear, the former Bedford surgeon and "Grand Amateur" astronomer who had gone out to the Cape to take over the Directorship of the Royal Observatory in Cape Town. Maclear also appears to have acted as the Herschels' family doctor, for on one occasion he came around to Feldhausen to remove one of Margaret's troublesome teeth, without, in the 1830s, any anaesthetic. There were also jolly picnics and visits to a nearby beach, although on such outings John often took a rifle to deal with marauding lions. Snakes were also a pest, and on one occasion the fearless Margaret despatched a puff adder in the drawing room with a poker.[6] The Cape skies had a breathtaking transparency, but the scent of a lone man up on the 20-foot telescope's framework could sometimes attract packs of hungry feral dogs at night, obliging Herschel to take a pistol loaded with buckshot when off observing.

After almost five years of relentless observing, measuring, and cataloguing the strange inhabitants of the southern deep skies, the Herschels sailed home in 1838 on board the *Windsor Castle* Indiaman. The financially astute Herschel paid the return fare out of the profits he made when he sold Feldhausen. The young Queen Victoria had just ascended the throne and was about to be crowned, and Herschel, the scientific celebrity, received a baronetcy, making him Sir John in England, to add to his earlier (like his father's) Hanoverian title of Knight Guelph. By nature a quiet retiring man, Herschel was overpowered by his new Europe-wide celebrity. While appreciative of the respect and admiration heaped upon him, he needed time to make sense of the fundamental discoveries he had made in the southern hemisphere, and relate them to those that he and his father had made in the north.

In 1833, the Herschels had taken their three children to the Cape, but they came home with three more. The old home, Observatory House, Slough, spacious as it had once seemed, was now getting cramped. Moreover, Isambard Kingdom Brunel's Great Western Railway – Brunel was another of Herschel's friends – had now reached Maidenhead, making it even easier for Herschel's admirers to drop in unannounced, to disrupt the calm thinking space he craved. Consequently, he quitted Slough and bought Collingwood House and grounds at Hawkhurst, then in the depths of rural Kent, for £10,500, while retaining ownership of the Slough property. Collingwood was a fine gentleman's residence, an ideal house in which to work, entertain friends, and bring up his ever-expanding family, especially after he had spent a further £1,400 on improving it.[7] The house would later become a private school, which I visited some years ago.

Nine years would elapse between the return from the Cape and Herschel's publication of his monumental *Results of Astronomical Observations... at the Cape of Good Hope* (1847). The data it contained, in conjunction with Herschel's other researches, would modify Sir William's earlier conclusions, and become one of the cornerstones of early Victorian cosmology.

THE HERSCHEL COSMOS OF 1850

At the Cape, John Herschel made a fresh series of meticulous zonal surveys of the Orion Nebula. At first, when comparing his Cape Orion surveys with those made in Slough in 1825–26, he believed that he would see new – perhaps evolving – structural detail within the nebula. If this were so, then it could suggest that the Orion Nebula was "close" to us, in cosmological terms, and presented an image big enough for an observer to detect small structural changes. But this would not turn out to be the case, for Herschel realized that from the Cape, at a latitude of about 33° south, the Orion constellation (now upside down) was much higher up in the sky than it was in Slough at 51½° north. Besides, the Cape air was much more transparent and purer than the often damp air of the Thames valley. So was the extra detail visible at the Cape simply the product of better seeing, or air transparency, conditions?

One significant conclusion Herschel drew from the Cape observations concerned the very structure of the galaxy (which in 1850 meant the Milky Way), occasioning a modification of his father's model for the "Construction of the Heavens". Instead of seeing the stars in a plane, or striatum, like Sir William, Sir John now came to see them as forming a circle or ring about a cosmological centre. Because the distribution of the nebulae in the northern and southern hemispheres appeared asymmetric, moreover (nebulae seemed to cluster around the galactic pole, in Ursa Major, Leo, and Virgo in the north, but were more evenly distributed in the south), John suggested that our solar system was not at the geometrical centre of the galaxy. Yet, as we have seen, in the south there were the nebula-packed Magellanic Clouds, of which there was no equivalent in the north.

Very importantly, in cosmological terms, Sir John began to wrestle with what kind of stuff the stars and nebulae might be made of. His rigorous training in Newtonian celestial mechanics at Cambridge had taught him to be sceptical about the existence of "luminous chevelures", "shining fluids", and other vague matter in the nebulae, for having no ostensible substance, how could such stuff respond to gravity, and how could it glow? His Newtonian training had taught him that gravity only acted between solid pieces of matter – be they the size of planets or of dust grains. He thus concluded that the nebulae must be *particulate* in their composition, or made of countless billions of discrete particles (somewhat analogous to individual particles in a sandstorm, or water droplets in a rain cloud), each obedient to the universal law of gravitation.

This particulate theory of nebulae found a parallel in the way in which dust in the air within a room produces shafts of sunlight, along with an astronomical phenomenon that, in our twenty-first-century light-polluted skies, is rarely seen: the zodiacal light. This is a band of light stretching up in the plane of the zodiac – or the sun's apparent path in the sky – as the sky darkens after sunset. Victorian astronomers – still in an age of candles and gas illumination – already realized that this faintly glowing band was caused by a fine dusty detritus in the plane of the solar system, the possible source of what Herschel called *meteorolites* (meteorites). Yet the zodiacal light was being reflected from individual particles, each bound by Newton's laws to move about the sun, and was not a vague shining fluid or flocky stuff.

Could not the nebulae, therefore, be composed of similar, physical-law-obeying *particles*, he suggested? Such thinking conformed to Herschel's argument that science must be inductive, or based on physical evidence, and had no place for vague flockiness, whereas it did for a particulate composition of the nebulae, even if they were only gas clouds composed of individual physical atoms. This inductive, particulate view was believed to be substantiated when Herschel's friend, Lord Rosse at Birr Castle,

Ireland, brought his giant 72-inch-aperture telescope into use in 1845. Its great mirror, with its enormous light-grasp, revealed that about a dozen hitherto "flocky" nebulae in the Messier's, William Herschel's and John Herschel's catalogues exhibited distinct structure and shape. In particular, the nebula in Hercules, number 51 in Messier's catalogue, displayed a spiral whirlpool structure: the first discovered of those spiral galaxies that would play a major part in the development of twentieth-century deep-space cosmology of Edwin Hubble and others. Clearly, gravitational forces must be at work in shaping these objects.

The size of the stars and their absolute brightness

While at the Cape of Good Hope, with its breathtakingly clear skies, John Herschel also turned his attention to what astronomers would come to call the *absolute brightness* of different stars. It was accepted, philosophically, that some stars might appear bright because they were relatively close, and others dim because they were remote. But how bright would individual stars be if, theoretically, they were all lined up at the same distance from us? What would their intrinsic or absolute, as opposed to their relative, brightness be? Here Herschel, the gravitational physicist, displayed his matching skill as a gifted practical technician and instrument designer, while also returning to his old love affair with light. This was shown in his invention of the astrometer or star-measurer: the original ancestor of all our modern photographic, electronic, and even digital starlight-measuring instruments.

To ascertain the absolute magnitude of the stars, Herschel needed an unvarying standard light source to serve as a comparison, alongside the star whose brightness he was measuring. His astrometer was basically simple in design, consisting of a 45° prism, a tiny hole of known width drilled through a metal plate, a metal arm, and a string-operated slider mechanism to move the prism and pinhole along the rod. The astrometer would be attached to the eyepiece end of a telescope, such as his favourite 20-footer with its 18.70-inch-diameter mirror.

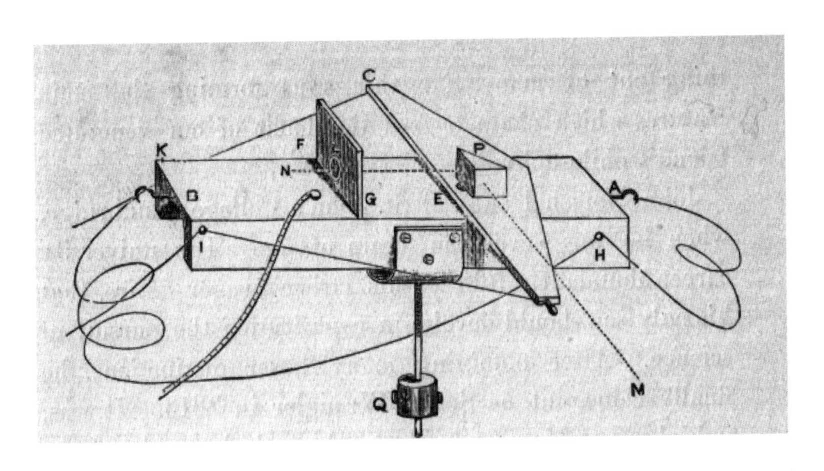

10.3 John Herschel's "astrometer", c. 1836. A ray of moonlight comes through a small hole drilled through a brass plate, to be reflected by a prism into the telescopic observer's field of view. The moonlight ray could then be compared with the brightness of a chosen star, such as Vega. By string-operated sliders, the prism and hole could be slid along an arm, until the "standard" ray of moonlight was of exactly the same brightness as the star. The distance and geometrical relationship between the holed brass plate and the telescope eyepiece enabled Herschel to calculate the proportionate brightnesses of different stars. (Herschel, *Results of Astronomical Observations Made… at the Cape of Good Hope* (1847) and R. S. Ball, *Great Astronomers* (1895), p. 249. Author's collection.)

On a night when the moon was appropriately placed in the sky, a selected star image, such as Arcturus or Vega, would be brought to focus in the telescope eyepiece. Then a single ray of moonlight would be directed from the prism, through the pinhole, and into Herschel's eye, in the same field as the star. This moonbeam would appear like a false star, its standard brightness (through a hole of, let us say, 1/25th of an inch across) capable of comparison with the star's natural brightness. The string and pulley mechanism made it possible to slide the pinhole plate closer to or further away from the eye, to make the moonbeam appear brighter or dimmer, until it appeared of identical brightness to the star.

Knowing the exact diameter of the pinhole, and its precise adjusted distance from the telescope eyepiece, Herschel was able to employ these geometrical proportions to establish which stars were *absolutely* brighter or dimmer. The sun's absolute brightness could be computed, as we knew its distance in miles with great accuracy. Then when Bessel in Königsberg measured the distance of 61 Cygni in 1838 – 10.3 light years (now remeasured at 11.4 light years) – this star's absolute magnitude could likewise be established. Therefore, by comparing absolute magnitudes against "standard distance" stars such as 61 Cygni, Herschel was able to slot many other stars into a scale of absolute brightness and possible size.

Vega and Arcturus were now shown to be vastly larger than our sun, while the puzzling planetary nebulae were also enormous, and at unimaginably vast distances from us. How did these bodies burn and generate their light, which must be truly prodigious to cross such vast tracts of space? Once again, we encounter the awe-inspiring cosmology of the late Romantic age, taken to an ever-widening readership, both by Herschel and his admirers, such as Professor John Pringle Nichol of Glasgow University, whose popular lectures and books, such as his *Views of the Architecture of the Heavens* (1843) and *System of the World* (1846), had a very wide appeal.

Sir John Herschel, the universal philosopher

In our present day, the term "philosopher" is usually applied to those pondering pure thought questions. Two centuries ago, however, it was understood much more broadly, designating a person of profound understanding, studying a range of matters extending from abstract metaphysics via physiology to unravelling the length, breadth, and depth of the stellar cosmos. In 1850, philosophy encompassed scientific research, while physics was natural philosophy.

By 1850, with his Cape of Good Hope discoveries behind him, it was Sir John Herschel who, more than any other British, Continental, or American savant, was coming to be recognized as *the* philosopher of the age. Every branch of intellectual endeavour had apparently been, in some way, illuminated by him. The acute mind that had studied the secrets of deep space in both hemispheres, which led to fundamentally new ideas about the nebulae and galaxy, had also advanced optics, chemistry, and physics. Nor had he neglected his beloved classical literature, for in 1866 he published a new translation of Homer's *Iliad*, in English hexameters. Thomas Hardy's fictional astronomer, Swithin St Cleeve, expressed his intention to make advances beyond Sir John's Cape of Good Hope researches in *Two on a Tower* (1882).[8] Yet in 1850, the great philosopher made the greatest mistake of his life, when he allowed the prime minister, Lord John Russell, to persuade him to accept the post of Master of the (Royal) Mint. Lord John's bait, of course, was that Sir Isaac Newton had held that post, between 1699 and 1727. Prestigious as the offer was, it inevitably plunged Sir John into a maelstrom of all those things he hated most: administration, committees, politics, and lack of mental space. For nearly five years, the job almost ruined his health, until, at last, a reluctant government allowed him to resign early in 1855.

We should also remember that Sir John Herschel's universal genius included a gift for friendship. Throughout his life, he made and retained friends. He hated and kept out of controversy, was not party political, and avoided making enemies. He was seen not only as *the* philosopher of the age, but as a true *Christian* philosopher, fascinated by and in awe of not simply nature, but creation. When he died aged seventy-nine in 1871, Sir John Herschel was honoured with a grand public funeral in Westminster Abbey: incongruously, one might think, for such an instinctively modest man, but a clear indication of the deep respect in which he was held by the world at large. Eight eminent men of science acted as his pall bearers. One of them was Charles Darwin, who claimed that, when young, he had been inspired by Herschel's 1831 *Preliminary Discourse on the Study of Natural Philosophy*.[9] Sir John was to be laid to rest near Sir Isaac Newton: his perceived ancestor as a truly "universal philosopher".

Chapter 11

There Must *Be Somebody Out There!*

Many people today assume that the idea of life on other worlds, of exoplanets, and strange, alien beings, is the product of twentieth-century science and science fiction. Yet even a cursory glance at the recorded scientific history reminds us that such ideas go back at least four centuries, and took hold with an enthusiasm that grew as the nineteenth century ran its course.

A fascination with "aliens"

It was probably the Assyrian–Greek Lucian of Samosata who invented "moon-men", for his ambiguously titled spoof *A True History* (second century AD) told of a ship blown out of the sea that, after seven days in the air, landed on the moon. Tales in this genre were generally written as political satire, the author describing rational moon-folk, who were like humans at their best but wiser, and provided a role model by which we could mend our own broken world. The idea of using an imaginary faraway place as the setting for political or social commentary on the present state of affairs was later used by Sir Thomas More in his *Utopia* (1516) (from the Greek *ou topos*, "no place"), and Sir Francis Bacon's *New Atlantis* (1626), although both More and Bacon set their wise and idealized communities not on an astronomical body, but on earth.

It was Johannes Kepler's posthumously published *Somnium* ("The Dream") of 1634, however, that really introduced Western readers to the idea of the wise and kindly alien. Fundamental to Kepler's *Somnium* were the imaginative possibilities for other worlds and their possible inhabitants, inspired by the telescopic discoveries after 1609, in the wake of which the moon, Venus, Jupiter, and Saturn were shown to be spherical worlds in their own right. If they were worlds, why should God not have populated them with intelligent living beings? It was the telescope, and the three-dimensional bodies it revealed, that opened up the serious possibility of life on other worlds, even though speculative theologians such as Cardinal Nicholas of Cusa had considered the possibility of other intelligences elsewhere in the universe in the 1460s.

There was one fundamental problem with alien beings, though: their theological status with regard to mankind. Were the moon-folk, Jupiterians, or Saturnians, who were clearly not descended from Adam and Eve, already in a state of spiritual grace, and saved, or were they, like us earthlings, sinful, damned, and needing God's saving grace? These and other spiritual questions do not especially bother modern "alienologists", but in the post-Reformation and Counter-Reformation of seventeenth-century Europe, they were weighty subjects to be pondered. The Revd Dr John Wilkins, whom I discussed in my book *Stargazers*, designed around 1640 a mechanically propelled flying chariot in which one might fly to the moon, and also made reference to the possibility of encountering Selenites, or moon-people. No-nonsense scientists like Robert Hooke and the Dutch brothers Christiaan and Constantijn Huygens (all Fellows of the Royal Society) spoke confidently of beings living on other worlds. The Huygens brothers, in Christiaan's *Cosmotheoros* (finished 1695, published 1698), considered that if these beings were similar to us, they would be technologically ingenious – and might already have devised telescopes for themselves!

By the eighteenth century, the invariably benign, human-resembling denizens of outer space were taken as read. Voltaire used his own kindly interplanetary tourist Micromégas, a giant from an "exoplanet" rotating around Sirius, and his equally gigantic friend from Saturn (who only got his lower legs wet when wading through earth's deepest oceans) in *Micromégas* (1752), to help teach us a few lessons; while William Herschel speculated on the existence of a cooler region inside the sun that might be capable of sustaining life. If the stars were suns like our own, then who was to say that planets might not be rotating around them, home to conscious beings of some kind?

The Revd Dr Thomas Dick of Broughty Ferry, Dundee

Early nineteenth-century Scotland, during that period often styled the Scottish Enlightenment, produced several men, often of a devout, Evangelical Protestant inclination, who accepted it as natural that God would have created whole races of beings on the planets and satellites of the solar system and the stellar universe beyond. These included the Revd Dr Thomas Chalmers of Glasgow, the geologist Hugh Miller, and the renowned Edinburgh physicist Sir David Brewster. One of their number was the Revd Dr Thomas Dick, born in Dundee in 1774, the son of a handloom weaver who, like so many in that age, rose to fame through hard work, self-education, and determination.

Able to read the Bible even before going to school, young Thomas Dick was captivated by a brilliant meteor (or what one might now style a fireball) that shot across the sky when he was nine. That meteor gave him a passion for all things celestial that would endure to his death, eighty-two years later. He entered Edinburgh University in 1794 and, as a poor youth, paid his own way by working as a teacher, using his earnings to finance his own higher education. He was next ordained by the Secessionist Church, married a lady named Elizabeth Aedie, and became Moderator of the congregation in Stirling.

Then things went terribly wrong. Some fifteen months after his marriage to Elizabeth, it was discovered that Thomas had been playing around with their maidservant, Betty Ker, and she was pregnant. He was found guilty of adultery, excommunicated, deposed, disgraced, and thrown out of his Moderatorship. Dick saw the hand of God at work in his humiliation, especially when both his wife's and Betty's babies died. As formal Christian ministry was now closed to him in the Scottish church, the deeply repentant Dick had to look elsewhere both for a livelihood and another form of ministerial outlet. He turned to his already proven talents as an educator and writer, becoming a schoolmaster and successful freelance journalist, especially on religious and scientific subjects. His *Christian Philosopher, or The Connection of Science and Philosophy with Religion* (1823) became a bestseller and, like his successive works, confirmed the intimate relationship between science and Christianity. Dick's books were destined to have a colossal impact on several succeeding generations, being translated into Welsh, German, and even Chinese, and influencing missionary explorer David Livingstone, American poet Edgar Allan Poe, and Welsh working-man astronomer and poet John Jones. Retiring from schoolmastering in 1827, Dick bought a house and built an observatory at Broughty Ferry, near his birthplace, Dundee. There he would complete over seven books and numerous articles.

Both implicit and explicit in Dr Dick's writings was the religious and statistical foundation of his argument for a "Plurality of Worlds", an inhabited solar system and stellar universe beyond. For – in accordance with arguments we have seen above – why should God have created so many potential celestial abodes if he had not wished them to be inhabited? Dick also took it as read, just like the Huygens brothers, Dr Wilkins, and Kepler, that these beings should be no less blessed with intelligence and curiosity than we are. But in an age in love with statistics as the nineteenth century was – especially the Scottish political economists after Adam Smith – Dick progressed beyond the old arguments of Divine Providence to produce some numerical calculations. Beginning with British population census returns, Dick calculated in *Celestial Scenery* (1838) that if the planets and satellites of the solar system contained a population density similar to that of England, there must be at least 21,895,000,000,000 beings living in the solar system.[1] God must have created many trillions more in the starry heavens. These books were not dismissed as flights of fancy, and by the 1830s, there was good reason to believe that the universe was a well-populated place.

In spite of the great success of his books, Thomas Dick does not seem to have made a corresponding fortune from their sale, although *Celestial Scenery*, for instance, was advertised in Edinburgh in 1838 as selling for 10 shillings and sixpence per copy, a sum equal to about four or five days' wages for a farm labourer. Fluent and persuasive as he was, he does not seem to have been a good businessman, and probably made bad deals with his publishers.

Hence it was a local subscription, a modest Civil List Pension, and a £300 gift from his admirers in the USA that kept Thomas and his second wife – another Elizabeth – in comfort over the last decade down to his death in 1857.

As wildly speculative as his celestial demographics might strike us today, we must nonetheless be aware that Dr Thomas Dick, LLD, probably did more than any other individual to alert readers on two, or perhaps three, continents to the likelihood that there was somebody out there. Even if one were cautious about large-scale cosmological demographics, there were eminently respectable astronomers seriously searching for signs of intelligent life closer to home, on the moon. The great north German lunar cartographer Johann Hieronymus Schröter believed the moon must be inhabited, while his friend and protégé, Franz von Paula Gruithuisen, was convinced that he had discovered a lunar city when examining what we now know to have been a complex and remarkably symmetrical lunar formation in 1822.[2] This cultural expectation led to a classic journalistic hoax, perpetrated in New York.

New York, August 1835, and the "Great Lunar Hoax"

As we saw in Chapter 10, Sir John Herschel was in South Africa, or else at sea, between November 1833 and June 1838. This made him conveniently remote for the enterprising New York journalist Richard Adams Locke, for it would be impossible to get a message from North America to Sir John, and receive a reply, in under five or six months in that age of sail. It tells us much, moreover, about Herschel's renown as a universal philosopher by 1835 that his internationally recognizable name should be used to front one of the greatest legendary scams in the history of journalism, under the title the "Great Astronomical Discoveries Lately Made by Sir John Herschel… at the Cape of Good Hope".[3]

On Friday 21 August 1835, when gripping news was thin on the ground, the enterprising New York *Sun* issued a short taster for things to come on its second page. Richard Adams Locke told his readers that a report had just arrived from a (non-existent)

Edinburgh periodical of wonderful discoveries made at the Cape by Sir John, passed on by Herschel's Scottish friend, the equally fictitious Dr Andrew Grant. The observations were made possible by "an immense telescope, of an entirely new principle". (As we saw above, the *real* telescope Herschel was using at the Cape was his well-known 20-foot reflector with its 18.70-inch mirror.)

The impact of the first instalment was left to sink in, and the eager public were all primed and loaded for the first of a sequence of revelations which would run from Tuesday 25 August to Monday 31, each one building on its predecessors. The tone of the *Sun* articles was perfect. Mimicking the sober tone of the non-existent Edinburgh learned journal from which the revelations supposedly came – a tone of cool-headed objectivity and technical jargon (especially about Herschel's fantasy telescope) – Locke landed his innocent readers as deftly as an expert fisherman might land an exotic catch. As Herschel could somehow increase the incoming moonlight by a new and entirely bogus optical physics, the familiar monochromatic lunar landscape displayed wonderful colours, both of geology and of rich vegetation, as he peered at the lunar surface and down its valleys as from a distance of 100 yards! And what discoveries did Herschel make?

The moon displayed dense and exotic flora, with beautiful giant plants, forests, a lunar cornfield, and rich grassy verdure. Then there were lovely animals, especially unicorn-like creatures resembling goats, and other animals resembling bison. The goat-unicorns capered joyously around like lambs. But the excitement became fever-pitched when humanoid creatures were espied, some covered with a coppery-coloured fur, with wings on their backs enabling them to fly. Crucially, they were both intelligent and benign, judging from their behaviour. These Lunarians, or Moonlings, were sitting in groups in the meadows as if enjoying a picnic, with mother Moonlings tending their babies. They clearly possessed language and a capacity for rational discourse, judging by their hand gestures and body language. This idyllic moonscape also contained architecture, for Herschel reported seeing perfectly cut obelisks, both in groups together and in chains across the verdant moonscape.

What is fascinating in our present context is that the existence of this Lunarian paradise was believed by so many. However, two academic astronomers, Denison Olmsted and Elias Loomis, who called at the *Sun*'s office to inspect the highly technical details from Edinburgh that the popular newspaper had not printed, smelled a rat when none could be found. Yet the *Sun*'s circulation shot through the roof, with some 19,360 copies being sold. The articles were reissued in pamphlet form, and these also sold thousands of copies. Richard Adams Locke did well out of it, his usual weekly wage of $12 being augmented by over $150 as a bonus.

European astronomers were much more sceptical of the *Sun* articles when the news arrived in England and France, as Herschel's genuine correspondents back at home had been told nothing about his lunar "discoveries". Besides, British and Continental astronomers knew that Herschel had gone to the Cape to study the deep stellar cosmos, not the familiar moon. Nor had he taken a special telescope of unprecedented power with him. Sir John's friend François Arago, Director of the Paris Observatory, was angry that Herschel's reputation was being used for a cheap publicity stunt. Yet, it was said, when Sir John himself finally received a bundle of American newspapers in South Africa, he laughed. Perhaps Herschel was flattered that of all the world's astronomical philosophers, he had been the one chosen by Richard Adams Locke to make this world-shattering discovery.

Aliens notwithstanding, the blatant hoax at the heart of the New York *Sun*'s tale of Lunarians damaged the credibility of American science. A decade later in 1846, for instance, the first reported case of painless surgery under ether anaesthesia (perhaps the USA's first world-changing scientific discovery) was received with scepticism, especially in England where "hoax" and "Yankee dodge" greeted the initial reports; until, that is, ether anaesthesia was shown to be transformative in its alleviation of pain and suffering.

All this apart, what is important is how tales about the discovery of an advanced culture on the moon was so readily believed even by educated people in 1835.

JULES VERNE: FROM THE EARTH TO THE MOON IN 1865

Like many people, I am sure, over the century since they were first published, I was captivated as a youth by the science-fiction novels of Jules Verne. Verne was in love with science, technology, and invention, and accumulated a great archive of articles and newspaper cuttings about all manner of scientific discoveries and technical contraptions of his own time. Then his imaginative alchemy took them further, to an adventurous crossing of Africa in a balloon, a journey to the centre of the earth, submarines, flying ships encircling the world in eighty days – and space travel. His stories were remarkably international in their characterization, with Englishmen, Germans, and Americans among his heroes, although a derring-do French gentleman adventurer would usually show up, such as the convivial, eccentric technophile M. Michel Ardan. Verne's novels invariably starred brilliant and daring private gentlemen, well educated, insatiably curious, independently well off, apparently unhindered by family commitments, and in command of their own time. Such scientific gentlemen – though often married – were a type in the days before science became professionalized by universities and governments in the late nineteenth century, especially in Great Britain and the USA, and I have styled them "Grand Amateurs". Edmond Halley and Sir John Herschel belonged to that breed, as to a large extent did William and Caroline Herschel, and we shall encounter more – including more ladies – over the next few chapters.

Jules Verne saw science and ingenuity as transcending national boundaries and constituting a brotherhood of the curious-minded, and this comes over at the start of *From the Earth to the Moon*, which was set in immediate post-Civil-War America. Yankees and Southerners are brought together by a love of guns, armour plate for "ironclad" ships, and ballistics, as the Baltimore Gun Club came up with a wonderful peaceful way of pooling their respective skills and passions. Could a giant cannon be cast that was capable of firing a great ball directly upwards? If fired at exactly the right second, as calculated by Newtonian gravitational dynamics, could that ball move in the right trajectory so as to hit the moon? Enter M. Michel Ardan, who has a proposal. Could the 9-foot-diameter cannon ball be modified, made into a pointed bullet-shaped projectile, and fitted out as a space capsule to convey the daring Frenchman to the moon? The old rivals Impey Barbicane and Captain Nicholl now join forces, and decide to accompany Michel Ardan on the lunar journey.

All the latest technology of the 1860s goes into designing the capsule, the interior of which resembles the well-padded interior of a gentlemen's club. To reduce weight, the projectile is made in the then scarce but lightweight aluminium, while a form of bottled air and gas provides for both respiration and illumination. Retro-rockets would even be used to reduce the descent to the lunar surface. Before a huge crowd, the giant Columbiad cannon, built at Tampa, Florida (ironically close to the later Kennedy Space Center), blasts the three into space with a charge of gun cotton ignited by an electric spark. Hydraulic buffers and a well-padded projectile interior ensure a safe launch into space. But then a near-graze with an unexpected meteorite upsets the trajectory, so that instead of the three landing on the moon, the projectile is flung into orbit around it. The story continues in Verne's sequel, *Around the Moon* (1870).

Through their large glass porthole, however, Ardan, Nicholl, and Barbicane obtain spectacular, close-up views of the lunar surface – and find no traces of life! Jules Verne's moon is a dead world. Having orbited the moon, they now make ingenious use of the retro-rockets, firing them at the exact time when they reach the computed balance point, where the lunar and terrestrial gravitational forces are exactly equal. The rockets' push delivers them not toward the moon, but into a terrestrial, home-bound orbit. The sturdy projectile luckily splashes down into the sea, off the American coast, at an incredible 115,200 miles per hour, but enjoying as they do the luck of the brave, they all survive without so much as a scratch. The USS *Susquehanna* has fortunately been cruising nearby, picks them up, and delivers them home to a heroes' reception.

Several commentators have drawn attention to the parallels between Verne's story and the Apollo 8's successful moon orbit in 1968. In this real-life space mission, just one century later, three men were launched into space from Florida in a small capsule, encircled the moon and observed the dark side, splashed down uninjured in the north Pacific to be picked up by a US warship, and were brought home to a global heroes' welcome, as we shall see further in Chapter 27. But by 1865, the weight of genuine scientific evidence was going against the existence of "Lunarians". What about life on Mars?

PITY THE POOR MARTIANS DYING OF THIRST: 1877

The planet Mars is some 48.8 million miles further out from the sun than we are (141½ million in total), and has a diameter of only 4,200 miles. Even by the 1870s, astronomers knew it must be a bitterly cold world, probably with a thin atmosphere. However, it did have observable seasons, for its white polar caps, like those of earth, increase and diminish as the north and south poles are tilted toward and away from the sun during Mars's two-terrestrial-year orbit. Since the seventeenth century, moreover, Mars had been known to have dark and light surface markings that might be caused by different land masses, or even vegetation.

Mars moves around the sun in a very elliptical orbit: it was Mars's unusual orbital behaviour that had first led Johannes Kepler in 1609 to his monumental realization that all planetary orbits were elliptical. One consequence of this ellipticity is that every two years Mars becomes relatively "close" in the sky, and depending on the Martian–terrestrial orbital cycles, it can become very close. In 1877, the Mars–earth proximity was only 35 million miles, which facilitated telescopic views of unprecedented detail.

It was Giovanni Schiaparelli of the Milan Observatory whose observations took the world by storm. Although only using, for that date, a relatively modest refracting telescope of 8 inches aperture, Schiaparelli reported seeing straight lines or, in the Italian, *canali* (channels) extending hundreds of miles across the Martian landscape. It had *not* been Schiaparelli's intention to imply that these lines were artificial or engineered channels, although the world's media began to talk of the discovery of *canals*, or engineered waterways, on Mars. So had the 1877 close opposition revealed that there were highly intelligent beings on Mars? And on their reddish arid world, were they engaged in massive engineering works, desperate to bring water from any lakes or seas, or even the polar caps, to irrigate possible agricultural areas?

The problem posed by the Martian *canali*, however, is that scarcely two astronomers agreed upon their routes or locations. Some "canals" appeared as single straight lines to one astronomer, but a parallel pair to another. Or were some individual "canals" as much as 50 or 100 miles wide? The wealthy Boston Grand Amateur Percival Lowell, a firm believer in the Martian canals, threw massive private resources into the search, including building the great Lowell Observatory at Flagstaff, under the pure skies of Arizona, in 1894. Mars was due for another close opposition in 1894 (though not as close as that of 1877), and using equipment on loan from Harvard as well as his own, Lowell hoped to confirm the existence of the canals. So committed was he to the Martian canals that in 1901 his Deputy Director, Andrew Ellicott Douglass, was sacked for expressing scepticism about their very existence, along with some of Lowell's other cherished ideas. Martian canals apart, Flagstaff, Arizona, would go on to be one of the world's great prime-sky location observatories.

The major problem with the Martian canals, however, was a general lack of consensus about their very existence. They were most frequently seen by committed believers, and tended to appear to best effect in telescopes of modest aperture and power. Edwin Walter Maunder, of the Royal Observatory, Greenwich, believed the canals came from optical aberrations in the telescopes, combined with ophthalmic and brain tendencies to connect incidental details into imaginary structures. Nowadays, the Martian canals are known not to exist, and I know modern astronomers, such as the serious London amateur astronomer John Dee FRAS, who has conducted telescope experiments that demonstrate the human eye's and brain's tendency to see straight lines and structures in blurred, complex images. But by the 1890s, the world had taken the Martians to its heart: clever, ingenious folk, no doubt peaceable and maybe more advanced than we are, struggling hard to conserve every drop of water on their desiccated planet. Herbert George Wells gave a new and terrifying twist to the Martian tale in 1898.

THE MARTIANS TURN NASTY

In many ways, the founding fathers of science fiction were Jules Verne and H. G. Wells. Both men, half a century apart, were science visionaries, whose novels exercised a profound influence on the collective imagination of the new century, with their prophecies of the impact of technological devices and of fantastic journeys. Wells's *The First Men in the Moon* (1901) tells how a couple of English experimenters produce an anti-gravity substance – named Cavorite after its inventor – which effectively propels their spherical spaceship to the moon. There they discover a bizarre subterranean world populated by super-intelligent 5-foot-high stick insects or Selenites, fat monsters whom they name mooncalves, and exotic vegetation.

By 1901, Wells's tale was already thin when it came to scientific evidence. Astronomers were fully aware by that date that the moon was a bitterly cold, airless world: yet who knew what lurked deep below the surface, or lived on the far, so-called dark side that we can never see from earth? His apocalyptic *The War of the Worlds* (1898) created the twentieth-century idea of superior, intelligent, aggressive, and utterly ruthless Martians. In Wells's brilliant imagination, the Martians are not only busy building canals and conserving their own precious water but, attracted by the blue of planet earth, determined to invade and take over the abode of primitive mankind. To add a true touch of horror, the advance party of invading spacecraft, which resemble great spherical objects walking on long insect-like tripod legs and contain the hideous, big-eyed, big-mouthed Martians, lands in Surrey. Very cleverly, Wells supplied familiar topographical and domestic detail, so that a resident of Woking or Horsell Common could go out and, in their imagination, see the exact locations where the vicious heat rays and other Martian devices had devastated all efforts to protect the populace. Here was Great Britain, at the zenith of her imperial greatness, being obliterated by a merciless alien foe, with obvious implications for Europe, the USA, and elsewhere. The latest high-power armour-piercing and explosive shells of 1898 could little more than scratch the great spaceships, which in retaliation could melt battleships with their heat rays and rip up great railway junctions like so much stubble.

All the glories of Victorian technology notwithstanding, it is Mother Nature who saves the day. The ruthless Martians are eventually "slain by the putrefactive and disease bacteria against which their systems were unprepared", such as the germs that cause the common cold, and against which they were powerless.[4] Novels, films, radio plays, and later television and digital technology would portray, adapt, and amplify Wells's story, to establish the twentieth-century love–dread affair with aliens.

So is there really anybody out there?

As things stood in 1898, one could still feel relatively sanguine about the prospect of life on other worlds. But as the twentieth century wore on, the increasingly pliable alien became the creature of the science-fiction writer, and no longer of the astronomer. Ever more powerful terrestrial telescopes, both optical and radio, then after 1960 space probes of increasing sophistication and range eliminated potential inhabitants from the solar system. The moon, visited by Buzz Aldrin and Neil Armstrong in July 1969, was clearly dead, while Venus had an acidic atmosphere that corroded away space probes. The "gas giants" Jupiter, Saturn, Uranus, and Neptune were shown to be turbulent snowballs of complex toxic gases, capable of generating surprisingly high temperatures due to a combination of rapid axial rotation, intense gas pressures, and internal and solar heat – resembling, perhaps, baked alaskas rather than frozen snowballs. Mars, alas, appeared lifeless, although geologically fascinating, with what looked like long dried-up riverbeds.

Only the possibly geothermal-warmed depths of the frozen oceans of some planetary satellites, such as Jupiter's Europa, retains a toe-hold for imagined aquatic life of some kind. Otherwise, as far as life in space is concerned, attention has shifted to possible earth-like exoplanets, rotating around stars. Even pulsating radio stars (pulsars), discovered by Dame Jocelyn Bell Burnell in 1967, whose pulses, it was hoped, might be radio signals from intelligent space-beings, were soon shown to be pulsing naturally. Thomas Dick's optimistic calculations for an inhabited solar system are no longer viable.

Since 1984, a privately funded American organization named the SETI (Search for Extra-Terrestrial Intelligence) Institute has been collecting and analysing electromagnetic data from space collected by radio telescopes, in the hope of finding something suggesting deep-space intelligences. But so far, the results have been zero. Then there is the Drake Equation, which attempts to compute the likely existence of intelligent life in space, though if you have no solid, evidence-based data to feed into the equation to start with, all you have left is numerological juggling, which tells us nothing, except that there are some earthlings who cannot abide being alone.

It has long struck me how so many devotees of deep-space aliens pursue their belief with an almost blind religious zeal, especially as many of them claim to be science-driven atheists, contemptuous of traditional religion. Some years ago, before an audience of 1,200 people, I ruffled a prominent alien advocate – who evangelically exhorted us to "reach out to our cosmic companions" – by suggesting that she was really engaged in a blind-faith secular religious activity. At least traditionally religious people, on the whole, such as Christians, honestly admit the faith component of their religion, and do not muddle-headedly claim it to be a species of science. I will, however, be returning briefly to life on other worlds in the last chapter of this book, where I add a few reflections about where astronomy might be heading in the twenty-first century. But as things stand at present, we seem to be alone. So it is perhaps best to leave the aliens to Hollywood!

Chapter 12

Mary Somerville: Mathematician, Astronomer, and Gifted Science Communicator

As scholarship in the history of astronomy has increased, and as the Royal Astronomical Society celebrated its centenary of the admission of women into the full Fellowship in 2016, we have become aware of the very significant contributions women have played in astronomical and wider scientific history. We have already seen the vital work performed by Caroline Herschel, while even before her, there was the seventeenth-century Pole, Elisabetha Catherina Koopman Hevelius, and in the eighteenth century Maria Gaetana Agnesi, who lectured on mathematics at Bologna. But as a gravitational, orbital mathematician, and as a high-level physical science and cosmological educator, Miss Mary Fairfax, or Mrs Somerville, would have a major effect on the age in which she lived.

Miss Mary Fairfax, the independent-minded admiral's daughter

Miss Mary Fairfax made her appearance in the world at Jedburgh, south Scotland, on 26 December 1780, shortly after her mother had waved her Royal Navy captain (later admiral) husband off to sea in the American War of Independence. The fortunes of war, and subsequent capture and imprisonment by the French, meant that Sir William Fairfax would not see his daughter for another eight years or so. She was to grow up at Burntisland, on the north shore of the Firth of Forth, across the water from Edinburgh. Her early outdoor lifestyle in lowland Scotland helped frame the iron constitution that would see her through ninety-two years. In addition to her toughness, she grew up to be beautiful, graceful, musical, artistic, and gifted at languages. But she developed a trait that worried her parents: for Mary had an irrepressible curiosity about numbers, equations, and all things mathematical. Was it not generally accepted that such studies softened girls' brains, and either drove them mad, or at least made them very peculiar, and hence, unmarriageable? The mathematics that especially enchanted Mary was that relating to the higher and most abstruse branches of the science, such as Newtonian gravitation, planetary orbital dynamics, and algebraic analysis. After negotiating with her parents, however, she was allowed a limited ration of mathematical reading time, and simply became hungry for more. Her sanity, charm, and femininity were clearly in no way impaired, much to her uncomprehending parents' relief. When Mary's own daughters edited their octogenarian mother's memoirs for publication in the late 1860s, they were at pains to emphasize that she had never been a bluestocking or female intellectual radical, and that her formidable intellect in no way diminished her grace and charm.

12.1 Mary Somerville in 1845, aged sixty-five. (Print in author's collection.)

Miss Martha and Miss Mary, however, were keen to suppress some of their mother's childhood reminiscences, which she had clearly written down with glee. Most notably, there was the story of a visiting army officer uncle on her mother's side, taking it upon himself to teach little Mary how to swear like a British trooper, and finding her a ready pupil. Her daughters edited this story out of their mother's memoirs, though it is there in the manuscript, and has been restored in Dorothy Mcmillan's edition of 2001.[1] In particular, there is Mary's childhood use of the swear word "B.....", which I interpret as that perennial military favourite, "bugger".

TWO CONTRASTING HUSBANDS

In 1804, the twenty-four-year-old Miss Mary Fairfax married her cousin, Captain Samuel Greig of the Imperial Russian Navy. He was descended from the Scottish Admiral Greig, RN, who had been invited over to Russia in 1763 when Czarina Catherine the Great was building the new Russian navy along British lines. As Czar Alexander I was fighting Napoleon Bonaparte on the British side in 1804, Mary met Captain Samuel Greig in London in a Russian consular capacity. Her mathematical activities were suspended, as Mary gave birth to a son, Woronzow Greig, and then a second, David, who died, and she played the part of a London naval-diplomatic housewife. Samuel had a low opinion of women's intellects, and did not encourage her. But in 1807 he died, leaving Mary an independent young widow.

She now returned to science with a vengeance, living at her parental home in Edinburgh, and paying for private lessons in higher mathematics from William Wallace, Professor of Mathematics at the University, and his brother John. By this time, she was at the cutting edge of European mathematics, studying British and Continental mathematicians, and especially the great Pierre-Simon Laplace. As we saw in Chapter 9, Laplace and his colleagues had taken the gravitational physics of Newton's *Principia* to new levels, and were analysing cosmological questions such as the possibly rotating nebulae and the dynamic stability of the solar system. The thirty-year-old Mary loved it, and before long Professor Wallace and his brother told her that they had taken her as far as they were able, and the rest was up to her.

In 1812 she married another, very different, military relative: the widely travelled senior British Army surgeon Dr William Somerville, MD, who was nine years her senior. "Somerville", as she always called him, was different because this medical FRS (elected 1818) was deeply proud of his wife's brilliance, and did all that he could to advance her studies and her scientific reputation. For the time, he must have been a husband in a thousand, although what doubtless made things easier was that their fields of expertise were quite distinct: mathematics and medicine. Dr Somerville became seriously ill with an infection during their honeymoon, but they were destined to enjoy forty-eight years of happy and deeply creative married life together.

Continental travel and international mathematical fame

Following the Battle of Waterloo in June 1815, which ended Napoleon Bonaparte's military tyranny in Europe, it became possible once more for the British to travel abroad. As in so many other pre-twentieth-century wars, military and political hostility did not necessarily stand in the way of scientific, literary, and cultural friendships between the combatant nations. The Royal Society never lost contact with its French counterpart, the Académie des Sciences, and when the Somervilles arrived in Paris in 1817, respect and warm friendship greeted them from the savants. William was a distinguished medical man, but it was Mary who became the celebrity. She was by this time the author of five short but deeply learned papers, published between 1811 and 1816, in Thomas Leybourn's *New Series of the Mathematical Repository*, while in 1811 she had won the Silver Medal for her solution to a complex mathematical question. In addition, the former Mrs Mary Greig's fame had spread to still war-torn Europe via correspondence with the wider scientific community, and by word of mouth.

In Paris, the Somervilles were entertained by Dominique François Arago, Director of the Observatoire, and shown the instruments, and by Jean-Baptiste and Madame Biot, and many other physical scientists. Mary was especially delighted to meet the elderly Marquis Pierre-Simon Laplace, whose as yet unfinished *Mécanique Céleste* (1796–1827) awaited its final volumes. As we saw in Chapter 9, it had been Laplace more than anyone else who had taken gravitational and planetary-orbit mathematical analysis beyond where Sir Isaac Newton had left it, to inspire new generations of creative Continental mathematicians. But Paris was not all science, and in her *Personal Recollections* Mary would recall attending parties and dinners, seeing the great tragedian François-Joseph Talma, and enjoying herself in the fashionable dress shops of the city.

From Paris, the Somervilles travelled south to Geneva, where they met up with old London friends, Dr Alexander Marcet, an Edinburgh-trained Swiss physician who had fled before French Revolutionary forces in 1794, and his English wife, Jane, herself a leading and very influential scientific writer. Mary found her mathematical fame had gone before her into Italy, and in Rome, this very Protestant Scotswoman even had an audience with His Holiness Pope Pius VII, whom she described as a "handsome, gentlemanly, and amiable old man".[2] The Somervilles did the sights of Italy, including the then new archaeological excavations at Pompeii. William's appointment, however, to the most plum job in the Army Medical Corps – the virtual sinecure post of Physician to the Royal Chelsea Hospital – in 1819 saw them splendidly set up in London. With a new home in fashionable Hanover Square, with easy access to the Royal Institution, the Royal Society, and all their scientific and artistic friends in London, and with a husband who was truly proud of his gifted wife, Mary's career was ready to take off.

Mary Somerville, Astronomy, and the Herschels

It was probably in the Paris Observatory, with Arago, that Mary first saw and perhaps used large state-of-the-art professional astronomical instruments. These would, of course, have included achromatic refracting telescopes, but more importantly, she would have seen transit circles, quadrants, meridian transit instruments, and precision screw-controlled telescope eyepiece micrometers. None of these instruments would have been used simply to look at the moon, stars, or planets. Instead, their purpose was to make critically accurate positional measurements of the sun, moon, and planets against the already meticulously mapped star constellations. This harvesting of geometrical data from the sky provided the mathematical astronomer with a bedrock of observed factual evidence upon which to base, and test, theoretical analyses. Such analyses included determining the independent proper motions of the stars in space (first noted by Edmond Halley in 1718) to establish the direction of the sun's motion in the galaxy, or they could be used to pin down the exact geometry of the immensely complex lunar orbit over its 18.6-year cycle.

Then there were the binary stars, first noted by Sir William Herschel in 1782: the close pairs (or triples, or quadruples) of stars that move about each other in elaborate geometrical minuets over decades, or even centuries. Very clearly, it was the force of gravity that connected them together by invisible ties, and which even enabled mathematical astronomers to calculate the orbital periods and individual star masses from their speed of motion and orbital eccentricities. There were deeper, philosophical questions, such as whether the present-day observed dynamics of the solar system could be used to determine if the planets had once been flung out of the sun, as Laplace had suggested, and whether the sun was a condensed nebula.

When one pushed the mathematical analyses of observed geometrical data further, one faced the problem of the solar system's long-term dynamic stability. Could the vast array of gravitational pulls and pushes within the solar system lead to its possible disintegration, either by gravitational collapse, or everything flying apart? At this birth-time of fossil geology, when the palaeo-anatomist Georges Cuvier was exhibiting the first skeletons of long-extinct aquatic creatures excavated in the Seine valley and elsewhere, one was forced to confront the question of the earth's vast geological history – first discussed by Edmond Halley, Robert Hooke, and others in the late seventeenth century. If giant fish had been entombed in a seabed which by 1815 had become the rocks of northern France, one was inevitably led to ask about long-term changes in the earth's climate and sea levels. Had the relative positions of the earth, moon, and sun changed over vast aeons of time, and how might present-day mathematical analysis of planetary data be used to cast light on ancient times? All of the above was but a part of that wonderful world of higher mathematics and its application to observed astronomical data that captivated the minds of Mary Somerville and her English, French, and other scientist friends. Precision observations of the heavens, combined with higher mathematical analysis, were seen as holding the key to the creation itself.

Back home in England, Mary would have learned more practical astronomy from their friends. The hospital surgeon astronomer Sir James South, on his estate at Campden Hill, near modern Notting Hill Gate, had a magnificent private observatory in the 1820s, and he gave Mary some practical teaching in using big observatory instruments to make observations of the celestial motions, such as measuring the positions of double stars with a precision micrometer. The Somervilles and the Souths shared several interests, especially medicine and astronomy, and Sir James would become the Founding President when the Astronomical Society of London (1820) obtained its Charter to become the Royal Astronomical Society in 1830. He had been a protégé of Sir William Herschel, and recorded several anecdotes about him, pertaining to his great skill as a big-telescope observer.

Although Mary would acquire an elegant 6-inch-diameter reflecting telescope made by James Veitch of Inchbonny, Scotland (an instrument that I examined and recorded on 26 October 1993), her astronomical interests tended more to the mathematical branch of the science rather than the observational. Even so, as her researches and her reputation grew, she fully understood the importance of instruments and their use. On their return from the Continent, Mary and William made a visit to Slough, where they met the elderly and iconic Sir William and Lady Herschel, though they were not able to meet Caroline, who happened to be away. But they did see the famous Herschel telescopes, and were impressed by the great 40-foot reflector that stood in the garden. But it was the young John Herschel who would play a major role in Mary's blossoming scientific career, and they would establish a friendship that ended only with John's death in 1871.

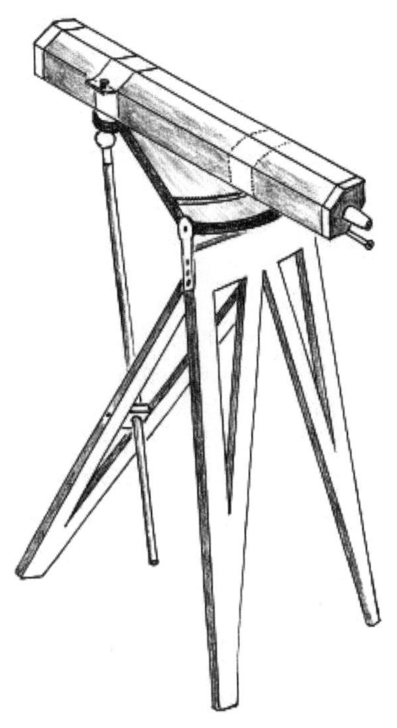

12.2 Mary Somerville's 6-inch-aperture speculum metal reflecting telescope, signed "James Veitch, Inchbonny". In private hands, the telescope was examined, measured, and drawn by Allan Chapman in October 1993. Its present whereabouts are unknown.

EARLY MATHEMATICAL AND PHYSICAL WORKS

In 1825, Mary read of the work of Professor Domenico Morichini of Rome, which claimed that the "chemical" or ultraviolet rays of the solar spectrum were capable of inducing magnetism into slips of steel. Assisted by Herschel and William Wollaston, who provided the required apparatus, she conducted a careful series of experiments over that summer, drawing the conclusion that "the refrangible rays of the solar spectrum have a magnetic influence".[3]

We do not know why, but her conclusions were incorrect, for UV does not induce magnetism. In 1825, the whole nature of light, the spectrum, static and current electricity, energy, and even gravity still posed mysteries for physical scientists. What were these strange forces that permeated the universe, from the sun and stars to objects in the laboratory? The significance of Mary's 1826 paper, announcing the results of her experiments of the previous summer, is that it was published in that massively prestigious organ of science, the *Philosophical Transactions* of the Royal Society, and sponsored by John Herschel. It was, perhaps, the first research paper written by a woman to be published in that journal.

Like her new friend Sir John Herschel, Mary Somerville was fascinated by light, and its effects upon different chemical and physical substances. In her optical thinking, she was firmly committed to Thomas Young's undulatory or wave model of light, arguing in her *On the Connexion of the Physical Sciences* (1834) in particular that a wave motion enabled one to explain far more observed optical phenomena than did the older Newtonian corpuscular model. In her *Connexion*, for instance, she describes simple experiments – borrowed from Young – that we have seen already, conducted with pinholes and slits cut into cards, and with fine stretched human hairs. She showed how interference patterns, dark lines, and colours could be produced with such simple table-top experiments, which could be easily explained by the wave theory of light, but not by the corpuscular theory. Voracious reading, simple experiments, observations, conversation, and epistolary correspondence lay at the heart of her work as a scientist and science communicator. She was also to write four monumental and deeply influential books.

MARY SOMERVILLE, THE PHYSICAL SCIENCES EXPOSITOR

Mary was a daughter of the Scottish Enlightenment of late-eighteenth- and nineteenth-century Edinburgh, and her call to major scientific authorship came from an Edinburgh friend. In 1827, Lord Henry Brougham, the Edinburgh-educated London lawyer, soon to be Lord Chancellor of Great Britain, and an old friend of the Somervilles, wrote to Dr Somerville at 12 Hanover Square, London. Brougham wanted Mary to write a book, and as it was not the convention of the day for a gentleman to write directly to another gentleman's wife, he channelled his request through her husband.

Brougham, as we shall see in Chapter 20, was the driving force behind the SDUK movement: the Society for the Diffusion of Useful Knowledge. This philanthropic movement aimed to educate the working classes, especially in science, technology, history, and literature, by funding accessible public lectures from high-powered people, mainly via the burgeoning Mechanics' Institute movement. It also used cheap, authoritative publications, such as the *Penny Cyclopaedia*, and specially commissioned books. Brougham wanted Mrs Somerville to write a book on modern, Laplacian, cosmology, for a general, basic-English-literate readership. She was enthusiastic, but her *Mechanism of the Heavens* (1831) shot far ahead of the intended mark and was unsuitable for the SDUK. It was a monumental treatise on mathematical astronomy and physics and, instead of providing a readable account of the "spirit" of Laplacian cosmology, contained a fundamental creative rethink of the whole science.

Their mutual friend, the Edinburgh publisher John Murray, took the manuscript of *Mechanism.* Its most accessible section was the "Preliminary Dissertation" covering the first seventy pages or so, a masterly overview of mathematical science, its scope, and usefulness, extending from recently discovered ancient Egyptian chronologies and sky-depictions such as the "Tentyris" (Dendera) zodiac down to gravitation theory. Gravity clearly enchanted Mary, and she provides a readable account of its manifold presence in astronomy, cosmology, and physics. She cites Thomas Young's calculation that at the centre of the earth gravitational attraction would be so great that "steel would be compressed to one-fourth and stone into one-eighth of its bulk".[4] She also praises the work of her heroine "Miss Caroline Herschel, a lady so justly celebrated for astronomical knowledge and discovery".[5] Words, however, can only take one so far, and she reminds her reader that a full knowledge of physical astronomy can only be attained through a knowledge of higher mathematics. Only then will the sheer beauty, harmony, and perfection – which Mary saw as a manifestation of the mind of God – be fully grasped.

After the "Preliminary Dissertation", she then moves to the truly tough part of *Mechanism*, explicating the intricate beauties of Laplacian cosmological physics, adapting, reworking, and going beyond it. That is why this book, clearly authored by "Mrs Somerville", as shown on its title page, caused such a sensation; for here were mathematical physics and astronomy of the highest order, written by a woman. Giving full acknowledgment to the recently deceased Baron Laplace and his *Mécanique Céleste*, Mary dealt with mathematical axioms: the motion and behaviour of solid and fluid bodies. She went on to the heavens, beginning with the mathematical complexities of the lunar orbit. Jupiter's satellite's orbits were dealt with in detail across 110 pages in Book IV, followed by the mathematics of planetary perturbations: how the sun and planets exert mutual dragging and accelerating gravitational pulls upon each other, to constantly modify and self-correct their elliptical orbits.

Mechanism sold at 30 shillings per copy – more than a skilled artisan's weekly wages – and the original print run was 750 copies. It received glowing reviews in the influential *Quarterly* and *Edinburgh* reviews, but perhaps the greatest accolade came from the Revd Dr William Whewell, soon to be Master of Trinity College, Cambridge, who wrote to Dr Somerville that a copy of his wife's book was now in the Trinity Library to help in the teaching of undergraduate mathematicians.[6] *Mechanism of the Heavens* became the second high-level science book written by a woman, after that of Maria Agnesi, to be used in university teaching.

On the Connexion of the Physical Sciences, Physical Geography, and *On Molecular and Microscopic Science*

The success of *Mechanism of the Heavens* left John Murray, Mary's publisher, and others wanting more. This led in 1834 to a work of staggering range and detail, for while *Mechanism* provided an overtly mathematical treatment of astronomy and physics, her *Connexion* delivered an equation-free physical sciences masterpiece accessible to the generally educated English-language reader. It ranged across all the then known branches of astronomy, electricity, optics, magnetism, and gravitation, even including astronomical data extracted from his translation of Egyptian hieroglyphs by the Somervilles' friend, Thomas Young.

Starting close to home, Mary discussed the electrical and magnetic causes of the aurora borealis as then understood – citing Faraday, Arago, and others – then dealt with the earth's own rotation. She also discussed the latest theories about meteors and the great shower of November 1799. Could these annual November showers, appearing in the constellation of Leo (now styled the "Leonids"), be caused by "a zone composed of millions of little bodies, whose orbits meet the plane of the ecliptic towards the point which the earth occupies every year, between the 11th and 13th of November"?[7] – an explanation for meteor showers virtually identical to the one accepted today. She paid much attention to comets, and how modern instrumental and gravitational studies had shown that they were not hot bodies. She tells how light from the brilliant comet of 1811 had been focused upon a thermometer bulb, yet had no thermal effect whatsoever. Comets must be flimsy bodies, for the comet of 1770 had "grazed" the earth within six lunar distances (about 1.5 million miles), yet analyses of the Greenwich observations, the most accurate in Europe at that date, had failed to record the slightest disturbance of the earth's daily rotation.

Section XXXVI, beginning on page 381 of the third (1886) edition, treats of the fixed stars, dealing with the binary systems of γ (gamma) Virginis, 70 Ophiuci, and others studied by her Grand Amateur friends, including Sir John Herschel, Admiral William Henry Smyth, and Sir James South. The stellar heavens were full of puzzles, for not only were there the misty nebulae and star clusters, but some appeared round, others elliptical, while others, in their misty glow, displayed a centre resembling "a candle shining through horn" (or as we today might say, through frosted glass).

Mary Somerville's *Connexion* was a late Georgian and Victorian bestseller, going through nine editions up to 1858, and selling some

15,500 copies at 10 shillings and sixpence each (between two and three times a maidservant's weekly wage, if she were lucky). There would be a 9-shilling posthumous edition coming out in 1877, some five years after Mary's death. With each edition *Connexion* got longer, and one can trace several new scientific and technological discoveries in successive editions, for Mary always brought each new printing right up to date. These included discoveries in the chemistry of early photography, the physics of electric batteries, spectroscopy, chemical analysis, nebular studies, and planetary theory.

Mary was delighted to hear how her treatment of planetary perturbation theory in the sixth, 1842, edition had first stimulated John Couch Adams to undertake his researches that led to the discovery of Neptune in 1846. Adams himself said this to Dr Somerville on a visit to Sir John Herschel's Collingwood House in Kent in 1848.[8]

The *Connexion* would be followed by two more magisterial syntheses of physical science. Her *Physical Geography* (1848) explicates the most recent discoveries in geology, and what we today would call geophysics. Here she discusses continent, mountain, and island-chain formations, along with vulcanism, while fossils supply us with a key to the earth's vastly ancient history. Carbohydrate chemistry, sunlight, and oxygen provide the physical building blocks for living matter. Nowhere, though, does she mention Noah's Flood.

Then, at the extraordinary age of eighty-nine, widowed and living in Italy, she produced her two-volume *On Molecular and Microscopic Science* (1869). Here, with an intellectual range and flexibility that belied her great age, Mary addressed the latest ideas regarding the elemental character of matter: in the laboratory, the sun, or the depths of cosmological space. From an astronomical perspective, however, the significance of *Molecular and Microscopic Science* lies in her treatment of very recent discoveries in spectroscopy, to which we will return in more detail in Chapter 17. The last thirteen years or so of her life saw the birth of both solar and stellar physics and the unravelling of the chemistry of astronomical bodies. It is amazing that this great cornucopia of wholly novel chemical and astronomical data had been so thoroughly mastered, lucidly argued, and presented, by a self-educated woman in her eighties.

NATURAL LAWS, RELIGION, AND HER FINAL VOYAGE

One point comes over time and again in Mary Somerville's four great books: that mathematical and physical science in all its integrated branches spoke of great laws. Science was about reason, not about the fantastical. While she showed little interest in dogmatic theology or conventional religious observance, she was convinced that "Surely it is not the heavens only that declare the Glory of God – the Earth also proclaims His handiwork".[9] In many ways, she saw science as a divine gift to mankind, an ongoing revelation. She concludes the *Connexion* by stating: "This mighty instrument of human power [or mathematical science] rests upon a few fundamental axioms, which have existed eternally in Him who implanted them in the breast of man when He created him after His own image."[10]

Calmly contemplating her own death, the admiral's daughter saw death as "a solemn voyage, but it does not disturb my tranquillity", for while "Deeply sensible of my utter unworthiness... I trust in the infinite mercy of my Almighty Creator"[11] to bring her safely to port.

On 28 November 1872, it is said that she had been correcting proofs, then the following morning the ninety-one-year-old glided peacefully out of her earthly anchorage.

Chapter 13
Sir George Biddell Airy of Greenwich: Astronomer Royal to the British Empire

Astronomy is an awe-inspiring science, posing all manner of profound questions about cosmological vastness and the invisible laws that bind it all together. On the other hand, since earliest times, it has also possessed a major utilitarian dimension, for by the motions of the heavens we construct our calendars, keep accurate time, and find our way across the great expanses of both land and sea.

It was toward this end that in 1675 the early Fellows of the Royal Society proposed to a very willing and nautically minded King Charles II that a national astronomical observatory might be established, for the benefit of both the Royal Navy and our merchant sea captains, both close to the king's heart. This Observatory would remap, or catalogue, all the stars to a new level of accuracy, made possible by rapid advances in instrument-making technology. Then it would make an exact, ongoing record of the motions of the sun, moon, and planets against the stars. The accumulated data might then, it was hoped, enable sailors on the high seas to exactly calculate their position in longitude and latitude with regard to an agreed point on the earth's surface, such as the king's new Royal Observatory at Greenwich. Its first Director, or Astronomer Royal, would be the Revd John Flamsteed, appointed under a royal warrant dated 4 March 1675.

Over the next 160 years, up to 1835, there would be five Astronomers Royal – Flamsteed's successor, 1720–42, being Edmond Halley, whom we met in Chapters 2 and 3. Although Flamsteed had been obliged to provide his own instruments, Edmond Halley, as we have seen, began the tradition of getting the government to provide state-of-the-art instruments, a custom that would continue into the twentieth century. These precision pendulum clocks, quadrants, transits, transit circles, and telescopes, would, between 1675 and 1835, represent the technological benchmarks of their day, while the data they harvested from the sky, when published, would go on to establish international points of reference in exact mathematical astronomy. A host of intended practical benefits would flow from them, from the finding of the longitude using the moon (lunars) in the 1760s (alongside John Harrison's quite separate chronometer method) to discoveries in geodesy, geomagnetism, and geophysics.

The Royal Observatory was not concerned with cosmology or philosophical issues regarding the structure of space, although it did observe, record, and publish orbital data pertaining to scores of comets. Publicly funded Greenwich was primarily a utilitarian institution, providing data and standards for maritime safety, cartography, and the efficient running of Great Britain, and any other country that chose to make use of its public observations. By the time that the Astronomer Royal John Pond retired on health grounds in 1835, Great Britain was not only the world's most advanced industrial nation; it was also the global superpower, and Pond's successor would, in many respects, transform the Observatory's role to suit the needs of the Victorian age.

SIR GEORGE BIDDELL AIRY (1801–92): EARLY LIFE AND ACHIEVEMENTS

All five of Airy's predecessors as Astronomers Royal had been gentlemen of independent means. Flamsteed had been the son of a rich merchant of Derby, who had spent £2,000 of his own money on equipping the Observatory, Halley was comfortably off, the Revd Dr James Bradley (in office between 1742 and 1762), was married to a Gloucestershire heiress, while his successor, the Revd Dr Nevil Maskelyne, was known to be so independently wealthy that when, in 1810, friends petitioned for an increase in his relatively modest annual salary of £350 – a sum that had not been increased since 1752 – Spencer Perceval, Chancellor of the Exchequer, turned his request down on the grounds that the now elderly Dr Maskelyne did not require any more public money.[1]

13.1 Sir George Biddell Airy. (Print in author's collection.)

The thirty-four-year-old Professor George Biddell Airy was a different case altogether. His enterprising yet financially modest father was of farm labourer stock, and while the maternal Biddells were better off, young George had nothing to inherit except brains, ambition, and driving energy. His uncle Arthur Biddell, who took young George under his wing, was a successful Suffolk land agent and modest landowner, well connected in East Anglia, in with the county gentry, clergy, and MPs, and with contacts in Cambridge University. George showed early promise, and from a place at Colchester Grammar School, where he excelled in mathematics and classical languages, he won a scholarship to Trinity, Cambridge's great mathematical college, Newton's own alma mater, and a rival of its next-door neighbour, Sir John Herschel's St John's.

Upon going up to Trinity in 1819, George found that Uncle Arthur's sartorial advice was out of date. The undergraduates were no longer wearing eighteenth-century knee breeches, but long trousers. Even so, Airy made friends quickly, and his Cambridge chums, mostly older than he, such as John Herschel, Charles

Babbage, George Peacock, Augustus de Morgan, and William Whewell, would remain friends for life, as would the Trinity geologist and Prebend of Norwich Cathedral, the Revd Professor Adam Sedgwick.

Airy's undergraduate career was simply brilliant, like Herschel's more than a decade before, and he won the top First of his year and the prestigious Smith's Prize in mathematics. His brilliance was openly acknowledged by his fellow undergraduates when, at the start of his second year in August 1820, young Mr Rosser and Mr Bedingfield paid Airy £14 per term apiece to coach them in mathematics. The ever-thrifty and independently minded Airy was proud that he was "now completely earning my own living", and no longer needing assistance from family or friends.[2] Upon graduation, he was quickly elected to a bachelor Fellowship of Trinity, and became an influential teacher. But having fallen in love with Miss Richarda Smith, whom he had met on a journey to Derbyshire, he was delighted to be appointed to the Plumian Professorship of Astronomy and the Directorship of the new Cambridge University Observatory, which enabled them to marry in 1830.[3] But he was already being headhunted by important government and Admiralty persons in London, as an obvious successor to the increasingly ailing John Pond, for the post of Astronomer Royal. Airy was appointed to the post in the latter part of 1835, and left Cambridge with his wife, moving into their new residence at the Royal Observatory.

With relation to both his Cambridge professorial and Greenwich appointments, he made it clear that he was, relatively speaking, a poor man. Unlike most senior Cambridge dons, Airy, a layman, had no additional ecclesiastical income from a sinecure benefice. Nor did he possess independent means, and his wife, though a lady, had no private wealth. He saw himself as an employed, working astronomer, and not a Grand Amateur. His ability as an astute financial negotiator, however, led to acts of "brinkmanship" both before Cambridge University's governing Senate (of which he was a voting member) and King William IV's government, and he secured hefty increases in salary before accepting both appointments. As Astronomer Royal, he would receive £800 per annum, plus a house;

some years previously, he had secured an annual Civil List Pension of £300 to be settled on Richarda, giving the Airys £1,100 in joint income, along with a free residence.

In consequence, Airy saw himself as a new type of man of science, whose job it was to serve the nation full time. Yet even with £1,100, an income close to that of Anthony Trollope's fictional Dean of Barchester Cathedral in *Barchester Towers* (1857), who received £2,000 p.a. before the sum was reduced to £1,200 by the Ecclesiastical Commissioners,[4] the thrifty Airy still saw himself as only modestly well off. When the newly appointed Astronomer Royal was offered a knighthood in December 1835, for example, he respectfully turned it down, on the grounds that he lacked "the necessary considerable fortune" to warrant such a title, seeing himself as just "the devoted servant of the country".[5] Several future administrations would offer Airy knighthoods over the decades, but it was not until 20 June 1872, by which time his beloved Richarda was in declining health, that he finally accepted the title of Knight Commander of the Order of the Bath, conferred upon the seventy-one-year-old Sir George by Queen Victoria at Osborne House, Isle of Wight.

Although married to a clergyman's daughter, and with a younger brother, William, an Anglican priest, Airy made it clear in his *Autobiography* that he possessed no clerical calling. He had an active interest in religious matters, however, was a loyal member of Greenwich parish church, and had a "veneration of our noble liturgy".[6] But he disliked the theological and doctrinal controversies then racking the Church of England, and was encompassing in his own churchmanship. Biblical history fascinated him, and in September 1865 he published a piece upon the possible location and features of the land of Goshen in Genesis, where Joseph settled his father and brothers when they fled from the famine in Canaan.[7] Airy also contributed to the support of John William Colenso, Bishop of Natal, who was removed from his see when he expressed reservations about the doctrine of eternal damnation, and the precise historical accuracy and authorship of the biblical books of Genesis, Joshua, and Judges. In his religion as in his politics, Airy favoured a sincere yet tolerant approach.

New instruments, chronometers, time, and the electric telegraph

Over three decades following his appointment in 1835, Airy obtained official grants that enabled him to fundamentally upgrade and improve the Greenwich Observatory instruments, and widen the scope of the Observatory to suit changing needs. While his Cambridge degree and prizes had shown him to be a top-class pure mathematician, he was an instinctively practical man, fascinated by the great engineering and electrical advances of that rapidly advancing age. The railway engineers Isambard Kingdom Brunel and George Stephenson, William Simms the instrument maker, and the Ipswich agricultural engineer Arthur Ransome were among his friends, and Airy was ideally qualified to apply sophisticated techniques of mathematical analysis to engineering and optical structures. Low-friction pivots used to balance heavy telescopes, elegantly counterpoised structures, and the famous "Airy Points" in load-bearing beams all enabled the post-1835 Greenwich instruments to obtain enhanced levels of accuracy in precision measurement. Airy saw the Greenwich achievement not exclusively in a British context, but in an international one, as each of his designs and innovations were minutely described and often illustrated with engineering drawings for publication. Queen Victoria's Astronomer Royal was as well known and respected in Paris, Berlin, St Petersburg, or Cambridge, Massachusetts (Harvard) as he was in Cambridge, England.

In addition to telescopes and other precision angle-measuring instruments at Greenwich, Airy's principal instrument was the great transit circle of 1851. This consisted of an 8-inch-aperture refracting telescope, some 12 feet long, set upon a pair of trunnions like those of a cannon, and capable of rotating, in a perfectly balanced engineered system, between a great pair of stone piers. The telescope was adjusted to the plane of the meridian, so that when rotated around 360 degrees, it described a perfect meridian, or north–south arc across the sky: the Greenwich, or "Prime Meridian" of the world.

13.2 The 1851 Greenwich meridian transit circle. (Œ. Dunkin, *The Midnight Sky* (1891), p. 156. Author's collection.)

Airy and his staff would then observe and record the exact position of every star, planet, the moon, and the sun, as they made the nightly, or daily, crossing of the meridian. The great transit circle could measure these positions to an angle smaller than a tenth of a single arc second of a degree, which in terrestrial terms would correspond to the angle displayed between the two sides of a ten-pence coin at a distance of over 30 miles. These observations – scores of thousands of them – would then be "reduced" (or calculated) and analysed by the "Computers", or mathematical calculators on the Greenwich staff, prior to publication and international distribution. They could then be freely used by astronomers in Vienna, Harvard, or Madras to provide tables for sailors of any nation on the high seas, or their exact astronomical coordinates could be used to make maps of any scale: maps of the whole earth, the Hawaiian Islands, or the Canadian great lakes.

In addition to observational astronomy, however, Airy's Greenwich developed other vital functions, serving not only Great Britain, her nation and empire, but also the wider world. Firstly, it ran a valuable service in monitoring and certifying the quality of marine chronometers made in London. These precision timepieces became vital adjuncts to accurate navigation and cartography, and fine Greenwich-certified chronometers by Charles Frodsham, John Fletcher, and many others might be purchased on the open market for a Royal Navy ship, or by a London merchant captain – or by a visiting Dutchman, Russian, or American.

Secondly, after 1838, Greenwich founded its "Magnetic and Meteorological" – or "Mag. & Met." – department to monitor both the weather and small variations in the earth's magnetic field. Both classes of data were not collected in isolation. Within Great Britain, James Glaisher, Superintendent of the Mag. & Met. under Airy, organized a small army of volunteer Grand Amateur co-observers across the country who sent their weekly observations to Greenwich: observations made with accurate, privately owned thermometers, barometers, hygrometers, wind-gauges, and in some cases early magnetometers. British Mag. & Met. work was shared with equivalent observers in Germany, France, Russia, the USA – wherever a field station could be set up. Airy, Glaisher, and their colleagues across the world hoped that by amassing and analysing a vast amount of meteorological and geomagnetic data, one might learn how to predict the weather, or how sunspots affected the terrestrial magnetic field: all of paramount importance to those in peril on the sea.

It was also Airy, more than anyone else, who established GMT – Greenwich Mean Time – as a national and global timekeeping point of reference, using astronomical data harvested by the transit circle and its smaller predecessor. Even when Airy came into office in 1835, the famous Greenwich Time Ball, high on the Royal Observatory roof, was hoisted each day and dropped exactly at 1 p.m. It served as a time check for all the sea captains whose ships, in those days, lined both banks of the River Thames for several miles downstream from the City of London and its docks.

The burgeoning of the new railway system in the 1840s also

created timekeeping problems, for the further west one travelled (such as from Paddington station, London, to Penzance in Cornwall), the more time one lost. Local noon in Penzance falls 23 minutes later than it does in London, owing to the earth's rotation. This east–west time difference was not so much of a problem in stagecoach days, but as Brunel's trains on the new Great Western Railway (alias "God's Wonderful Railway") could go at up to 60 miles per hour, then discrepancies and problems arose.

The solution to a "mean" or shared time problem across Great Britain could also be solved by another new and wonderful technology: the electric telegraph, whose wires were spreading like tentacles across Great Britain, Europe, and the eastern seaboard of the USA by the early 1850s. As electricity travels down a wire at 186,000 miles per second (the speed of light), if a switch were activated in Paddington station, a bell in Penzance at the other end of the 300-mile-long wire would ring as near to simultaneously as made no practical difference. In tandem with the electric telegraph, ingenious horologists developed electric-controlled clocks in the 1840s, so that a signal sent into the national telegraph network from Greenwich Observatory could by means of electromagnets control the time on every single clock on the national telegraph system, from Penzance to Manchester, and on to Aberdeen; the first "GMT" signals.

Some people objected to the Astronomer Royal intruding his time-signals into their towns or cities. The Dean and Chapter of Christ Church Cathedral, Oxford, for instance, insisted that the clock striking the bell Great Tom in Tom Tower should still display local, Oxford, time, which is 5 minutes slow of Greenwich. Then it was conceded that the face of the clock should have two minute hands: one showing Oxford time, the other "God's Wonderful Railway" time – handy if one were rushing to catch a train. The double fingers on the face in front of Great Tom have long since gone, but one of the glorious eccentricities of Oxford lies in the fact that even today, Christ Church Cathedral services, "by long custom", begin 5 minutes later than GMT time: six o'clock Evensong, for example, still begins at 6:05 p.m. (and the same applies to dinners and other Christ Church events). We enjoy explaining this to visitors.

Electric telegraphic timekeeping, in addition to its domestic applications, had important cartographic uses. One could, for instance, employ it to establish the exact east–west separation of astronomical observatories from each other, such as those of Cambridge, Oxford, Liverpool, and Edinburgh. Time signals could also be sent to both naval and mercantile ports, so that sailors could check and regulate their longitude-finding chronometers prior to sailing. Liverpool's Bidston Hill Observatory served this function for the bustling merchant port. Then after the first successful submarine electric cable was laid linking Great Britain to France and beyond, in 1851, signals from Airy at Greenwich enabled astronomers as far apart as Paris, Madrid, Berlin, and St Petersburg to coordinate their respective observations. As overland or submarine wires were further laid to India (1870), the USA (1858, then 1866), and Australia (1872), it meant that astronomers, meteorologists, and even newspaper editors could now tap into the telegraph systems of the world. By 1880 there was nothing technically impossible in the idea of an astronomer in San Francisco exchanging signals with a colleague in Moscow.

13.3 The Royal Observatory, Greenwich, c. 1870, showing the Time Ball and public electric clock on the gate. It was from here that the electric time signal was distributed through the telegraph network after 1852. (E. Dunkin, *The Midnight Sky* (1891), p. 153. Author's collection.)

Sir George Biddell Airy did not invent the electric telegraph, or even the mechanics of telegraphic time-signalling, but his creative imagination, astute awareness of the rapid technical and economic progress through which the mid-nineteenth century was passing, and position as the senior government scientist in the already global British empire meant that he was in a key position to be an innovator in the public service. The global telegraphic network that came into being between 1845 and 1880, with its capacity for instant communication, has come to be dubbed "the Victorian internet".[8]

Airy the scientific civil servant

Airy belonged to a new type of British astronomer – the conscientious well-salaried professional – and personified a new type of civil servant. A senior government post in the early nineteenth century was generally seen as something of a sinecure. It was not good manners to expect a gentleman to do a full day's work, as might be expected from a lowly copying-clerk, let alone a factory hand. Clergymen (especially cathedral dignitaries), Oxford dons, hospital consultants, lawyers, and Treasury or Foreign Office mandarins worked at their own speed and in their own way. Airy's financially independent predecessors at Greenwich had operated in a similar manner, although this did not mean that the work – be it amputating a leg in St Bartholomew's Hospital, advising the prime minister about a diplomatic strategy, or observing the moon from Greenwich – was not done in a thoroughgoing fashion. But Airy, in many ways, saw himself as a gentleman *employee* of HM Government. This was epitomized early on, when Airy applied for a period of holiday leave, and the puzzled minister of state dealing with the matter told him to do as he saw fit, for he was in charge at Greenwich.

There, as in Cambridge, he was an immensely hard worker, in organizing the working routines of the Observatory and its staff, and taking on the extra burden of advising the government on a whole array of science and technology topics, ranging from the Railway Gauge Commission of 1845 to the calibration of new gas meters, imperial observatories in India and South Africa, and even the design of sewers. But Airy was no workaholic, and defended both his own leisure time and that of his staff, while making sure that wages were promptly paid in an age when official stipends sometimes fell into arrears. Airy always had a sharp eye where the value of money was concerned: both his own and that of HM Government.

By temperament, however, he was a domestic man. Gentlemen's clubs were not for him, and after fulfilling his Observatory, learned society, or government-meeting duties, his real pleasure was to spend time with Richarda and their children. Holidays with the family were also precious, either at their small cottage retreat in his native Playford, deep in the lush Suffolk countryside, or in the Lake District, especially in the area around Cat Bells. So while Airy clearly enjoyed his work, it was not by any means his whole life.

Airy and the discovery of Neptune, 1846

Not long after Sir William Herschel's discovery of the new planet Georgium Sidus, or Uranus, in March 1781, it was found not to be following the gravitationally computed orbit predicted by the mathematicians. Were Newton's laws in error – Heaven forbid! – or could there be yet another planet further out than Uranus that exerted a gravitational effect upon it, once one had taken into account the known pulls of the sun and inner planets? Two men, in England and in France respectively, independently addressed this problem in the early 1840s. The young Englishman John Couch Adams of St John's College, Cambridge, claimed, as he later told Dr William Somerville, that he had been inspired to work on the Uranus-disturbing planet after having read Mrs Somerville's *On the Connexion of the Physical Sciences*. The Frenchman was the brilliant Urbain Jean Joseph Le Verrier, in Paris. It says something about the acknowledged quality of the published Greenwich Observatory data that both astronomers used it.

Adams had a computed position for the Uranus-disturbing object by October 1845, but on two occasions he called personally, but unannounced, on Airy at Greenwich, and was not able to see him. On the first occasion, Airy happened to be away on Railway Gauge Commission business; on the second, he and his family were on tenterhooks as Richarda was about to give birth to Osmund, their ninth child, after a difficult pregnancy. Some modern scholars have blamed Airy for not paying attention to Adams's unannounced calls, but by consulting the manuscript "Astronomer Royal's Journal", now in Cambridge University Library, I was able, in 1986, to establish what Airy was doing on the days of Adams's unannounced visits.[9] The brilliant twenty-six-year-old Adams was not only given to cold-calling upon busy men, but when Airy wrote to him in Cambridge, requesting vital computational data, Adams never replied. However, after Le Verrier made his own calculations public a year later, Airy encouraged his Cambridge Observatory successor, the Revd Professor James Challis, to begin a search for Le Verrier's "planet" using the large Cauchoix "Northumberland" refractor. But this was in the summer of 1846, when skies in the latitude of Cambridge had long morning and evening twilights, and pinpointing a small unknown speck of light was not easy.

Le Verrier, in France, was having even less luck in initiating a search at the Paris Observatory, though, being older, ambitious, and much more worldly wise, he sent his results to Professor Johann Galle in Berlin, whose staff detected "Le Verrier's Planet" over the night of 23 September 1846, after which the news reverberated across Europe. (Le Verrier subsequently claimed the right to christen "his" planet Neptune, after the Roman sea god.)

So why did Airy not begin a full-scale search for Adams's planet at Greenwich? A variety of theories, some downright fanciful, have been proposed by historians, such as Airy being a snob, too "lofty" to bother with poor little Adams. Knowing what we do about Airy, this is clearly rubbish. Others suggest that Airy lacked patriotism, being willing to concede Le Verrier's priority, which again is ridiculous. All of these theories are countered by the fact that Adams bore no grudge against Airy and always spoke well of him.

I would suggest two eminently practical reasons – railway engines and a wife about to go into labour notwithstanding – for Airy not initiating a search at Greenwich. Firstly, he was dubious about the method employed by Adams to come to his position: using known planetary data to try to back-calculate and find a planetary needle in the celestial haystack. Airy had written to Adams on 5 November

1845 to request further information on this point and apparently received no reply. Secondly, and more importantly, was the way in which Airy conceived his role as Astronomer Royal, and how he interpreted his official duties. As he saw it, these were to observe already known physical data, such as solar, lunar, and geomagnetic criteria. This had to be analysed, tabulated, and published as a service to the nation and the world of learning. Most definitely, he did *not* see the Royal Observatory as a research institution, combing the heavens to ferret out new facts about the cosmos. To do so would not, in his way of thinking, constitute a proper usage of taxpayers' money. Airy, like his admired politicians Peel and Gladstone, was a small government thinker in the noble independent Victorian liberal tradition. Taxpayers' money should be used only to provide essential services, such as to pay for the defence of the realm, law and order, and safety on land and sea, and not to fund pure scientific research, lacking any practical application.

It is puzzling that Adams, in the search for Neptune, failed to publish his calculated coordinates in *The Times* newspaper or in the notifications of the Royal Society or the Royal Astronomical Society. Had he made his data public in this way, as he had done previously on another matter, his priority would have been assured, no matter who subsequently *saw* the planet first. When news of the Berlin Observatory sighting of "Neptune" first appeared, with its coordinates, in *The Times*, the Liverpool Grand Amateur William Lassell immediately saw Neptune, not as a star but as a distinct *disc*, on 2–3 October in his mighty 24-inch-aperture telescope. On 10 October Lassell noticed that Neptune was accompanied by a satellite, subsequently christened "Triton".

The Astronomer Royal and his staff

By the time of Airy's appointment as Astronomer Royal in 1835, the Greenwich Observatory had an official, full-time staff of around six men, and a varying number of casually employed "Computers" who did the routine calculation work. None of the warrant observers were university-trained. Most would have begun employment at the Observatory at sixteen or seventeen, upon leaving a good school, and with a demonstrated ability at mathematics. A sort of apprenticeship system operated, in which the most able Computers were promoted to permanent warrant rank when a senior officer died or retired. Directly beneath the Astronomer Royal was the chief assistant, who directed the lower assistants, who would work their respective shifts at a number of major angle-measuring instruments. Then, along with the Computers, these warrant assistants undertook the time-consuming "reduction" calculations, correcting for errors, and preparing the polished observations for final publication. Whether observing with the instruments, or doing the calculations, it was all routine work, primarily in the service of the Royal Navy.

In 1835 Airy inherited from John Pond, his predecessor, a colourful chief assistant, Thomas Taylor. A former Royal Navy officer or official, he joined the staff in 1807, probably in the wake of Navy cutbacks following Nelson's victory over the French in the Battle of Trafalgar, 1805. As one might expect from an ex-Royal-Navy man, Taylor was a competent practical astronomer, but also something of a rogue. He was sometimes drunk, given to various rackets, sold unpublished Greenwich observations to almanac-makers, and brought in some of his own sons as assistants. Airy made sure he was liberally pensioned off at sixty-three: no doubt to his delight.

Over the next couple of decades, Airy built up a body of men upon whom he could rely. Edwin Dunkin, for example, was fluent in French, and rose as high as he could go on the Observatory staff. James Glaisher, astronomer, pioneer meteorologist, and geomagnetic physicist, was, with Airy's support, elected FRS in 1849. After the sacking of Taylor, Airy began the Greenwich custom of appointing a young high-flying Cambridge mathematics graduate as his chief assistant. The longest serving was the Revd Robert Main MA, who was glad to resign his bachelor Fellowship in Cambridge, so that he could marry his sweetheart and set up home in Greenwich, while also helping out in local churches.

Why did the ordinary warrant assistants, clearly well-educated men, never go to university? Evidence suggests that several were not Anglicans at a time when Oxford and Cambridge, King's

College, London (1829), and Durham (1832) were primarily Anglican foundations. Some of Airy's staff were Methodists and Congregationalists, while the Breen family, several members of which worked as assistants or Computers, were devout Roman Catholics of Irish descent. Airy has been criticized for working his staff hard, but he backed their claims for salary increases (warrant salaries were between £150 and £300 per annum, roughly equivalent to those of senior government and commercial clerical staff). Airy also gave liberal holiday allowances for that time, sometimes up to a month with full salary, and often a decent pension at sixty-five or seventy.[10]

He also encouraged his staff to continue in their professional reading, the Observatory possessing a good library. Several of the staff became Fellows of the RAS Airy showed concern, too, for the manual and technical staff attached to the Observatory, such as park keepers and handymen, often petitioning the government for pensions for them when a loyal old servant retired. He may have been strict, but he was scrupulously fair. A warrant assistantship was a secure job for life, and very few of his warrant staff sought employment elsewhere, either dying prematurely or retiring.

On several occasions during his forty-six-year reign as Astronomer Royal (1835–81), Airy spelled out in official documents what he saw as the role of the Greenwich Observatory: to observe the known heavens and related physical phenomena to the very highest standards of accuracy, then publish that data for the good of the Navy, the nation, and the wider world. It was to provide a practical service from which all might draw benefit, be it safety at sea, or time signals for the railways. But it was not the job of the Royal Observatory to undertake speculative fundamental research, in an attempt to discover new astronomical phenomena. That was the job of those rich and highly motivated astronomically minded "barristers, brewers, peers, and engineers" who constituted the British Grand Amateur community, and whom we shall now meet.

Chapter 14

Barristers, Brewers, Peers, and Engineers: Paying for Astronomical Research: the British "Grand Amateur" Tradition

The conviction of the Astronomer Royal, Sir George Airy, that in Great Britain fundamental research in astronomy was the domain of private gentlemen (and ladies) and the financially independent universities, rather than taxpayer-funded state institutions, reflected a deeply held attitude in Victorian Britain. It applied not just to astronomy and the other non-utilitarian sciences, but also to music, art, literature, scholarship, and culture in general. These were all eminently praiseworthy activities, objects of national pride, and even fitting subjects for the bestowal of public honours, but it was not deemed to be the function of the state to fund them.

Funding Astronomy in Great Britain: the roots of a tradition

This private initiative and financing approach to science and culture did not apply in Continental European countries, however, although in many ways it did in the USA (which even after 1776 inherited many British attitudes and practices). In Germany, France, Austria, and Russia, if you were a scientist, a musician, or an artist, the way to advancement was to become attached to a royal or princely household. William Herschel's brother Jacob, for example, was employed as Vice-Concert Master in Hanover when William and Caroline were rising to independent fame in England. In a noble household, largesse might be showered from above: a largesse deriving from the high taxing of the general population who, in personal terms, saw very little in return for their taxes. One has only to think of the careers of Haydn and Mozart in music, the Cassini dynasty in astronomy, or the numerous painters, architects, and savants who clustered around the courts of Louis XIV and XV, Prussia's Frederick the Great, Austria's Joseph II, and Napoleon Bonaparte, to see how the system worked.

Following the efforts of the Congress of Vienna to re-establish *ancien régime* Europe after the French Revolutionary and Napoleonic wars, high culture became a major vehicle for national identity, with new universities, science research institutes, observatories, laboratories, and other high-prestige institutions being established or refounded. Invariably they were paid for by public or plundered money, be it Bonaparte's new École Polytechnique in Paris, or Czar Nicholas I's magnificent Pulkowa Observatory of 1839. This post-1815 world of European high culture in many ways created the structures of the modern academic profession, with their hierarchies of students, postgraduates, junior and senior lecturers, and, at the very top of the tree, great professors, who also acted as institute directors. In this academic world, the possession of the new PhD research degree was the arbiter between the academic sheep and the goats, letting down ladders for the successful and pulling them up against academic lesser fry. And there was no place for women in this world.

In Britain, things were very different, primarily as a result of very different political and economic circumstances. England's unique elected Parliament had, since the thirteenth century, enjoyed a constitutional capacity to rein in a monarch with absolutist tendencies, through its crucial leverage over finance and tax-raising. Even the aspiringly absolute Tudor and Stuart monarchs found it impossible to go to war or raise and disburse taxation-derived funds without Parliament's assent. King Charles I's attempts to raise taxes and appoint military commanders without Parliamentary consent had played a major role in triggering the outbreak of the English Civil War in 1642, and, following the abdication of King James II in 1688, royal power was fundamentally restructured. Parliament now unequivocally had the whip-hand, and while the Hanoverian Georges might be cheered along by patriotic crowds, their power over political and financial policy was severely circumscribed by the freeholder-elected Members of Parliament, the basis of Great Britain's constitutional monarchy.

Among other things, this election-based character of government and public finance facilitated, as we have seen, an immense expansion of independent middle-class wealth in the eighteenth century that made elected politicians fearful of raising taxes unnecessarily, or woe betide them at the next election. Nelson, Wellington, and their brave fighting men may have won the battles that finally led to the overthrow of Bonaparte's military dictatorship by 1815, but the ships, guns, foodstuffs, and other equipment that resourced the soldiers, sailors, and civilian population depended upon an agricultural, industrial, and financial infrastructure that was unique in history. A king or a duke might be rich in relative terms, but in Great Britain, a City of London financier, an industrialist, a mine owner, a brewer, a manufacturing engineer or a textile magnate could be equally so. Many of this latter category were enterprising self-made men, and when some, on the grounds of their great wealth and public service, were ennobled during the Victorian age, they might jokingly be said to have been given a "Beer-age".

Astronomy, as a high-capital investment science, where increasingly large telescopes became the only way of making major new discoveries, was, therefore, a politically and economically related science. Given the above-mentioned circumstances of English taxation and patronage, it is clear that no government would even dare to disburse tens or hundreds of thousands of pounds of public money to fund purely intellectual, as opposed to practical, astronomical research. So while the practical astronomy of Sir George Airy's Greenwich might rightfully make demands upon the public purse, "useless" research into the nature of the nebulae, gravitational physics, or spectroscopy was the domain of the Grand Amateurs.

They were "Grand" because they addressed the cutting-edge questions in cosmology, solar, and solar system studies; yet they were "Amateur" from the Latin word *amat*, "he/she loves". As Sir John Herschel, Admiral William Henry Smyth, and others claimed, they did astronomical research for *love*, out of their own pockets, not because the state paid them to do so.

The Grand Amateur astronomical world

Nearly all of the Grand Amateurs were men of middle-class background, and most had either made their own fortunes or had greatly augmented inherited commercial, landed, or professional money. All were well educated, driven by intellectual curiosity, a love of technology, and a strong sense of public service, and invariably wanted to make the world a better and a wiser place. Many, but not all, were actively Christian across a spread of denominations, including Anglican and free-church clergy, and all were firm believers in progress. They also tended to be sociable, often meeting at the newly founded (1824) Athenaeum or other clubs when in London, or at each other's town or country houses. Many made their money in the industrial powerhouses of the north, most notably Liverpool and Manchester, while others made it in London, via the City, medicine, or the law. Some, like the above-mentioned Admiral W. H. Smyth – a London shopkeeper's son who ran away to sea and rose up through the Navy ranks – made it in part from prize money, the fortunes of war, and a good marriage to the daughter of the British Ambassador to the Royal Court of Naples.

Central to their world was the Astronomical Society of London, which we shall examine shortly, while many Grand Amateurs were also Fellows of the Royal Society, re-echoing its founding definition as laid down by Thomas Sprat (later Bishop of Rochester) in 1667, that the Fellows were "*Gentlemen*: free and unconfin'd".[1] They paid their own way, and this proud independence was epitomized in 1852 in the Liverpool brewer Grand Amateur William Lassell FRS, who was about to travel with his family and privately designed and built giant reflecting telescope to Malta for a year or so. Lassell requested his friend Lord Rosse, then President of the Royal Society, to give him a Presidential letter of introduction to the British Commander at Malta so that they could enter the social life of the island's naval and diplomatic community. Lassell emphasized to Rosse, however, that he sought "no pecuniary aid from any quarter" for his forthcoming expedition to work beneath the clear skies of Malta: he would fund it out of his own pocket.[2]

Unlike on the Continent, Grand Amateur research was not especially university-based, for while many of those who were old Oxbridge men, and some resident dons, such as Airy's independently wealthy friend the Revd Richard Sheepshanks, Fellow of Trinity College Cambridge, became active members of the Royal Astronomical Society, their astronomical world tended to be outside formal academia. In 1840 Oxford and Cambridge Universities were anyway much closer to the Grand Amateur spirit than were the centralized and hierarchic German and French universities. And in many respects, they still are.

In Oxbridge, the functional bedrock of the university was the self-electing, self-governing resident bachelor dons, or Fellows, members of their own individual collegiate societies. Professors, in whatever discipline, were the poor relations, being obliged to live out in their own accommodation, though sometimes granted the right to marry. Yet whether one were a college Fellow or a professor, teaching duties were light, research optional, and most enjoyed quite separate sources of income, such as a church benefice, a medical or legal practice, private investments, or landed rentals. They were nobody's employees.

By the 1830s, however, England's first new universities were founded, with University College London (1826), King's College, London (1829), and Durham (1832). Even here, especially in Durham and King's, both with strong Anglican ecclesiastical connections, Oxbridge attitudes toward employment and gentlemanly values were rarely far away. Their freedom from state control did not, however, impede a university observatory's astronomical research capacity. Oxford's Radcliffe Observatory of 1771, built and equipped from the private Radcliffe Trust benefaction, was described by the visiting Dane Thomas Bugge in 1777 as the finest observatory in Europe. Cambridge University's 1823 Observatory acquired an enhanced major research facility in 1833, when the scientifically minded Hugh, Duke of Northumberland, endowed it with the great refracting telescope which still bears his ducal name. The Durham University Observatory was built upon the foundation provided by the astronomical instruments of the Revd Thomas Hussey after 1840. Durham's first Professor of Astronomy, the

Revd Dr Temple Chevallier, a fine mathematician and astronomer, was also an ecclesiastical and academic pluralist of truly Trollopian dimensions, and an anti-Catholic bigot who believed that the friendly approaches of the priestly scholars of nearby Ushaw Jesuit College were driven by nefarious motives.

Independent Grand Amateur attitudes and values pervaded British scientific thinking. In this world, the private gentleman-scientist was king, be he Sir John Herschel in astronomy, or Charles Darwin in biology. *The Origin of Species* was conceived, researched, and written entirely upon the strength of Darwin's large private income, without its author ever needing to undertake paid employment, and he paid his own bills during the HMS *Beagle* voyage, on which he was effectively a passenger.

These circumstances did not incline British science to be inward-looking or xenophobic. British Grand Amateurs were always visiting their professorial friends and colleagues in Paris, Berlin, Königsberg, or Pulkowa, especially as the growing network of steamships and railways made Continental travel increasingly easy and comfortable. Continental scientists were also always coming to Britain, attending Royal Society and Royal Astronomical Society meetings, and doing the Grand Amateur visiting circuit of observatories and country houses. Following the rumpus surrounding the disputed discovery of Neptune in 1846, Sir John Herschel invited John Couch Adams and Urbain Le Verrier to a congenial scientific and social gathering at his country mansion, and the two astronomers soon developed a healthy regard for one another. Convivial dinners, good drink, music, and entertainment have a remarkable capacity to dispel animosities. Honours were exchanged, as medals, fellowships of European academies and the English Royal Societies, and honorary degrees passed both ways across the English Channel, and the Atlantic.

The week-long British Association for the Advancement of Science jamboree meetings, held annually since 1831, also became showcases for contemporary scientific research, in all scientific disciplines from physiology to cosmology, with European and American scientists invariably attending and delivering papers. Nothing better illustrates the genial mingling of British Grand Amateurs, the Astronomer Royal, and visiting distinguished

Continental professionals than an incident following the British Association meeting in Glasgow in 1850. At that date, the railway across the Scottish lowlands had not yet been completed, so that people travelling to the meeting from the south went by train as far as Liverpool and then took the fast passenger steamer from Liverpool to Greenock, on the Clyde, near Glasgow. On the way back, in early September 1850, several people stayed at Starfields, the Liverpool observatory and mansion of William Lassell. Professor Wilhelm Struve of the Pulkowa Observatory, St Petersburg, was one of them, and he caught his train back to London leaving behind a freshly laundered shirt. Sir George Airy, the Astronomer Royal, dropped into Starfields on the way to Greenwich, and Maria Lassell gave him the shirt, to return to Struve in London.

This little incident says much about the astronomical world of 1850, in which a self-made, brewery-owning Liverpool Grand Amateur could ask the Astronomer Royal to deliver laundry to a visiting Germano-Russian professor, whom Lassell addressed in the intimate "My dear Struve" in his accompanying letter.[3] But what were the major astronomical contributions these Grand Amateurs were making?

The Liverpool brewer and the Manchester steam-engine builder

Judging from his surviving notebooks and letters preserved in the Royal Astronomical Society archives, William Lassell (1799–1880) was bitten by the astronomical bug at an early age. A devout and lifelong member of the Congregational Church, he obtained his earliest schooling in his native Bolton, Lancashire, where his father owned a timber and building business. Following his father's early death, he was taken by the widowed Mrs Hannah Lassell to live with the rest of the Lassell family in Liverpool, where they were engaged in business. Not being an Anglican, he was not eligible for Oxford or Cambridge, but received a sound education at the renowned Rochdale Dissenting Academy and in Liverpool. Then he became a partner in, and later owner of, the 20 Milton Street Brewery, Liverpool, which provided the wellspring for his ample private fortune, combined with the support of his wife Maria King, formerly of Toxteth, Liverpool. She came from another commercial family, and had two astronomical brothers, Alfred and Joseph.

14.1 William Lassell. A photograph of c. 1850. (By kind permission of Liverpool Astronomical Society.)

By the late 1820s, Lassell was casting and figuring high-quality speculum mirrors, and his 9-inch-aperture, 112-inch-focal-length reflector of 1836–37 became a popular prototype for future developments. Although its aperture was modest by that date, its optics were superlative, and the high purity of the tin and copper of the speculum alloy enabled Lassell to impart a particularly brilliant and lasting polish. Perhaps more significant, however, was his radical new mount for the 9-inch. Instead of the usual wood pulleys and rope stands of the Herschels, Lassell positioned his new telescope upon a perfectly balanced and precision-engineered mount in cast and wrought iron, set in the equatorial plane. This new mount enabled him to direct his 9-inch to any part of the sky with ease, while its finely balanced and roller-bearing tracking mechanism enabled him to keep an object exactly in the centre of his field of view from horizon to horizon.

Lassell's 9-inch was the ancestor of every large, equatorially mounted research reflector that came thereafter, facilitating an ease of usage unknown to the Herschels. The 9-inch design would then be upgraded to produce Lassell's great 24-inch equatorial of 1844–45, and the 48-inch after 1857: truly great monuments in the history of large-scale precision engineering, now in the service of astronomy. These instruments, tragically, no longer survive other than in pictures and in engineering drawings, but these were of sufficient quality to enable Merseyside Museums and Liverpool John Moores University to build an exact working replica of Lassell's 24-inch telescope in 1996, for which I had the honour to be involved as historical consultant.

14.2 William Lassell's 48-inch-aperture reflecting telescope, depicted here in its original location in Liverpool, which Lassell took to Malta in 1861. (*Phil. Trans.* 1867; also Simon Newcomb, *Popular Astronomy* (1898), p. 132. Author's collection.)

What made the 24- and 48-inch telescopes truly splendid, however, was the large steam-powered industrial-size figuring and polishing machine, in which geared wheels exerted an exact control over the motions of the figuring tool. Optical lathes went back to the seventeenth century, but in Lassell's machine we find something on the scale of Victorian railway locomotive engineering being applied to something as delicate as watchmaking.

Finding that the damp air of Liverpool frequently lacked the transparency to enable these optical giants to be used to full advantage, Lassell and his family undertook two quite lengthy sojourns in Malta, the first in the 1850s, with the 24-inch, and the second in the 1860s, with the 48-inch. His favourite objects of study were Saturn and the other planets, followed by the nebulae, all fields in which he made discoveries. As his above-cited letter to Lord Rosse emphasizes, his entire astronomical enterprise was paid for out of his own resources, and he must have spent several tens of thousands of pounds, in the money of the 1850s, on astronomical instrumentation and overseas expeditions. He epitomized the British Grand Amateur tradition.

One wonders, however, how much technical advice and even heavy industrial assistance Lassell may have received from his longstanding friend, the Edinburgh–Manchester ironmaster James Nasmyth.

14.3 James Nasmyth. (*James Nasmyth Engineer: An Autobiography* (1889), frontispiece. Author's collection.)

Nasmyth's own great fortune derived from his engineering factory in Patricroft, north Manchester, where he used his world-changing invention, the steam hammer, to mass-produce railway locomotives and huge marine engines.

Back in his cultured Edinburgh childhood, Nasmyth had developed a passion for astronomy, inspired partly by visits from his artist father's friend, the great Edinburgh optical physicist Sir David Brewster. When this youthful astronomical passion was joined to his mature heavy engineering resources, Nasmyth was ready to join the Grand Amateur league. In addition to any assistance given to Lassell, Nasmyth designed and built a splendid and original large reflecting telescope of his own. Not only did the instrument contain a superlative speculum mirror of 20 inches in diameter, but it was mounted upon an ingenious "altazimuth" stand, by means of which the whole instrument rotated around a fixed eyepiece. When Nasmyth sat on his ground-level observer's seat, he could rotate the instrument to locate objects in any part of the sky, and track their motions by means of a precision system of gears controlled by a pair of handles. He styled it his "comfortable telescope", and it was perfectly adapted to undertake the lunar, planetary, and solar research that interested him.

14.4 James Nasmyth's "comfortable" or "trunnion vision" reflecting telescope.

From his seat, Nasmyth was able to turn this 20-inch-aperture telescope to any part of the sky, by means of two geared adjusting wheels. The eyepiece fitted into the hollow trunnion axis that supported the tube, and was always at the right level for the observer's eye. (*James Nasmyth Engineer. An Autobiography* (1889), p. 338. Author's collection.)

The moon was Nasmyth's especial object of study, though where he particularly advanced knowledge was in his fascination with the lunar geology. As the owner of a great iron foundry, Nasmyth was struck by the visual parallels between the geological features of the lunar surface and the cracking and buckling of the slag that formed upon molten iron. His studies convinced him that the moon had once been an internally hot body, the cooling of which produced the long cracks radiating from craters such as Tycho, the "volcanic" craters, with central peaks and surrounding rampart walls, the flat lunar lava plains, and the buckled and twisted mountain ranges. As a dexterous craftsman, Nasmyth even replicated these geological features in his workshop and studio. He was also interested in the sun, and made some of the first reported observations of the complex, unstable granulations structure of the solar surface seen at very high magnifications. What were these packed lenticular formations upon the solar surface, and how did they relate to sunspots in the solar burning process?

Within Lassell's and Nasmyth's local Lancashire Grand Amateur circle was the physician and Congregationalist minister William Rutter Dawes, then living at Ormskirk, between Manchester and Liverpool, and later at Haddenham, Buckinghamshire. Dawes, who enjoyed the advantage of having married the widow of a wealthy solicitor friend, was also interested in the planets, although his real passion lay in the complex physical motions of double and compound star systems. Unlike Lassell and Nasmyth, however, Dawes was not a telescope builder so much as a commissioner of large and powerful refractors, such as his magnificent 8-inch instrument of 1855 by Alvan Clark of Boston, USA. As Sir John Herschel and other Grand Amateurs in England, and Friedrich Wilhelm Bessel and others on the Continent realized, these compound star systems provided vital evidence for the action of Newton's truly universal laws in the very depths of space. By

1840 these northern gentlemen were in no way isolated, as the burgeoning railway system made it easy to meet friends and attend learned society meetings in London and elsewhere. Similarly, fast steam packets sailing between Dublin, Liverpool, and Bristol made it almost as easy for Irish Grand Amateur astronomers such as Lord Rosse to do likewise.

The Irish nobleman who discovered the "whirlpools" of deep space

William Parsons, known first as Baron Oxmantown, then, after inheriting the title in 1841, as the third Earl of Rosse, of Birr Castle, was a profound admirer of Sir William and Sir John Herschel, and determined to use the resources of his Irish earldom to extend their work on the nebulae of deep space. Educated at Trinity College Dublin, and Magdalen College, Oxford, the young lord was fascinated by applied mathematics and engineering as well as astronomy. When he was forty, in 1840, Lord Rosse, after many years of metallurgical experimentation, completed a reflecting telescope with a superlative 36-inch-diameter mirror that showed the heavens as never before and spurred him on to build the largest and most powerful telescope of the nineteenth century.[4] He knew all the leading astronomers and engineers of the day, many of whom were personal friends and often visited Birr Castle. He became an excellent landlord, training up men on his estate to perform complex tasks in metallurgy and engineering. All the Rosse telescopes were designed and built on-site using local skills and labour, under the direction of his blacksmith foreman William Coghlan. During the devastation of the Irish Potato Famine of 1845–47, Rosse liberally spent his own money to devise projects, many not astronomically related, by which his tenants could still earn wages and enjoy an income. This Church of Ireland (Anglican-connected) nobleman also gave money to build both Protestant and Roman Catholic churches on his estate.

14.5 William Parsons, 3rd Earl of Rosse. (R. S. Ball, *Great Astronomers* (1895), p. 273. Author's collection.)

Just before the potato crop failed, in 1845, Lord Rosse brought his iconic giant reflecting telescope into commission. It had a pair of speculum mirrors of an unprecedented 72 inches, or 6 feet, in diameter, of 52 feet focal length. At any one time, one mirror was mounted ready for use in the great wooden tube resembling a factory chimney and moving between a pair of great stone walls, set in the plane of the meridian, while the other was in the castle workshops being repolished. The great telescope could move sideways (or in right ascension) through a 15° angle, while in the vertical it could encompass an angle of over 130°, between the southern horizon and the northern pole. It also had a pair of great mirrors, each weighing around four tons. Such paired mirrors, often styled A and B, were the norm with the big speculum reflecting telescopes, for the optical surfaces tended to tarnish and lose reflectivity and hence definition. When mirror A had lost its optimum lustre, it would be removed from the tube, and taken to the workshops for a repolish, while mirror B would be wheeled out on a special trolley to replace it, until it, in turn, tarnished. In this way, clear observing nights need not be lost through want of a shining mirror.

14.6 Lord Rosse's great telescope at Birr Castle, Ireland. It had a 72-inch-diameter mirror of 52 feet focal length. (J. N. Lockyer, *Stargazing* (1878), p. 295. Author's collection.)

It was while Lord Rosse was observing Herschel's objects visible from the latitude of Birr, on an April night in 1845, that the new "Leviathan of Parsonstown" made its classic discovery. (In the nineteenth century, the Irish market town of Birr was named Parsonstown, after the earl's family name of Parsons.) Rosse noticed that when examining nebula number 51 in Messier's Catalogue, the hitherto diffuse glow of light now displayed a clear internal structure: appearing as a distinct spiral with arms branching out from a bright centre, which might well indicate an anticlockwise rotation. This galaxy, in the constellation of Canes Venatici, would become immortalized as the Whirlpool, for that is what it resembled.

In all, the "Leviathan" telescope revealed internal structural shapes in about a dozen nebulae, including the Crab nebula in Taurus, the Ring nebula in Lyra, the Dumb-bell in Vulpecula, and the Owl in Ursa Major – all so-named from the visual shapes which they suggested by their wispy structures. This was a tiny fraction of the several thousand nebulae known by that date, but it represented a momentous hurdle crossed by astronomy. Spiral and other shapes suggested that gravitational forces were at work, and – very likely in the case of the Whirlpool – complex internal motions were taking place over vast periods of time. Yet if an instrument as large as the Leviathan of Parsonstown could still resolve so few of the several thousand then-known nebulae, how could we discover anything about those nebulae that, even in that vast 72-inch mirror, remained mere smudges of light? Things looked sadly limited in 1850, though between 1860 and 1890 the new sciences of astronomical spectroscopy and photography would transform Western cosmological knowledge, as we shall see in Chapters 15–17.

14.7 Lord Rosse's drawing of the "Whirlpool" nebula, observed with the 72-inch mirror telescope at Birr Castle, 1845. Rosse's great telescope was the first to reveal internal structural detail inside a nebula. (J. P. Nichol, *Thoughts on… the System of the World* (1846), pp. 22–23, pl. VI. Author's collection.)

In 1836, Lord Rosse had married Mary Wilmer Field, a Yorkshire heiress (he had been born in York in 1800), and a lady of remarkable intelligence, energy, and enterprise. Among her other achievements, Countess Mary would become Ireland's most celebrated art photographer. Fully supporting her husband's astronomical researches, she would also become the mother of several highly gifted children. Their eldest son, Lawrence, the fourth Earl of Rosse, would further develop astronomical and physics research at Birr Castle, and perform pioneering experiments to determine the surface temperature of the moon. His younger brother Charles Parsons would design and build the first marine turbine engine, which would, at a stroke, revolutionize marine engineering. The Parsons (Rosse) boys would be tutored in science by the young graduate Robert Stawell Ball. Ball would become a lifelong friend of the Rosse family, hold senior professorial observatory director posts in both Trinity College Dublin and Cambridge, and become the celebrity astronomy lecturer and writer of the late Victorian age. We shall meet Sir Robert, as he became, in Chapter 20.

I know Birr Castle, and am indebted to the seventh earl, Brendan, and to Countess Alison for graciously granting me access to the castle archives. It was also the seventh earl who, in the 1990s, set about the gargantuan process of restoring the great telescope and its fittings. While its 72-inch mirror has long been in the Science Museum, London, I have learned much about this magnificent monument to Grand Amateur astronomy by the direct, hands-on business of operating its restored winches and windlasses, climbing up the great wall ladders, and standing on the eyepiece-level observing platform, with a 52-foot drop below.

14.8 Allan Chapman operating the windlass to elevate the 52-foot tube of the restored Rosse telescope, c. 2000. (With thanks to the photographer, John Fitzsimons, of Sligo, Ireland.)

In addition to Lord Rosse, Victorian Ireland was a hotbed of high-powered Grand Amateur scientific research. There was Edward Joshua Cooper of Markree Castle, County Sligo, who, along with his assistant Andrew Graham, performed cutting-edge astrometric and asteroid work with a large 13.3-inch-aperture, 25-foot-focus refracting telescope. John Birmingham was a devout Roman Catholic county gentleman of Galway in Ireland's far south-west, famous as a variable and supernova star observer, while the friends who met and researched at William Edward Wilson's Daramona House Observatory, County Westmeath, would, among other things, pioneer the photometric measurement of "absolute" star brightness in the early 1890s. Ireland possessed three academic observatories, Armagh, Trinity College Dublin's Dunsink – both archiepiscopal foundations – and Cork, while the formidable and long-lived Revd Professor Thomas Romney Robinson of Armagh had a fifty-nine-year formative influence on many Irish astronomers, including Lord Rosse.

Dublin was also home to one of the world's great optical engineering and telescope building firms: that of Sir Howard Grubb. The Grubb optical factory was even visited by royalty in 1876, when Dom Pedro, the scientifically minded emperor of Brazil, made an informal visit. Upon leaving, the emperor accidentally took Sir Howard's fine silk top hat, leaving his own battered old one behind, though he later returned it, with fulsome apologies.

The Royal Astronomical Society: a Grand Amateur creation

On Wednesday 12 January 1820, a group of fourteen astronomers met for dinner at the Freemasons' Tavern, Lincoln's Inn Fields, London, to discuss the formation of an Astronomical Society of London. All were gentlemen in comfortable circumstances, including the young Sir John Herschel, the clergyman Revd Dr William Pearson, the surgeon James South, and the City financier Francis Baily, along with military officers and senior Cambridge dons. Only two men at that dinner could probably have been described as "professional": the visiting Peter Slawinski, professor at the University of Vilna, Russia, and Dr Olinthus Gregory, a mathematics professor at the Woolwich Military Academy. The Society was rapidly constituted, with the elderly Sir William Herschel as its non-attending President, and, along with its originally associated dining club, it thrived. In 1830 it obtained its Royal Charter – transforming it into the Royal Astronomical Society – and rapidly became the principal British astronomical society, as it continues to be. To this day, the Dining Club always refers to its offspring, the RAS, as "the Other Place".

However, while the succeeding Astronomers Royal and professors of all the British universities would enter into the Fellowship, the RAS was an overwhelmingly Grand Amateur creation. In this respect, it mirrored the Royal Society itself, most of whose Fellows were still private scientists in 1850, with the addition of a few professors; and most of the astronomers in the Royal Society were RAS Fellows as well.

In addition to Lassell, Rosse, and the Grand Amateurs discussed above, the Royal Society and RAS contained many more. Take, for example, the later career of the above-mentioned Captain, later Admiral, William Henry Smyth, who retired from active service to Bedford in the 1820s and began serious double-star and other astronomy. Through the Bedford legal Quarter Sessions Court, Smyth met a rich Cambridge-educated landowning barrister, Dr John Lee, QC, of Hartwell House, Aylesbury, who was also a passionate observatory-building Grand Amateur astronomer. The two men quickly became friends, and Hartwell House became one of the great social centres of British astronomy, as the surviving Hartwell Albums and Scrapbooks, now preserved in Oxford's Museum of the History of Science, testify.[5] Everybody who was anybody in British, Continental, or American astronomy felt free to drop in to Hartwell, for Grand Amateur astronomy was very much of a fellowship. The astronomers often came with wives and families in attendance, enjoying not just the science, but the entertainments and socializing as well.

If the nature of the nebulae and the physics and chemistry of the sun and stars still remained a mystery in 1855, the next generation of Grand Amateurs in Great Britain and the USA would find the golden key to resolving them. They spent their own gold and time in developing the crucial new technologies that would unlock the hitherto unimagined wonders of deep space, by means of spectroscopy and photography.

Postscript: Grand Amateur astronomy today

Over the course of the twentieth century, astronomy and cosmology became professionalized both intellectually and technologically, as the cost and complexity of fundamental research soared. Yet much of the once cutting-edge technology of the Grand Amateurs, such as big mirror making, big telescope engineering, astro-photography, and spectroscopy, was simply adopted and developed further by the rising breed of paid professionals. Serious amateur astronomy, however, is nowadays more active than ever, and on a global scale. While no modern amateur expects to solve the mysteries of cosmology as did their Victorian brewer or barrister ancestors, yet today's Grand Amateurs are still willing to spend substantial sums on large top-of-the-range, commercially made telescopes, linked in with sophisticated computer software, visual imaging, processing, and analytical technology. We will be discussing them in more detail in Chapter 28.

Chapter 15

The Camera Does Not Lie: The Birth of Astronomical Photography

Since time immemorial, the only technology by which one could record the visual details of the natural world was to draw, paint, or engrave them. While the development of perspective drawing based on the *camera obscura* in fifteenth-century Europe, and its subsequent refinements, enabled an artist to faithfully delineate the world on a flat surface, everything still hinged on the accuracy of the human eye, hand, and brain, to make sense of it. From ancient times onwards, artists had scratched or drawn images of the sun and moon, while from Thomas Harriot's surviving drawings of 26 July 1609 and onwards, astronomers had been mapping the face of the moon with increasing precision using telescopes. As telescopes got better, they then recorded such things as the belts and spot of Jupiter, Saturn's rings, the polar caps and Syrtis Major on Mars, and supposed dark or light areas on the crescent Venus. Yet all of this was still hand and eye work, albeit aided and refined by the use of new measuring instruments such as the telescope eyepiece micrometer. No matter how much an astronomer might have been helped by new drawing technologies by 1800, however, a skill in draughtsmanship remained a prerequisite of good astronomical recording.

On the other hand, the action of light upon a variety of substances, such as in leather tanning, had been known for centuries. But it would need the rapid advances made in both optical and chemical knowledge after 1800 to fundamentally transform our ability to record all manner of natural phenomena, be it a human face, a lunar crater, or even a nebula.

Monsieur Louis Daguerre, Sir John Herschel, and Mr William Henry Fox Talbot

In the nineteenth century, it was often the case that an Englishman and a Frenchman would make a parallel discovery or invention.

Think of Adams and Le Verrier and the discovery of Neptune in 1846. And here we have a Parisian theatre scene painter and a Wiltshire country gentleman and FRS in 1838–39 inventing photography. Science and technology were advancing so rapidly by that time, and information flying so fast between countries and continents, that it is easy to understand how two ingenious long-standing rival nations, only separated by 21 miles of sea, could be home to parallel discoveries.

Photography's genesis, however, was in the 1810s, when Joseph Nicéphore Niépce, a French country gentleman, began to experiment with "heliography", or sunlight-drawing. In 1816, he obtained a picture of his garden by using a lens to project an image on to a silver-salts-coated plate, after an 8-hour exposure. He later went into partnership with Louis-Jacques-Mandé Daguerre, the Parisian artist, who later so improved the now deceased Niépce's process that, by 1839, Daguerre was able to take pictures on a silver-sensitized metal plate from a 25-minute exposure. This length of time would be greatly reduced when Daguerre accidentally discovered how exposure to mercury fumes could develop a latent image on a plate that had only been exposed for 1 or 2 minutes.

Daguerre's new process of "light-drawing", after being presented to the Académie des Sciences, then published, was followed by an English translation entitled *The History and Practice of Photogenic Drawing*, by Dr J. S. Memes.[1] It then caused a sensation across Europe, and quickly became all the rage. Having one's portrait taken photographically was vastly cheaper than visiting a professional painter, and pictures of ordinary people suddenly became commonplace. Professional likeness-takers sprang up everywhere, as did the first amateur photographers, for the basic kit of camera and chemicals was cheap and easy to use, and a handy man or woman could easily make up one for themselves. Many of the early professionals were already practising artists, and many studios, such as that of Hill and Adamson of Edinburgh, became famous. The Parisian optician Charles Louis Chevalier designed the first specifically photographic lens in 1839, his achromatic meniscus, aspiring to get a flat field of view and a shorter exposure time by sharply focusing more of the otherwise scattered light:

itself a new adaptation of an already established astronomical lens design. Chevalier also became one of the first men to sell cameras commercially. Then in 1840, Joseph Petzval brought out his fast lens, aimed at reducing exposure times for portrait sitters, and the new technology began to leap ahead.

Even so, exposure times still required the sitter to be neck-clamped and unblinking for anything up to a couple of minutes on a dull day, and it was impossible to record movement. Daguerre's needle-sharp photographs of the boulevards made Paris look like an eerie, deserted ghost city, as no person or vehicle stayed still for long enough to register on the plate. As a daguerreotype was a one-off image on a piece of sensitized metal, the only way to do a duplicate picture was to re-photograph the original subject.

William Henry Fox Talbot's early experiments with light-drawing during the 1830s largely derived from his frustration at not being able to draw. Even his drawings made with the aid of his friend Sir John Herschel's invention, the camera lucida, look wooden. So this Wiltshire country gentleman, owner of the Lacock Abbey estate, astronomer, friend of the great railway engineer Isambard Kingdom Brunel, Member of Parliament, and FRS devoted himself to optics and chemistry. In August 1835, Fox Talbot succeeded in obtaining a photographic picture of the great window in Lacock Abbey on a piece of finely woven paper that he had treated with silver chloride.

These prints, where the light areas appeared black, were what John Herschel would soon style negatives. By treating the paper with delicate oils, to make them almost transparent, Fox Talbot was able to flatten them against new pieces of sensitized paper, re-expose them to sunlight, and obtain positive prints, in which the black and white areas were reversed to produce a normal picture. Due to the inevitable texture of the paper, many of his prints were not as needle-sharp as Daguerre's, yet they possessed the immense advantage of being capable of multiple duplication, without the need for re-photographing. Then when his friend Herschel – newly returned from the Cape of Good Hope – advised the use of hyposulphite of soda (hypo) as a way of fixing his images permanently, so that they would not darken in sunlight (see Chapter 10), all the basics of pre-digital photography were in place.

Fox Talbot's invention earned him the Royal (1838) and Rumford (1842) Medals of the Royal Society, while his *Pencil of Nature* (1844) was the first book ever to be illustrated with actual photographs. Fox Talbot, who had learned his classics at Harrow, would christen his process "calotype", from the Greek *kalos*, "beautiful". Once the essential process and technique of photography were in the public domain by 1838 – to use a lens to form an image upon a silver-salt-sensitized surface, and then render it permanent – possible uses and improvements began to fly thick and fast. The indefatigable Sir John Herschel was quick to recognize the potential of the new photographic technology for astronomical and other scientific recording. While he never photographed an astronomical body, he did use the Fox Talbot process to obtain a stunning silhouette of the great wooden framework of his late father's 40-foot telescope of 1789 – the first ever photograph of a scientific instrument. This was just before the long-disused telescope was demolished and the Herschels moved to Hawkhurst, Kent. By 1840, therefore, two viable photographic processes were available: Daguerre's, where critical, high definition was needed, and Fox Talbot's softer-focus one, when multiple copies of a picture were required.[2]

Dr John William Draper of New York: the first astronomical photographer

Sir John Herschel was not the only scientist to realize photography's enormous potential for recording natural phenomena. Early photographs through the microscope – photomicrography – were relatively easy, as the specimen, microscope, and camera were all stationary, so even long exposures posed no problem. But nothing in the heavens is stationary, and even though sophisticated clock-driven equatorial mounts were in widespread use in 1840, enabling the telescope to track an object across the sky, they often lacked the faultless precision demanded by photography. If, for example, one were trying to photograph a bright star or a lunar crater, then even the slightest mechanical straying of the telescope and its mount resulted in a useless blur.

Photography was first announced in the New York newspapers on 20 September 1839, and Dr John William Draper was experimenting with the daguerreotype process by December. Probably in July 1840, Draper took the USA's oldest surviving photograph. This was of his beautiful sister Dorothy Catherine, after a 65-second exposure. There she sits in her large bonnet, her clear eyes looking unflinchingly ahead in what must have been an act of real self-discipline. Before this portrait, however, Draper had reported to the New York Lyceum of Natural History that he had photographed the moon, and that the lunar *maria* or "seas" were very distinct. This modest start would trigger a landslide in astronomical understanding.

Draper, who was born in St Helens, Lancashire, England, and whose mother had settled the family in the USA, was an active amateur astronomer. Professionally, he was a doctor who held the Chair of Chemistry at New York University, and in 1850, he used daguerreotype "photomicrography" to illustrate a physiology textbook. He also did important laboratory photo-chemical research, and in 1843 used photo-chemical change to measure light intensity. His American-born son Henry would also become a professor of medicine, though his enduring fame would come through his work as an astronomical photographer and a spectroscopist, as will be shown in Chapter 17.

As things stood in the 1840s, daguerreotype was the process of choice when it came to recording fine scientific detail, although it was much too slow and required exposure times that were far too long to be anything more than a novelty, astronomically speaking. But the astro-photographic fuse had been lit.

The "miracle" of the "wet collodion" photograph, 1851

The ideal vehicle for photo-sensitive recording was not metal or paper, but glass, for a glass plate gave perfect transparency and an unlimited potential for printing off duplicate images. Sir John Herschel and others had experimented with glass, probably as early as 1839, but there seemed no practicable binding agent whereby the emulsion could be firmly bonded to the glass, so as not to wash away during processing in water-based fluids. As happens so often in science, however, it was a technological cross-over that provided the solution in this case, from a mutual borrowing between 1840s explosives technology and medicine. Gun cotton, which is far more powerful than black gunpowder, had been accidentally discovered when the Italian chemist Ascanio Sobrero mopped up some of the newly invented deadly nitroglycerine with a housemaid's pinafore which was lying nearby. The maid, no doubt, was alarmed to find that when flicking her dried-out pinny, it vanished in a flash and a puff of smoke. Christian Friedrich Schönbein (also the discoverer of ozone) in Germany and Louis-Nicolas Ménard in France made chemically equivalent, though less domestically alarming, discoveries around 1845–46.

Then in 1848 Dr J. P. Maynard of Boston, USA, found that when he dissolved the gun cotton (nitrocellulose or pyroxylin) in a mixture of the newly discovered surgical anaesthetic sulphuric ether and alcohol, he obtained a thick transparent fluid. When poured out onto a flat surface, it produced a tough, completely transparent film. This was collodion, and by 1850 one of its uses – like the ether anaesthetic employed in its manufacture – was medical, for it produced an excellent surgical styptic, and could stop bleeding. As early as 1849, Frederick Scott Archer in England was finding a photographic use for collodion, for when it was impregnated with silver salts, such as silver chloride, it formed a photographic emulsion that, on drying, made a close, waterproof bond with a glass plate. Thus was born the "wet collodion" process, which Scott Archer announced in 1851. Not only could one produce as many prints as one desired from a single glass plate, but the images were also needle-sharp. It was also a more light-sensitive or fast process than either Daguerre's or Fox Talbot's, sometimes allowing exposures as short as five or six seconds in strong sunlight. Like all the early photographic processes, however, the wet collodion required an element of laboratory back-up.

The polished glass plate first had to have one side covered with a wholly smooth and homogeneous coat of clear collodion: itself an art requiring considerable dexterity. It then had to be immersed in a bath of silver chloride or bromide solution, to impregnate the collodion with silver. The wet plate had to be manipulated into the camera by means of a light-proof wooden dark slide, and exposed. The plate was then developed in a dish of glacial acetic acid and iron sulphate to produce the negative. Whether one wanted to photograph a human face, a landscape, or the moon, the same technique was employed. The whole operation had to be performed while the plate, and its emulsion, were still wet: hence the name of the process. Awkward as early photography might have been, things had come a long way since Niépce's first "heliographs" of 1816.

Warren De La Rue: the Guernsey-born paper manufacturer and pioneer of astronomical photography

In 1840 Warren De La Rue made a visit to James Nasmyth's engineering factory and foundry at Patricroft, Manchester, to consult with the great engineer about a process for making white lead, for chemistry and electricity were leading interests for this successful businessman. Then, upon seeing Nasmyth's telescopes and observatory, and the 13-inch-diameter mirror that Nasmyth was currently finishing, his mind was turned to other things. In addition to whatever was agreed about white lead manufacture, De La Rue ordered a 13-inch-diameter speculum mirror blank for an astronomical telescope, which he subsequently ground and figured, and the two men began an astronomical friendship. Though a Guernsey man, De La Rue had a fine house and grounds at Canonbury, north London, where he set up an observatory.

Here, the very technically minded De La Rue did several very significant things. Firstly, he designed and built a new kind of mount for his 13-inch mirror.[3] Instead of mounting his mirror inside an iron or strong wooden tube, De La Rue fabricated an open tube of iron rods securely held in place with iron rings. Such a design had several advantages. It was lighter in weight than rolled sheet iron would have been, thus simplifying the engineering logistics of the mount and clock-drive mechanism. Also, in warm weather, it would easily disperse any thermal inequalities between the surrounding air and the air inside the tube, thereby reducing any fluttering of the image due to air currents. This would become a great advantage when De La Rue used the telescope for photography, for while the eye of a trained observer could to some degree compensate for image-flutter, the inert photographic plate demanded total stability if the resulting image were not to be blurred.

One wonders whether Warren De La Rue's open-tube design influenced William Lassell when he came to build his great 40-foot-focus, equatorially mounted (non-photographic) instrument later in the 1850s, for such a design could have proved invaluable for observing in the warmer climes of Malta. Open or skeleton-frame tubes, especially for large-aperture telescopes, would become the accepted norm in the great twentieth-century professional observatories of Mount Wilson (1906 and 1917), Mount Palomar (1948), and elsewhere, as the professional astronomers and telescope builders picked up ideas from the prior achievements of the Grand Amateurs.

In 1852, De La Rue devised a glass plate holder that was positioned at the prime focus point of the 13-inch mirror: at the point where the eyepiece would normally go. De La Rue's Cranford establishment would become the first observatory to also incorporate a photographic darkroom. The object of his particular attention was the moon, and he would go on to produce a stunning series of lunar photographs, some of which were published commercially in stereoscopic form. By the mid-nineteenth century, the moon would become an object of intense interest, especially to the Grand Amateurs, for a variety of reasons. James Nasmyth was not alone in seeing parallels between the twisting slag that formed upon molten iron in the foundry and the complex topography of the moon. Geology was also advancing rapidly by the 1840s, and – considering the popularity of Laplace's hypothesis for the stellar ejecta origin of the solar system – could not the atmosphere and oceanic-erosion-free lunar surface tell us something about the ancient earth?

15.1 The moon, photographed by Warren De La Rue, 27 August 1860. (One of a set of stereoscopic views of the moon. Author's collection.)

Having the time and the independent resources to examine the lunar surface at high magnification over long periods, many Grand Amateurs (often active field geologists) began to monitor the surface of our satellite as no one had ever done before. Surely vulcanism must have played a major part in the formation of the craters, with their high rampart walls and central Vesuvius-like mountains (a possibility first proposed by Robert Hooke in 1665)? Having a much lower gravitational attraction than the earth (being much smaller) and lacking any atmospheric resistance, could not the young moon have blasted out plumes of magma which fell back downwards to produce crater walls? Photography as an impartial recording medium for scientific data – independent of the eye, hand, and brain of the observer – was perceived as possessing enormous potential.

Then around 1866 lunar observers were agog when Julius Schmidt announced what seemed like the disappearance of the small crater

Linné (named after the great eighteenth-century Swedish botanist Carl Linnaeus) in the lunar Sea of Serenity. Was measurable change taking place on the moon even today, and challenging the accepted view that our satellite was a dead and changeless world? Photography, with its capacity for cold, clinical recording, clearly pointed the way forward.

Yet no matter how beautifully engineered a telescope, its mount, and clock-drive mechanism might be, it was still extremely difficult to obtain needle-sharp astronomical photographs, because trains of clockwork gears and their regulating escapements inevitably contained tiny irregularities due to slight metallic inconsistencies, lubrication, and even air resistance. These tiny irregularities would be unthinkingly compensated for in the brain of a visual observer, but the brutal honesty of photography exposed blurred images.

Warren De La Rue hit upon a solution. Instead of just letting the tracking machinery run when exposing a wet collodion plate for even a few seconds, why not also guide the telescope manually? If the main optical system of a large telescope were being used to expose a photographic plate, why not further equip the instrument with a high-powered "finder" telescope? Using this method, the astronomical photographer would set the cross-wires of his finder telescope upon some prominent object on the lunar surface. By means of fine-screw adjusters, it was then possible to give the whole telescope a series of gentle tweaks during the exposure so that the image falling on to the wet collodion plate remained in exactly the same place, thus ensuring a sharp photograph. This method of personalized tracking would be used by astronomical photographers until the digital age rendered it partially – but not wholly – obsolete.

The first "custom-designed" photographic telescope

As we shall see in Chapter 16, the physics of the sun was becoming a subject of growing astronomical interest, especially in the wake of the discovery of the periodicity of sunspots in the 1840s. No heavenly object was more amenable to photographic study than the brilliant sun, where even for a wet collodion plate one might secure a good image with an exposure time of a fraction of a second.

In conjunction with the independently funded Royal Society and British Association for the Advancement of Science, Warren De La Rue designed and commissioned from the optician Andrew Ross the first specifically photographic telescope.

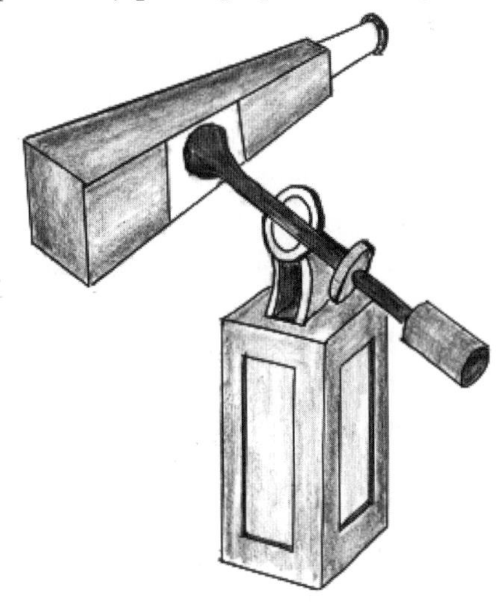

15.2 Warren De La Rue's photographic telescope, 1854. Instead of having an eyepiece, the long wooden pyramidal tube terminated in a photographic plate-holder. (Drawing by Allan Chapman.)

Its 3½-inch-diameter lens of 50 inches focal length, mounted in a wooden pyramidial tube, was corrected for violet light rays, as the spectral sensitivity of early photographic emulsions was different from that of the human eye. The new telescope, with a plate-holder instead of an eyepiece, was set up at the primarily Grand Amateur geomagnetic Observatory at Kew, Richmond, in the late 1850s. It stood upon a cast-iron pillar, with a fine clock-drive, and a spring-loaded roller blind shutter, to obtain very brief exposures of the solar surface. As the solar storm of 1859 (see the next chapter) had demonstrated that the sun exerted a magnetic influence upon the earth, it was deemed useful to photograph the sun on a systematic basis, to look for solar flares and other surface changes.

Along with Sir John Herschel, De La Rue argued that the sun should be photographed daily, and the photoheliograph was to play a major role in this project. In 1860, De La Rue accompanied the British expedition to Rivabellosa, Spain, to photograph the total solar eclipse visible on 18 July, and succeeded in obtaining some decisive photographs. Perhaps most significant were those of the solar prominences which, while observed visually at the Turin eclipse of 1842, and that of Scandinavia in 1851, were still objects of conjecture. De La Rue's photographic research back in England further enabled him to confirm the visual observation that when seen edge-on upon the solar limb, sunspots appeared to be physical depressions on the sun's body.

After Spain, De La Rue took the photoheliograph to his new observatory at Cranford, Middlesex, after which it went to Kew, to be replaced after 1874 by the 4-inch Kew Photoheliograph, which would take a daily photograph of the sun. These pioneering photographic telescopes are now preserved in store at the Science Museum, London.

Although Oxford University's Radcliffe Observatory was already a century old, academic and trustee politics led to the university establishing a new observatory around which the twentieth-century university's Science Area would grow up. In 1873, De La Rue gave his splendid 13-inch telescope to this new observatory, which would be directed by his independent astronomer friend, the Revd Dr Charles Pritchard – a teacher who had previously educated the sons of Sir John Herschel. Photography would play a major part in the Oxford Observatory's research programme, and a daily solar photograph would be taken almost to the end of the twentieth century, latterly by Dr Madge Gertrude Adam, who died aged eighty-nine in 2001. I had the honour of conversing with her on many occasions, and she was a mine of information.

James Nasmyth's *The Moon* (1874): Photographing the Moon at Second Hand

James Nasmyth was the son of Alexander who, it has been suggested, was the true founder of the Scottish landscape painting tradition, and James was also an accomplished artist with pen, brush, and plaster modelling. Fine as De La Rue's lunar photographs were, the engineering technology of the 1850s could not yet up take crystal-clear, high-magnification images of single lunar craters and other features, which could reveal the geological minutiae of internally stratified craters such as Copernicus. Nasmyth, therefore, devised a novel alternative strategy. Using his great 20-inch-aperture reflecting and other telescopes, he employed his meticulous artistic skill to make detailed drawings of certain key craters and other lunar features. He then used these drawings as the basis for a series of highly detailed scale plaster models. Next, he photographed the models with his wet collodion camera, using angles of illumination which replicated the real crater as seen at different lunar phases, and from low landscape-level angles that would show the crater as it would not be seen until the space age. Even 150 years on, it is easy to mistake one of the plates in *The Moon* for a real crater, taken at a modern observatory. Several of Nasmyth's original plaster models still survive in the store of the Science Museum, London, and they are exquisite, both astronomically and artistically.

15.3 James Nasmyth's photograph of a plaster model of the lunar crater Triesnecker. In the early 1870s it was still not possible to obtain high-definition photographs of single lunar features. Instead, Nasmyth fell back upon his skill as an artist, using drawings as a guide for making meticulously detailed plaster models of specific formations. For publication, he then photographed the models at appropriate angles of illumination. (J. Nasmyth and J. Carpenter, *The Moon* (1874), pl. XI. Author's collection.)

THE "DRY GELATIN" PLATE AND NEW POSSIBILITIES

During the 1870s, photography passed through its next evolutionary stage. In England, Richard L. Maddox began to experiment with gelatin as a replacement for wet collodion as a vehicle for the sensitive silver salts in the photographic emulsion. By 1871, he had developed a dry sensitive emulsion, though inconveniently, it was very touch-sensitive. Once the idea of the dry plate began to circulate, Charles Harper Bennett made crucial improvements whereby the plate became more stable and less sensitive to handling, and could be made more light-sensitive, or *faster*. Finally, in 1879, the Eastman Company in the USA began to manufacture fast dry plates on an industrial scale, and a whole new range of photographic possibilities opened up, from holiday snapshots to the photography of galaxies. Such plates could be bought, stored, exposed, stored again, and developed weeks or months later.

Photography was yet another example of how, once a new technology had made something possible, it came to develop, quite literally, at the speed of light! It opened up whole new vistas for the astronomer, especially as a photographic emulsion could store up light, unlike the human eye, which depended on a constant stream of individual photon impulses acting upon the nerves in the observer's eye. So could a photographic plate, by storing up light during a long exposure, reveal detail in a distant object that the human eye could never detect physiologically? The plate would become a godsend for deep-space photography, and would come to play a major role in our understanding of the distant nebulae.

ISAAC ROBERTS: PHOTOGRAPHER OF THE GALAXIES

An early individual who came to recognize this potential was an ex-farm-boy from North Wales. Isaac Roberts was born on his parents' Denbighshire farmstead in 1829. As a young man, he travelled to Liverpool to complete his training in the building trade, at a time when Liverpool was Great Britain's boom town, with its expanding docks, factories, great warehouses, and private homes. Isaac became a highly successful master builder, contractor, and civil engineer, having augmented his practical daytime on-site experience with night school study. Like so many Grand Amateurs, he was driven by a determination to succeed, intellectual curiosity, and commercial acumen.

Like his fellow-Liverpudlian of the previous generation, William Lassell, Isaac Roberts was fascinated by the heavens. In 1879 he bought a splendid 7-inch-aperture refractor, stand, and clock-drive from the master-optician Thomas Cooke of York, which he set up in his private observatory at Maghull, Liverpool. He later teamed this refractor with a 20-inch-aperture silvered-glass mirror reflector by Sir Howard Grubb of Dublin, both telescopes being eventually mounted upon the same clock-driven equatorial mount. This formidably powerful optical combination enabled Roberts to use the large refractor as a precision guide telescope for the 20-inch reflector, which could also be used as a camera.

Then in 1885, astronomers on both sides of the Atlantic became fascinated by the appearance of a "supernova", or new star, in the Andromeda nebula, which was M31 in Charles Messier's catalogue of 1781–84. At this time, of course, the nature of this long, lenticular nebula, which had already been exhaustively studied and drawn by William Cranch Bond and his son George Phillips Bond, using the great 15-inch visual refractor at the Harvard Observatory, was a puzzle. Was it made of stars, or of glowing gaseous matter? The supernova suggested the presence of stars. Isaac Roberts began to expose a series of meticulously guided plates of the Andromeda nebula: a task much more difficult than photographing the moon, even with the new fast plates, for the nebula is scarcely visible to the naked eye, and can only be studied telescopically. In a plate exposed on 10 October 1887, he detected detail in the Andromeda nebula which seemed to confirm what the Bonds had seen visually: that M31 consisted of a bright nucleus surrounded by circular arcs. But a 4-hour guided plate of 29 December 1888 showed things clearer still.

In 1890, Roberts left Liverpool and built a new observatory under the clearer rural skies of Crowborough, Sussex, where he came to a new interpretation of the Andromeda nebula. From a 90-minute guided exposure taken on 17 October 1895, he now concluded that

M31 consisted not of curved arcs, but a pair of spiral arms curving outwards from the nucleus. Superior plates taken in 1899 and 1902 confirmed this spiral structure. It did contain stars, but they appeared to be conjoined with faintly glowing bands of gaseous or dusty material. The Andromeda nebula, therefore, seemed to be a spiral, like Lord Rosse's "Whirlpool" (M51), yet whereas we see the "Whirlpool" face-on, Andromeda was sideways-on to a terrestrial observer.

Between 1887 and 1899, when Roberts published his *Photographs of Stars, Star-Clusters and Nebulae*,[4] one can trace the development of the power of photography to enable us to interpret the universe. Yet while the camera might not lie, its images could confuse if they lacked decisive clarity or seemed at odds with what we could see visually at the eyepiece, and the eye of the astronomer had not yet learned how to interpret them. By the 1880s, however, the combination of big telescope engineering and rapidly advancing photographic chemistry had also set others in pursuit of the nebulae, including, most notably, the Newcastle-upon-Tyne Grand Amateur engineer Andrew Ainslie Common. Working from Ealing, west London, with a magnificent 36-inch-aperture reflector and the new gelatin plates, Common began to photograph the Orion Nebula (M42) in January 1880.

15.4 M31 and M32 in Andromeda, photographed by Isaac Roberts on 29 December 1888 after a 4 hour 7 minute exposure. (I. Roberts, *Photographs of Stars, Star-Clusters, and Nebulae* (1893), pl. 44.)

As these developments were taking place in Great Britain, similar ones were in process among independently funded astronomers in the United States. Lewis Morris Rutherfurd of New York City set up a splendid private observatory equipped with an 11½-inch refractor in 1856, and was soon obtaining fine photographs of the moon, planets, and star clusters. In 1868, Rutherfurd obtained a 13-inch refractor, equipped with a corrector-plate to convert it for photographic work. Likewise, the elderly John William Draper and his physician-chemist Grand Amateur astronomer son Henry would pioneer astronomical photography in America. Henry Draper and Isaac Roberts were in a friendly competition to obtain the best photographs of nebulae before Draper's sudden death in 1882. (For more on this, see Chapter 17.)

In addition to its application to the moon, planets, and nebulae, however, photography would become the invaluable handmaiden of spectroscopy, that wholly new optical technology which would go on to burst the old bounds of "traditional" eye-at-the-telescope astronomy, and transform the science beyond recognition from the 1860s onwards.

Chapter 16

Unweaving the Rainbow Part 1: Sunlight, Sunspot Cycles, and Magnetic Storms

In 1835, the Positivist philosopher Auguste Comte specified what he considered to be an eternally unknowable category of knowledge: what the sun and stars are made of.[1] There was no conceivable way in which a piece of a vastly distant body radiating light and energy could ever be taken into the laboratory for physical and chemical analysis. Comte should have been more cautious in his pronouncement, however – as should all those people over the years who confidently declared what we could or could not ever know. A mere quarter-century after 1835, the great French philosopher's pronouncement would be shown to be spectacularly mistaken.

It must be admitted, however, that the birth of solar and stellar physics came about in a way which Comte could not really have envisaged in 1835. To his generation, astronomical advancement resided in precision angle-measuring followed by advanced mathematical analysis, while physical and chemical discovery was rooted in meticulously executed laboratory procedures, and never were the twain likely to meet. Then things began to accelerate with surprising speed by the late 1840s, especially after the *annus mirabilis* of 1859, when a coalescence of heat, light, magnetism, and spectroscopy first showed mankind how we might unweave the rainbow.

Understanding the Sun, our Nearest Star

Unseasonable weather is not new, and from the fourteenth century onwards diary keepers recorded the weather's perceived vagaries, such as William Merle of Merton College, Oxford, and Driby, Lincolnshire, who did so in some detail between 1337 and 1344.[2]

Some centuries later, astronomers began to ask if these vagaries could be caused by changes in the sun's heat output. The discovery of sunspots after 1611 and their careful recording thereafter had led to speculations about the possible connection between sunspots and the weather. Then as we saw in Chapter 9, Sir William Herschel published an analysis of wheat prices in Windsor, based upon those recorded in Vol. 1 of Adam Smith's *The Wealth of Nations* (1776), and variations in sunspot appearances. Working on the premise that wheat was cheap in years of abundant harvests and dear in times of dearth, Herschel attempted a correlation between sunspots and good and bad agricultural yields. It appeared that the more sunspots there were, the better the harvests.

The rapid expansion of palaeontology and fossil geology after *c.* 1800 had indicated that the earth's climate had swung through great extremes, with exotic tropical jungle growing and then forming the coal measures of what today are South Wales and Newcastle, and great ice-sheets once covering now temperate Europe. Geologists, physicists, and early geophysicists were beginning to realize that the sun might not be the permanently stable and regular energy source it had once been assumed to be. More than anything else, however, sunspots would turn out to be the key to solar physics, and between 1826 and 1843, the German apothecary and amateur astronomer Samuel Heinrich Schwabe of Dessau made a daily sunspot record whenever the sky was clear. Schwabe's seventeen-year record demonstrated a distinct cycle, or periodicity, of sunspot appearances, extending just over ten years (we now know it to be eleven years). His discovery was first published in *Astronomische Nachrichten* in 1844,[3] but it was only when Alexander von Humboldt drew attention to it in Volume 3 of his monumental *Kosmos* in 1851 that he began to receive major recognition. Yet what might actually *cause* that periodicity, the discovery of which won Schwabe the Gold Medal of the Royal Astronomical Society in 1857, and an FRS in 1868, remained a mystery.

Then what were the "pink flames" or "protuberances" seen by an international group of astronomers in Turin, Italy, during the eclipse of 1842? The flames were seen and recorded again by William Lassell in the Scandinavian eclipse of 1851, and finally photographed in Spain in the eclipse of 18 July 1860. Were they optical distortions occurring during eclipse totality, were they something rising up from the moon, or were they on the sun? Were they connected with sunspots? By the 1850s, the nature of the sun, its endless supply of radiant light and heat, and its relation to terrestrial climatic, physical, biological, and growth processes were very much under discussion across Europe and the USA. Did the sun have its own "weather"? Hervé Faye in France suggested that sunspots were the product of violent cyclonic systems on the sun. All this new research must be seen in relation to the discovery of infrared and ultraviolet radiation after 1800, and Young's and Fraunhofer's discovery of puzzling black lines in the solar spectrum, already mentioned in Chapter 9.

The great solar storm of 1859

Between 28 August and 4 September 1859, a violent storm erupted on the solar surface: one which would probably have been undetected on earth only a few decades previously, the detection itself being a product of new technologies. At 11:18 a.m. on 1 September 1859, at his large and well-equipped private solar observatory at Redhill, Surrey, Richard Christopher Carrington – another Grand Amateur, whose wealth derived from brewery profits – witnessed a dramatic development on the solar surface. A large white arch between some dark sunspots flared up. This was a facula. Measuring its size as a fraction of the known solar diameter, Carrington calculated that the facula was 35,000 miles long, or four and a half times the diameter of the earth. The facula's existence was confirmed by Mr Hodgson, of Highgate, London, another amateur, who also observed it. White faculae had been seen previously, but the significance of the great solar storm of 1859 was its unexpected collateral impacts elsewhere. One was the appearance of spectacular and sustained aurora borealis and australis displays in both hemispheres, while an American ship heading for California in the Pacific reported sights suggesting that both the sky and sea were on fire. These phenomena had been reported going back to late August 1859.

Most alarming of all was the sudden, dangerous behaviour of the new electric telegraph technology, the wires of which already criss-crossed Europe and the eastern part of the USA. Telegraph keys began to chatter uncontrollably, even when the batteries were disconnected, and some telegraphers even received bad electric shocks. It was soon realized that the storm being observed on the solar surface was inducing electric currents in thousands of miles of pole-suspended telegraph wires which might earth themselves through a telegraph key – or through a telegrapher's body. Michael Faraday had discovered electromagnetic induction in London's Royal Institution's private laboratory in 1831: a natural process whereby a moving magnetic stream could generate an electric current in a wire. The solar storm of 1859 demonstrated that, in addition to all the other optical and thermal forces which the sun radiated, it gave off powerful magnetic energy.

Carrington's Redhill Observatory and the Observatory at Kew were, after Schwabe, some of the first locations to monitor the solar surface on a daily basis, although Carrington's observatory was much bigger and better equipped than Schwabe's more modest enterprise. Here, the Cambridge-educated Carrington observed and recorded daily solar data at a very high level of magnification and detail, and analysed the solar surface physically. He began giving a running number to the sunspot cycles, which remains in use today, enabling analysts to compare different cycles over long periods. He also refined Schwabe's figure of 10+ years for a cycle to about 11 years.[4] Very significantly, his observations and analyses showed Carrington that the sun's surface is not fixed, like that of the earth or moon. From meticulous timings and averages, he discovered that the solar equator rotated slightly faster than those latitudes nearer to the solar poles. Sunspot cycles also tended to begin in those solar regions over 35 degrees above and below the solar equator, then gradually migrated toward the equator as the cycle progressed. This suggested that the sun, instead of being a solid sphere, was a viscous or semi-fluid body. All of this posed questions about how the sun burned, and what went on inside it to produce such an endless stream of diverse energies. These questions would have to await the new physics of the early twentieth century – atomic physics, relativity, and quantum theory – before coherent and experimentally verifiable answers would be forthcoming.

"Rice grains", "granules", and the solar surface

Under almost all observational conditions, the non-spotted parts of the solar surface seemed bland: a smooth, white expanse. In June 1860, however, things began to change when the ever-resourceful James Nasmyth (the ironmaster we met in Chapters 14 and 15) reported a discovery. Having retired from the heavy engineering trade at the age of forty-eight in 1856 as a very rich man, he had bought a country mansion and estate in Kent, near Sir John Herschel. Enjoying the clearer skies of rural Kent, Nasmyth had relocated his observatory, and had commenced a run of very high magnification observations of the solar surface. He was examining the structure of sunspots and how they appeared not just as depressions in the solar surface, but as actual holes or pits going down to the interior. Could the spots not really be dark, but merely less light-radiant than the blinding white solar surface, seeming dark by contrast?

It had been known since the seventeenth century that sunspots possessed a dark umbral central region and a lighter surrounding penumbral one – suggesting a descent into a cavity – some changing sunspot groups being much greater than the entire size of the earth. In 1860, Nasmyth was examining these spots at very high magnification when he noticed something new: instead of being bland and smooth, the solar surface appeared to consist of countless elongated units packed together, resembling grains of rice or granules. As he was observing in June, the sun was at its highest in the sky, thus minimizing atmospheric distortion, but even then the rice grain structure was only visible under conditions of pure atmospheric transparency and stillness. Yet the rice grains were really there, as other observers began to substantiate, and Sir John Herschel styled Nasmyth's achievement "a most wonderful discovery".[5]

16.1 "Rice-grain" or "willow leaf" granulations on the sun's surface, drawn by James Nasmyth. (*James Nasmyth Engineer. An Autobiography* (1889), p. 370. Author's collection.)

SOLAR KNOWLEDGE BY 1860: A RÉSUMÉ

Between 1800 and the summer of 1860, our knowledge of the sun had advanced enormously. As shown by a range of scientists extending from geologists to optical physicists, the sun was no longer the changeless, eternal source of popular belief. Laplacian nebula ideas suggested that it might be a ball of gas gradually spinning and cooling across vast aeons of time. It had clearly gone through stages, in which it had produced profound fluctuations in the terrestrial climate, varying between jungle temperatures in the latitude of modern Newcastle-upon-Tyne and ice sheets across the latitude corresponding to the modern French Riviera. The sun was also a source of various energy bands undetectable without modern instruments, such as ultraviolet, infrared, and magnetism; while its surface was "fluid" and went through eleven-year cycles of energy output that might affect terrestrial weather and food production. These were only the very short cycles, for who could tell when the long-term archaic cycles of the dinosaurs and the ice ages might return? All of this was doing the rounds of the scientific community worldwide before the spectroscope began to reveal the sun's chemistry and inner physics, and to demonstrate conclusively that it was in fact a star.

Chapter 17

Unweaving the Rainbow Part 2: Cosmologists and Catholic Priest Pioneers of Astrophysics

We saw in Chapter 9 how the dark and coloured lines visible in the solar spectra were objects of puzzlement and fascination to nineteenth-century chemists and physicists. By the mid-1850s, however, the potential for the laboratory use of the spectroscope for chemical analysis was developing rapidly. But it would be in Heidelberg that things would first coalesce to make astronomical spectroscopy possible.

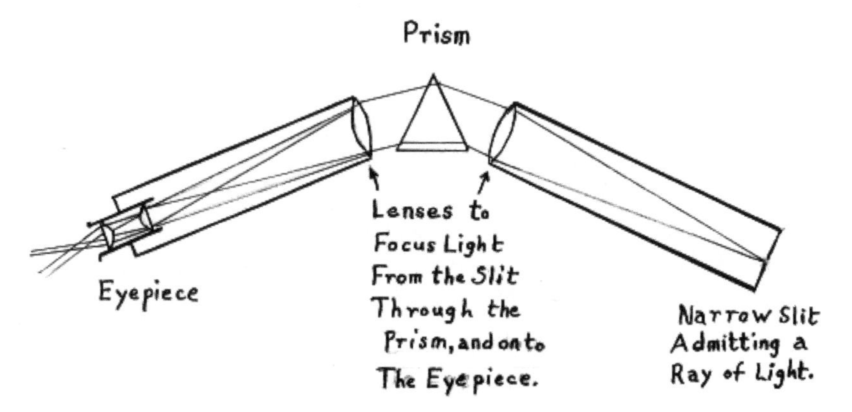

17.1 A basic spectroscope. A ray of light is admitted through a fine slit, and focused by a lens through a dense glass prism to disperse it into a spectrum. It is then focused down a small viewing telescope. If this viewing telescope is attached to a graduated scale, such as a surveyor's theodolite, or is fitted with a precision eyepiece micrometer, it can be used to measure the exact angular dispersion of the colours and their ensuing black Fraunhofer lines. (Drawing by Allan Chapman.)

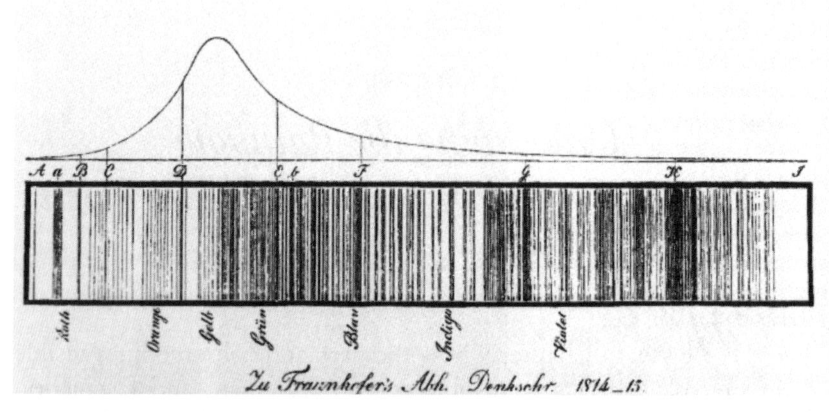

17.2 Josef Fraunhofer's solar spectrum of 1814. Note the two adjacent thick black lines that, it would later be discovered, correspond to the element sodium. (J. N. Lockyer, *Stargazing* (1878), p. 392. Author's collection.)

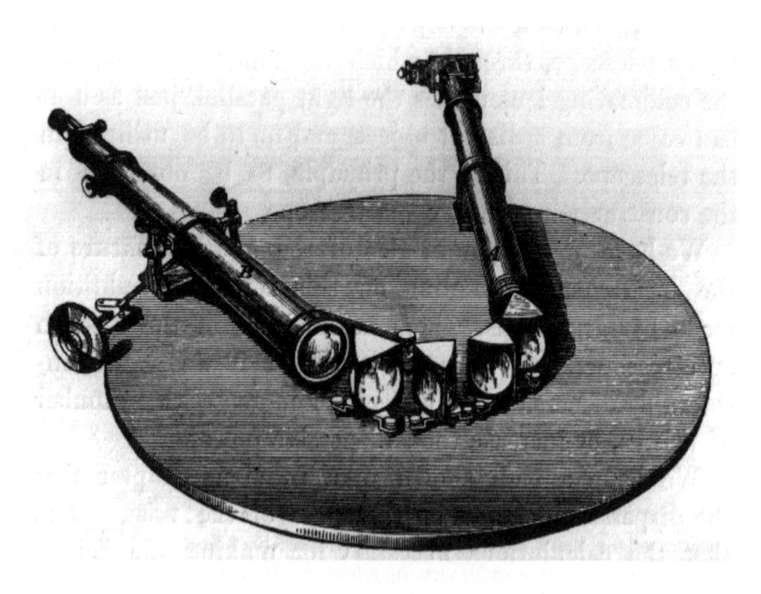

17.3 A four-prism spectroscope. These prisms, made from very dense glass or quartz, would disperse the starlight focused through the telescope over a large angle. Each prism would disperse the light yet further, making the resulting spectrum larger and its Fraunhofer lines more conspicuous and easier to measure. (J. N. Lockyer, *Stargazing* (1878), p. 396. Author's collection.)

AN AFTERNOON WALK IN HEIDELBERG IN 1859

The birth of solar and stellar physics after 1859 came about in a way that Comte could never have reasonably envisaged in 1835. Yet such unexpected turnabouts, as one new set of facts and technologies encounters another, are the very stuff of scientific discovery, in whatever branch. This is why scientific research is so exhilarating and so unpredictable. For men of Comte's generation, as we saw above, astronomical and chemical advancement resided in quite separate domains, at least until they metaphorically nodded to each other in the autumn of 1859.

Robert Wilhelm Bunsen and Gustav Robert Kirchhoff both belonged to the highest traditions of German academic university-based science and scholarship. Bunsen (who was also the inventor of the laboratory gas burner that immortalized his name) was one of Europe's leading chemists by the late 1850s, while Kirchhoff was eminent as an electrical physicist. Shared interests in electricity, electro-chemistry, and radiation physics brought them together scientifically. Bunsen in particular was interested in developing the use of incandescent flame tests as a way of analysing complex chemical substances. If salt, copper oxide, or magnesium were incandesced at high temperatures (such as in a Bunsen burner flame), they emitted brilliant colours unique to their elements.

By the 1840s, of course, these flame colours were accepted pieces of chemical knowledge, but what would connect laboratory knowledge with the sun, and open the way to chemical spectroscopy, was yet another pair of discoveries made independently by an Englishman and a Frenchman. The Paris Observatory physicist Jean Bernard Léon Foucault and his King's College, London colleague William Allen Miller both discovered that when sunlight was passed through incandescent sodium compounds, such as common salt, the yellow sodium flames became dark. The two bright yellow lines produced by incandescent sodium salt occupied exactly the same position in the solar spectrum as did the two black lines which Joseph von Fraunhofer had styled the "D" lines in his own map of the solar spectrum back in 1814 (the "Fraunhofer lines" mentioned in Chapter 9). Others interested in this newly discovered phenomenon included Sir David Brewster in Edinburgh and the Irish-Cambridge physicist George Gabriel Stokes.

Yet why should an incandescent element emit both coloured and black spectral lines? Foucault realized that when a more intense energy source, such as that of sunlight, or even the light emitted from a battery-powered electric arc, encountered a lower energy source (such as in a normal flame), energy was absorbed, as well as emitted. So if one incandesced salt in a laboratory spectroscope, one saw the brilliant yellow sodium "D" lines. But if a beam of focused sunlight was then passed through the flame, the yellow sodium lines would suddenly turn black upon absorbing sunlight. This discovery, as further amplified by Miller, Brewster, and others, suggested a powerful new analytical tool for the laboratory chemist and physicist: though not yet for the astronomer.

Kirchhoff and Bunsen drew all this observational data about coloured and black spectra together. This has been styled "Kirchhoff's Law": when a higher energy source (such as sunlight) encounters a lower energy source (an incandescent chemical in the flame of a Bunsen burner), a reversal takes place, and the coloured lines become black, while remaining in exactly the same places with regard to the other lines in the spectrum. Thus, one had an *emission* spectrum when the light was coloured and an *absorption* spectrum when energy absorption caused it to go black. This phenomenon was outlined in a letter of 15 November 1859 sent by Bunsen to his old Heidelberg chemistry student Henry Roscoe of Manchester.[1]

Before analytical spectroscopy could really advance, however, it was first necessary to obtain pure chemical reagents, or reactive agents, for all the then known elements, to establish their precise coloured emission and black absorption positions in the spectrum to serve as benchmarks. Robert Bunsen in Heidelberg was, among other things, involved in this work, in conjunction with Kirchhoff: a field where chemistry, physics, electrochemistry, and optics began to coalesce. While it was understood by the late 1850s, therefore, that the intense energy in sunlight affected incandescing elements to produce black-line absorption spectra, no one so far was seriously connecting this purely laboratory research with astronomical bodies, although Foucault had thrown some broad hints in that direction.

The train of events that would lead laboratory spectral analysis onto the creation of solar and astro-physics proper appears to have been set in motion by a combination of chance, following up a hunch, and then sheer inspiration. Working together one evening in their Heidelberg laboratory during the autumn of 1859, Bunsen and Kirchhoff noticed the bright light of a fire burning in Mannheim, almost twenty kilometres away down the Necker Valley toward the Rhine. They focused their laboratory spectroscope on to the distant flame (easier to do in an age of almost zero light pollution) and obtained a reading. They detected the elements barium and strontium in the resulting spectra: a chemical presence dramatically confirmed when they subsequently discovered that quantities of paint had been present in the blaze. Soon after, Bunsen and Kirchhoff took a stroll around Heidelberg's beautiful Philosopher's Walk when a realization struck them: if we can detect the chemical contents in a fire in Mannheim, might we not use the spectroscope to discover the chemical composition of the sun?

Sir William and Lady Margaret Huggins discover gaseous nebulae from a south London garden

William Huggins was heir to a successful City of London silk business. His passion, however, was not for commerce, but for science. He never went to university – he was obliged to abandon hopes of Cambridge in order to go into the family firm – but after a first-class school education, he enjoyed ample means to develop a Grand Amateur scientific career, joining London learned societies, travelling on the Continent, meeting people, and doing work in microscopy, chemistry, and optics. Already the owner of a collection of fine scientific instruments and books, in 1858, aged thirty-four, he bought a superlative 8-inch-diameter telescope object glass by the American optician Alvan Clark, from his medical Grand Amateur friend William Rutter Dawes for £200. Huggins then commissioned Thomas Cooke of York to mount it, to produce a first-rate equatorial refractor. When his father retired from business around 1854, the bachelor William sold off the firm for a profit, and the family moved to Tulse Hill, south London, which then bordered on the Kentish countryside, with its clear, unpolluted skies.

Settled in Kent, in a fine house and with the 8-inch telescope in his garden, Huggins began serious visual observations of the moon and planets and made double star measurements. This was relatively routine work, and he began to look around for more ambitious astronomical projects. Sometime around 1861, he read of the discoveries of Bunsen and Kirchhoff in Heidelberg, which came to him like "a spring of water in a dry and thirsty land".[2] The thirty-seven-year-old bachelor of independent means with a fine refractor at his disposal had found his life's vocation. With his friend and neighbour, Professor William Miller, Huggins had built a powerful eight-prism, high-dispersion spectroscope that could spread the spectra across a wide angle, thus making it easier to measure the positions of the lines with great accuracy. What especially interested him, however, was not the sun, but the stars and nebulae of deep space, for if they were but suns removed, as it were, to vast distances, then should not their physics and chemistry be similar? Huggins's new spectroscope, when working in conjunction with his 8-inch telescope, produced distinct stellar spectra.

First, however, just like Bunsen and Kirchhoff, he had to set about laying a foundation in laboratory chemistry. Pure chemical reagents had to be obtained or prepared in his own laboratory. Absolute chemical purity was essential, for if a reagent, be it of a metal or a gas, was in the least bit contaminated by another substance, then a false reading would result, with misleading lines in the spectrum. Once his basic set of pure reagents had been prepared, he then had to incandesce them in a high-energy source and map their respective positions manually with a precision micrometer in the eyepiece of the spectroscope. In 1864, Huggins published a paper in the Royal Society's *Philosophical Transactions*[3] giving the spectral line positions of 24 pure elements. This, and later element analyses, would provide the chemical bedrock for all of his subsequent astrophysical researches, by providing a comparative yardstick.

In many respects 1864 would be an *annus mirabilis* for Huggins and spectroscopy. Not only did he collect and publish stellar spectral data showing the diverse chemical and physical characteristics of individual stars, but he also examined various nebulae. Most of these yielded spectra suggestive of densely packed star fields, which seemed to substantiate the idea of Sir John Herschel and Lord Rosse that the nebulae were stellar or particulate in their structure. On the night of 29 August 1864, however, Huggins's spectroscope disclosed a most curious phenomenon in the Cat's Eye nebula in the constellation of Draco the Dragon (called the Cat's Eye because of its odd resemblance to a feline eye). Its spectrum did not indicate the normal black absorption lines of a stellar composition. Instead, all that he could detect were one or two emission lines, suggesting that it was made up of glowing gas.[4] So had Sir William Herschel been right after all when he suggested that some wispy nebulae were composed of "luminous chevelures" or glowing hair-like stuff rather than stars?

The glowing stuff in the Cat's Eye did not display a spectrum that resembled any known gas. Could the nebulae, therefore, contain gaseous material chemically and physically unique to themselves? The name nebulium came to be attached to this substance. Several physicists worked on this puzzling spectrum over the following decades, but it was not until 1927 that Ira Sprague Bowen announced his discovery that nebulium was the complex spectrum of oxygen in a doubly ionized state. Physical ionization was scarcely understood in 1864, but was so by the 1890s. Huggins would go on to undertake other groundbreaking nebula studies, such as that of Orion in 1865.

His status as an astrophysical pioneer was beyond doubt by 1868 when he announced his work on the star Sirius. Recognizing the 1841 acoustic work of the Austrian physicist Christian Doppler – who noted that the pitch of a sound (such as a train whistle) increases as it approaches, then drops as it passes away – Huggins applied the same thinking to light. As both sound and light are wave phenomena, would not approaching and receding stars have their spectra displaced to the blue and the red end respectively? This led him to announce that Sirius was receding from the earth at the velocity of 24.9 miles per second: a figure which he later amended in detail but not in substance. Huggins's Sirius work would provide one of the key techniques of the twentieth century cosmologist, especially Vesto Slipher, Edwin Hubble and Father Georges Lemaître, as we shall see in Chapters 24 and 25.

By the optical standards of the day, however, Huggins's 8-inch refractor was relatively modest, and in 1870, he was granted the private Oliveira Bequest of the Royal Society to build a new pair of large aperture telescopes. These included a 15-inch-aperture refracting telescope and a Cassegrain reflector, both mounted upon the same great equatorial axis by Sir Howard Grubb of Dublin. This novel configuration of two telescopes on one axis would soon be used by Isaac Roberts (see Chapter 15), allowing a big refractor to be used, if necessary, to guide a large spectroscopic or photographic reflector. The new telescope was intended to extend Huggins's probe into deep space yet further, and over the decades ahead, he would do much to establish and demonstrate the capacity of astrophysics and chemistry, leading to his knighthood in 1897.

Then in 1875, something remarkable occurred: the fifty-one-year-old bachelor got married. The bride was an attractive twenty-seven-year-old Irishwoman, who had admired Huggins since she was a girl. Margaret Lindsay Murray was the daughter of a wealthy Dublin banking family, and had a passion for science which stretched back to her childhood. As a girl, she had read of William Huggins's work, and, like so many people in the 1860s, was captivated by the fact that, all of a sudden, science was opening up a realm hitherto believed to be permanently sealed up in mystery: the chemistry and physics of the sun and stars. In many ways it was a strange romance, originally sparked by reading scientific books and papers before she even met Huggins. Their marriage, while producing no children, was one of the great astronomical partnerships of history: like those between Elisabetha Catherina and Johannes Hevelius, Caroline and William Herschel, Mary Anna and Henry Draper in the USA, Richard and Sallie Proctor, and, in 1875, Isaac Roberts and his professional astrophysical second wife Dorothea Klumpke.

The future Lady Margaret Huggins would introduce several crucial innovations into the Tulse Hill research schedule: innovations having a parallel to those developed by Mary Anna Draper at the Hastings-on-Hudson, New York, observatory of her husband Henry. Margaret would show her accomplishments in the ever-necessary chemical laboratory side of astrophysics, and as an efficient time manager. Right back to the early 1860s, when William had first begun to measure stellar spectra, it had been a long, tedious job. It could take several nights to manually record a single star spectrum – cloudy skies permitting – using the eyepiece micrometer. Being aware of all the recent innovations in photography, Margaret Huggins started the practice of photographing stellar spectra on the new fast plates. So on a clear night, doing their best to keep warm, she and William could take guided photographs of several star spectra in a few hours. Keeping warm meant wearing heaps of clothes, for it is impossible to heat a working astronomical observatory. If one tried, the rising heat would cause the air in front of the telescope to flutter and produce a dancing, blurred image.

After working all night, and hopefully having exposed a series of dry gelatin plates and obtained the long narrow spectral bands with their distinct Fraunhofer lines, William and Margaret could then develop the plates later. Thus they accumulated a library of glass spectral plates which could also be measured during the daytime or on cloudy evenings. Photography liberated Huggins from the slow slog at the eyepiece micrometer in a dark, freezing observatory. They also were working entirely in the Grand Amateur tradition. Exactly contemporary with the Hugginses, however, was an Italian Jesuit astronomer-priest who was also inspired by Bunsen's and Kirchhoff's discoveries, who went on to establish the first spectral classification system of the stars.

Father Angelo Secchi of Rome: the Jesuit pioneer of astrophysics

Pietro Angelo Secchi, Member of the Society of Jesus, belonged to an almost three-century-old tradition of eminent Jesuit scientists, and stands as a contradiction to the popular myth that the Roman

Catholic Church suppressed science. Born in 1818 in Reggio, Italy, he and his Jesuit brothers had been forced to leave Rome during the violent secularist revolution of 1848. He travelled to the USA, where he taught astronomy and mathematics at Georgetown University, Washington DC, and also became a familiar face in England where, among other things, he taught at the famous Jesuit School at Stonyhurst, Lancashire, with its still famous astronomical observatory. He returned to post-revolutionary Rome at the behest of Pope Pius IX and succeeded Francesco de Vico as Director of the Papal Observatory in the Vatican City. Here, he pioneered early astronomical photography, and was in Spain in 1860 to observe and photograph the total eclipse in July, although it was the news from Heidelberg that led him to truly new scientific pastures.

Like Huggins in England, whom he knew through Europe's and America's growing scientific network, Father Secchi was struck by the great variation to be found in stellar spectra. There was also a sense of wonder – no doubt of religious awe – in the realization that the stars, which for millennia had been thought to be essentially the same, now displayed such spectral diversity while still containing the same basic chemical ingredients. Between 1863 and 1867, Angelo Secchi began to observe and classify 4,000 stars, and found that he could arrange them into four main classes. Class 1 displayed strong hydrogen lines and included white and blue stars. Class 2 displayed a variety of lines, including yellow. Class 3 displayed spectra in which the red lines were sharp, yet became fuzzy toward the violet. Class 4 contained contrasting stars, where the lines were sharp at the violet end and became fuzzy toward the red. Secchi's list was completed with the addition of the carbon stars, and his mature list would be announced to the world at the 1868 meeting of the British Association for the Advancement of Science, held in Norwich.

The Stonyhurst College Jesuit Observatory

Stonyhurst College is a distinguished Roman Catholic public school near Clitheroe, Lancashire. It was established in 1794, after the violently anti-Christian armies of the French Revolution invaded Flanders, and forced the school for Roman Catholic English gentlemen at St Omer to flee to a new home in England. Although many Victorian public schools would come to boast fine astronomical observatories, Stonyhurst, with its Jesuit priest astronomer masters, would be the only one to develop a full-scale internationally recognized geomagnetic and astrophysical observatory within its grounds, originally directed by Father Walter Sidgreaves between 1863 and 1868. Although founded in 1838 as a meteorological and geomagnetic observatory, solar and stellar physics would come to rank high among the Observatory's research priorities, along with photography. It would become the home for a succession of distinguished Jesuit astronomers. In addition to Angelo Secchi, who was there briefly, an early solar physicist of distinction was Father Stephen Perry who, in addition to his Stonyhurst work, travelled abroad to observe eclipses. On the way home from one such eclipse to the Îles du Salut in 1889, his ship, HMS *Comus*, put in to the notorious Devil's Island French penal colony. The courageous Father Perry insisted upon going ashore during a virulent epidemic to take the Holy Sacrament to the neglected prisoners. He caught a fatal dose of dysentery fever, and was buried at sea.

Going into the early twentieth century, there was Father Aloysius Cortie and other distinguished solar and stellar physicists. Although no longer involved in front-rank astrophysical work, the Stonyhurst Observatory still operates, as I know from personal experience; as does the Vatican Observatory. In the 1930s, however, the increasingly polluted climate of the Eternal City led to its being relocated to the grounds of the pope's summer residence at Castel Gandolfo, some 25 kilometres from Rome, where I had the honour to be welcomed by Secchi's professional successor, Father George Coyne. Yet even the sweet air of Castel Gandolfo, overlooking the blue lagoon of an ancient volcanic caldera, is no longer pure enough for modern astrophysical research. The principal Jesuit research observatory is now located under the pristine desert skies of Tucson, Arizona, USA.

In some ways, perhaps, one can understand why Jesuit Fathers took solar and stellar physics so much to heart, for it had a certain measure of theological significance, showing that the sun, stars, and nebulae were all made up of the same basic elements as are our own bodies. One could think of astrophysics, if so inclined, as containing evidence for a unified Divine Design. Yet much of our thinking about the chemistry and physics of the stars after 1861 was coloured by what we were learning about our nearest star, the sun, and what the spectroscope was teaching us about it.

The sun and the spectroscope

After painstaking correlations had been established between chemical reagents burned in the laboratory and the black Fraunhofer lines in the solar spectrum during the 1860s, several major puzzles remained. Why, for example, if there were only 90-odd known chemical elements, were there *hundreds* of black Fraunhofer lines seen in a ray of sunlight? What were the physical mechanisms that produced coloured (emission) and black (absorption) spectra? What were the solar prominences and how did they relate to the sun's mysterious internal energy generation process? And what was the glowing corona with its projecting streaks of light, sometimes seen briefly during total eclipses but at no other time? Everything would be so much more straightforward if these tantalizing prominences and coronal glows and streaks could be studied at leisure, rather than being hungrily grasped at during the couple of minutes of eclipse totality. James Nasmyth and Charles Piazzi Smyth, Director of Edinburgh's Calton Hill Observatory, had in the 1850s experimented with eclipse-simulation devices, and had corresponded about them with Sir George Airy, the Astronomer Royal.[5] But no viable instrument could be devised, largely due to the sheer brightness of the sky when the sun was not eclipsed, and atmospheric light scatter.

Things changed dramatically, however, in 1868, when, once again, a Frenchman and an Englishman made an almost simultaneous discovery. The French physicist Pierre Jules Janssen had travelled to India to see the total eclipse of the sun visible on the subcontinent on 18 August 1868. He subsequently photographed a spectacular prominence which he calculated rose some 88,000 miles above the solar surface. Then following the end of the eclipse, Jules Janssen had one of those insights which are the stuff of experimental genius. He modified the slit of his spectroscope and adjusted it so that it formed a tangent to the solar limb. The spectacular prominence of the previous day had vanished, demonstrating how transient prominences were. But he later found that when moving his spectroscope slit around the sun's limb, other new prominences could be seen. The golden key to solar physics had suddenly appeared, enabling astronomers to examine prominences and other solar atmospheric phenomena whenever they chose, without the need for an eclipse.

Shortly afterwards, in London, Joseph Norman Lockyer, a civil servant and serious amateur astronomer who had been analysing sunlight through his own spectroscope for three or four years, suddenly and quite independently had the same inspiration. Applying his own spectroscope to the edge of the sun, Lockyer also was gratified to see prominences in the solar chromosphere. Jules Janssen did not fully report his findings until his return to France from India; Lockyer announced his much more quickly, so that, quite by chance, both reports came out simultaneously. Both sets of results arrived during a meeting of the French Académie only a short interval apart, and the brilliant double discovery was soon telegraphed around the world.

Far from the discovery generating Anglo–French acrimony, Janssen and Lockyer admired each other's work, and became friends. The French savants then had struck a commemorative medal to celebrate the joint discovery, with both men's portraits together. Lockyer's own copy is preserved in a glass case in the Norman Lockyer Observatory near Sidmouth, south Devon. (I was deeply touched on 6 May 2017, after I had opened an exhibition at the Norman Lockyer Observatory, and the Trustees presented me with a perfect electrotype copy of the original.)

17.4 Medal presented jointly to Jules Janssen and Sir Joseph Norman Lockyer by the French Académie, 1868. The obverse displays the portraits of both men, while the reverse side depicts what is probably an allegorical representation of Phoebus the sun-god and his chariot. (By kind permission of the Norman Lockyer Observatory.)

Lockyer would go on to be a true pioneer of solar physics, and in 1873, he published his controversial dissociation theory of why so few elements produce so many spectral lines. Could it be that temperature, pressure, and energy systems in the sun and stars cause the same chemical elements – sodium, hydrogen, etc. – to somehow dissociate and emit different energy bands? This theory later helped to make sense of Huggins's nebulium, which was shown not to be a new element, but a terrestrial gas in an excited state. In the wake of his post-1868 studies of the sun's atmosphere, Lockyer found something which produced a pale, yellow line, unconnected with any then known terrestrial element. From the name of the Greek sun-god Helios, Lockyer christened the source of this new spectral line helium. Then in 1895, the Scottish chemist Sir William Ramsay isolated the gas experimentally in his laboratory, thus demonstrating that it was not unique to the sun.

Lockyer was a powerful and persuasive figure, and succeeded in having a publicly funded solar physics observatory established on the Royal College of Science site, South Kensington (now Imperial College, London). Manoeuvring from his original post in the War Office, Lockyer would become Professor and Director of the South Kensington Solar Observatory. Here, as Great Britain's first salaried astronomical physicist, Lockyer exerted a profound impact upon scientific education and research. He was also an influential communicator of the new spectroscopic studies, and his *The Chemistry of the Sun* (1887) was a benchmark of where knowledge stood at that date. Then, when official policy caused the Observatory to be moved out of the smoky metropolis to the green fields around Cambridge, he resigned his chair and returned to his Grand Amateur roots. His second wife Thomasine Mary owned land at Sidmouth, in south Devon, and there, in 1911, the seventy-four-year-old Lockyer established a fine private observatory specializing in solar and stellar physics. The Norman Lockyer Observatory, with its array of five fine telescopes and facilities, is now one of the gems of British astronomy, being staffed by volunteers and operated by a charitable trust. Lockyer's archive is now lodged in Exeter University. The Norman Lockyer Observatory and Society are now mainly devoted to astronomical outreach and education work. The Observatory is in a glorious setting of rolling green fields above Sidmouth, and commands a dark-sky site with the English Channel directly to the south.

OUR AMERICAN COUSINS AND OUR IRISH FRIENDS

We saw in Chapter 15 how, in 1840, John William Draper, the New York academic physician and chemist, was the first to make a photograph of an astronomical body. He also experimented with using the daguerreotype process to photograph spectra in the 1840s, while his work on radiant heat and the refrangible rays that heated substances emitted would be of importance to Kirchhoff and the early solar physicists. His son Henry would become one of the pioneers of spectroscopy. This brilliant young man had already passed his medical qualifying examinations before reaching the legal age to practise medicine in the USA. It was in these years that his astronomical and physical interests really took shape, as he and his father made trips to Europe and established friendships with Lord Rosse, Sir John Herschel, the young William Huggins, and many French and German scientists.

In addition to ample money on the Draper side, Henry married (as his father had) an heiress, in 1867. Mary Anna Palmer was the highly intelligent daughter of a New York property magnate. Upon her marriage to Henry Draper, she and Henry became a team, like William and Margaret Huggins in England, with Mary Anna proving herself very skilful in the laboratory and with photographic recording. With a mansion on Madison Avenue, Manhattan, an estate at Dobb's Ferry, and a magnificent private observatory at nearby Hastings-on-Hudson, the Drapers were firmly placed in the American Grand Amateur league.

Learning the astronomical importance of the new dry gelatin photographic plates from William Huggins in 1879, Draper developed spectroscopic photography yet further. Before that, however, he had advanced stellar photo-spectroscopy when in August 1872 he obtained a landmark photograph of the spectrum of the star Vega – still at that date on a wet plate.

Henry Draper went on to photograph eighteen emission lines of oxygen in the solar spectrum and much else, and was making plans for an observatory in the pure, clear skies of the Rocky Mountains when in November 1882 he died of pleurisy. His wife Mary Anna, used their ample resources to establish the Henry Draper Memorial Institute at Harvard, which would become one of the world's powerhouses for spectroscopic and astrophysical research.

Ireland was also a major astronomical research centre in the nineteenth century. In addition to Lord Rosse, Edward Joshua Cooper, and Thomas Romney Robinson there were university and episcopally funded observatories, at Armagh, Dunsink (Dublin), and Cork. Margaret Lindsay Huggins, whom we have met, was an Irishwoman, and there were others. As far as solar and stellar physics are concerned, William Edward Wilson pioneered the new technology of astronomical photometry at his Daramona House Observatory, splendidly equipped with a 12-inch refractor and a tandem-mounted 24-inch silver-on-glass reflector by the ubiquitous Sir Howard Grubb of Dublin. Wilson's friends included a whole cluster of late Victorian Irish Grand Amateur and academic scientists, including the Dublin barrister William Stanley Monck and the physicists George Minchin and George FitzGerald (of subsequent "Lorentz-Fitzgerald contraction" fame). Of central importance to their work was the selenium-coated aluminium photo-electric cell, devised by Minchin in 1892. This photoelectric apparatus, built by Yeates of Dublin, turned light into electrical energy, making it possible to calibrate the energy output of different astronomical bodies. Between 1894 and 1901, Wilson and his friends made a series of increasingly refined measures of the temperature of the sun, while Monck's recognition of the mathematical relationship between star distances and luminosities would initiate a series of studies which in 1910 would become the ancestor of the Hertzsprung–Russell Diagram of stellar life-sequence classification.[6] The new technology of cinematography was also pioneered at Daramona House, but using time-lapse photo frames to record and analyse the structural changes taking place within sunspots.

Over the last two chapters in particular, we have seen how changes and technologies unimagined in 1835 would transform mankind's knowledge of the universe by 1900. Yet while photography and spectroscopy opened spectacular new vistas of astronomical understanding, and provided the keys to unlocking the chemical and physical secrets of the cosmos, many of the old problems remained. What was the relationship between the stars and the newly discovered gaseous nebulae? Was the universe stable or evolving? What other as yet undetected energy systems might there be out there, especially in the wake of James Clerk Maxwell's fundamental work on the connection between magnetism and electricity, Lord Kelvin's thermodynamics, Heinrich Hertz's discovery of radio waves, and Henri Becquerel's and Pierre and Marie Curie's discovery of atomic radiation in the 1890s? Were the nebulae all contained in one great Milky Way star system, or were they vastly remote island universes? And where had it all come from and where might the universe be heading?

These are the Big Questions, where cosmology impacted upon theology and philosophy, to which we will return in subsequent chapters. But it would not be until the new research techniques of the twentieth century that we would gradually edge toward answers.

Yet all the new technologies notwithstanding, these questions remain today. Over the next three chapters, though, we will examine how astronomy and cosmology excited mass public interest, and how this interest was catered for.

Chapter 18

The Revd Thomas William Webb and the Birth of "Popular Astronomy"

In earlier chapters we saw how so much fundamental astronomical research in Britain was driven by highly motivated self-funded private individuals with a passion for science. They staked out fresh fields of research, were stimulated to look deeper into space than their predecessors, and risked the time and the venture capital to devise new large telescopes and research technologies. Yet in addition to the trailblazing Grand Amateur researchers, another class of astronomer was emerging by the 1850s. They came from a diversity of backgrounds, stretching from day labourers to beneficed clergy, teachers, lawyers, and medical practitioners, and they shared a fascination with the heavens. These men and women were not able, or even motivated, to commit themselves to grand research projects, yet they wanted to know how to find their way around the night sky, to own and to learn how to use a telescope of some kind, and to be informed about what discoveries were being made. They began to found local, then national, astronomical societies, to create a market for accessible non-mathematical scientific books, and to attend lectures, sometimes commercially sponsored, aimed at aspiring star-watchers. They represented the birth of popular astronomy. One man who recognized and helped to meet their needs was a Herefordshire clergyman.

The Revd Mr Webb of Hardwicke, astronomer and popularizer

Reading astronomical books and owning telescopes were not uncommon when the forty-nine-year-old Thomas William Webb became vicar of the rural Welsh-border parish of Hardwicke, near Hay-on-Wye, Herefordshire, in 1856. The Society for the Diffusion of Useful Knowledge (SDUK) and its popular *Penny*

Magazine encyclopaedia in the 1830s, and John Pringle Nichol's books on the Herschel and Rosse cosmologies, were already taking the big ideas and discoveries to a wider readership, as we shall see in Chapter 20. Webb, however, was taking practical hands-on observational astronomy to new audiences. He created popular, amateur observational astronomy in a way that it would come to flourish worldwide down to our own day. Webb's *Celestial Objects for Common Telescopes* (1859) would become an iconic book, teaching the reader how to become an amateur astronomer: not a *Grand* Amateur, but someone who could take an informed delight in the sky and go on to enthuse others. Webb was one himself, for while his Oxford education, clerical profession, social connections, and liberal financial means suited him to the Grand Amateur league, had he so chosen, he looked at the sky for reasons other than scientific curiosity. His focus on popularizing astronomy rather than making great discoveries derived in no small way from his character. The son of an established clerical and gentry family, whose wife Henrietta also had connections with the south Welsh gentry, Webb was a devout Christian and a *gentle*-man, and service to his flock came uppermost in his priorities. While he had worked in other parishes in the Ross-on-Wye area before moving to Hardwicke, and held stalls in both Hereford and Gloucester Cathedrals, he was a modest, quiet man, not ambitious for high ecclesiastical office or worldly fame.

Webb's astronomy, like his parallel interests in meteorology, natural history, and even seismology, had more to do with contemplation than pure research, and dovetailed into his wider faith. His surviving manuscript observing books bring home to us how much he *enjoyed* looking at the moon, planets, star clusters, and nebulae through his telescopes, under the dark and unpolluted skies of rural Herefordshire – skies which remain spectacular in their transparency even today. All of this observing and recording must be seen as of a piece with his devoted ministry to his rural flock. He possessed a happy combination of traits which his fellow clergyman astronomer and protégé, the Revd Thomas H. E. C. Espin would later say led him "to walk with vigorous stride up the hills to see some distant parishioner… And then, when the cottage was reached… impart the sunshine of his life to theirs."[1]

Celestial Objects for Common Telescopes and Webb's telescopes

Nowadays we tend to use the word "popular" in a sense which implies elementary or simplistic. This was *not* the way in which the Victorians used it. To Webb and his contemporaries, "popularizing" meant explaining a subject without recourse to complex mathematics. His *Celestial Objects* is a meaty book by any standards. In it, he discusses astronomical history and the significance of great discoveries, and even assumes that his reader can understand a Latin quotation without the need for a translation; as one might expect from a Victorian solicitor, fellow-vicar, general medical practitioner, schoolmaster, or a lady who might want to take up the study of the night sky as a serious and committed hobby. To some degree this clearly genteel clientele would be those who could afford to buy a common telescope. Most of the observations and astronomical exercises, extending over a quarter-century, upon which Webb's book was based were done with just such an instrument, with a fine 3.7-inch-diameter achromatic object lens. Before the increase of relatively mass-produced telescopes after 1860, it would have cost a good £25–£30 purchased new, which was, roughly speaking, six months' wages for a working man, or almost a year's wages for a working woman.

After 1859, Webb began to work with bigger and more powerful telescopes than his albeit fine 3.7-inch refractor. In 1859, he acquired an excellent 5½-inch-aperture refractor by the Boston USA optician Alvan Clark, and then he began to observe with the new technology – the silver-on-glass reflecting telescope – as part of an interesting partnership with the nearby Hereford Bluecoats School Headmaster and telescope-maker, George Henry With.

It is hard to overestimate the significance of the new silver-on-glass reflecting telescope technology in the burgeoning of serious amateur astronomy after 1860, especially in the wake of Webb's book. It did three things at a stroke. Firstly, it greatly increased the aperture potential, making reflecting telescopes of

6-, 10-, and even 15-inches aperture relatively commonplace. (As we saw in Chapter 4, the bigger the optical aperture, the greater the "light-grasp", and hence the greater both the magnification and the optical resolution of detail.) Secondly, by using what was, by 1860, easily accessible and relatively cheap commercially manufactured plate glass as the basis for the mirror, it was vastly simpler to make, not requiring all the metallurgical furnaces and hot-metal casting facilities necessary for producing the older speculum mirrors. And thirdly, the cost of serious astronomical telescopes fell through the floor, as an optical capacity that in 1830 would have cost £150 could in 1870 be bought for less than half that price, and such a telescope put together at home by a skilful person for £20. If you threw in the tools, resources, and additional services needed to build a silvered glass reflecting telescope, all this launched it into the already growing Victorian hobbyist "Do it yourself" movement. And as a fourth advantage, a silvered optical surface was not only about 25 to 30 per cent more reflective than a speculum surface, but it was much slower to tarnish than the tin and copper speculum alloy. In practical terms, this meant that a silvered glass mirror of 6-inches aperture would perform as well, if not better, than a speculum one of 8 inches. Glass being much lighter in weight than speculum, inch for inch, a glass mirror presented fewer engineering problems when it came to mounting in a tube for use. So why was nobody manufacturing silvered glass mirrors back in 1720, when John Hadley presented his newly designed Newtonian reflector to the Royal Society? The answer is simple: it was not until the 1850s that French and German chemists devised a method by which a perfectly uniform thin coating of silver could be applied to a polished glass surface without its being easily rubbed off.

The true impact of the glass mirror reflecting telescope would not really be felt until after the first edition of Webb's *Celestial Objects* had been published. But when the telescope could be bought, or made at home, and used in conjunction with Webb's practical guidebook to observing the heavens, all was set to create a new popular scientific movement.

The "modest" amateur astronomer and the new reflecting telescope

Soon after the news of Léon Foucault's and Carl August von Steinheil's new technique for depositing a brilliant coat of silver upon a piece of glass circulated in the late 1850s, reflecting telescope construction changed for ever. The professional Paris Observatory commissioned a silver-on-glass reflector of 80 cm (31½ inches) aperture from Foucault in 1862, and with the exception of the unsuccessful Great Melbourne Telescope of 1869, the traditional speculum mirror reflector – beloved of generations of astronomers from Newton to the Herschels and on to Lord Rosse and William Lassell – was doomed to extinction. Silver and later aluminium-on-glass mirrors would dominate astronomy from the 1860s down to the great Mount Palomar 200-inch in 1948, and on to today. As mentioned above, however, it would be the amateur astronomer who would turn it into the instrument of choice by 1875.

Two manufacturers rapidly rose to prominence when it came to the commercial production of the necessary optics and ancillary fittings: the above-mentioned George H. With of Hereford, and George Calver of Chelmsford. Although With was a schoolmaster, his was a charity school for the poor, and his salary was only £85 per annum, plus a house. He also did other teaching around Hereford, as well as fulfilling duties as a lay reader in the cathedral, but one suspects that extra cash was always useful. Being a modest amateur astronomer, and enjoying practical tasks, George With tried his hand at figuring a glass mirror, and as William Herschel had done with speculum a century before, found that he was good at it. One made a glass mirror by a similar process to figuring a speculum one: a circular plate-glass blank, about an inch or more thick, was worked face-down upon a convex tool, using a range of coarse abrasive powders. This would give it a concave figure. By periodically washing away the abrasives, one could test the focal length of the concave mirror by focusing sunlight or lamp-light upon a white screen. Once the concave mirror gave the desired focal length, then softer powders were used, such as jeweller's rouge, French chalk, or talc (carborundum powder would later be used) until the concave curve exhibited the correct parabolic shape and focal length.

Next, the still transparent mirror would be tested with the reflected image of a watch face, or a page of fine print, and perhaps repolished and adjusted, until the resulting image was of a uniform clarity in all its parts. In 1858, Léon Foucault devised his knife edge test for mirrors, which soon ousted the earlier tests.[2] After the tests had been successfully done, the mirror was ready to receive its coat of silver, which would be applied by a newly invented process using volatile mercury and other dangerous chemicals as a catalyst, or flux. If the silvering were done in a well-ventilated place, and one tried not to breathe in the inevitable fumes, one ended up with a beautiful, shining parabolic optical mirror.

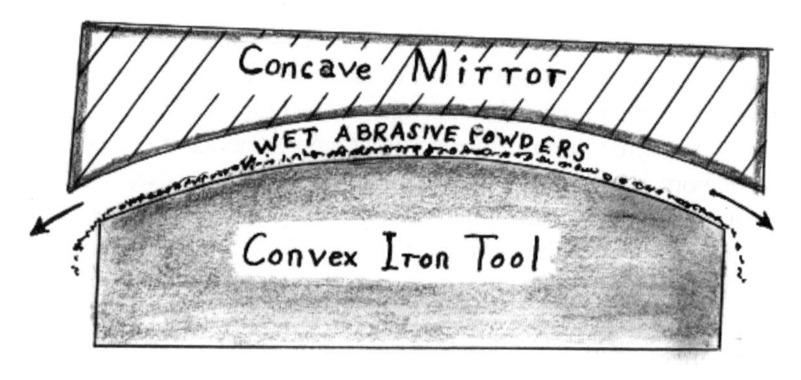

18.1 Figuring a concave mirror. See the explanation in the text. (Drawing by Allan Chapman.)

Such a mirror, starting with a blank glass disc bought by mail order from Chance Brothers of Birmingham or another good plate-glass manufacturer, could be ground and figured in the garden shed, or even in the kitchen. It required no furnaces or molten metal casting facilities: only commercially available tools, abrasives, and chemicals, skill, and patience – and deep breaths when it came to silvering. Or if one wanted to buy a finished mirror, schoolmaster George With was soon advertising his exquisite ready-made mirrors for sale: a large 10¼-inch, for example, was £38-10-00d, and smaller ones pro rata, and before long, his prices were coming down. Likewise,

George Calver was advertising his own mirrors. The efficient Victorian railway and Post Office parcels service made it easy for an astronomer in Wales or Leeds to purchase such ready-made mirrors from Hereford or Chelmsford, by cheque or postal order. Once the mirror had safely arrived, one might pay a local wood or metal worker to build a mount to a particular design, or one might "do it yourself". By 1865 metal rods, pieces of iron girder, sheet metal, and nuts and bolts were all commercially available, either from a good hardware shop, or more ambitiously, from an order placed with a Birmingham or Newcastle supplier.

Victorian clergymen-astronomer-engineers

One extraordinary thing about nineteenth-century clergymen, be they Roman Catholic priests, Congregationalist ministers, or, overwhelmingly, Anglican vicars, is how many of them loved science and practical technology. Their ranks included balloonists, still and experimental cinematographers, steam-engine makers, microscopists, cannon and gun inventors, amateur architects, sanitary and civil engineers, and lots of astronomers. The historical record makes us all too aware of those men who spent their energies locked into academic theological controversies, but we hear surprisingly little about those clerical technologists who happily combined their Christian faith and ministry with a love of contraptions of various kinds. As a breed, they were all highly educated, financially secure and comfortable, enjoyed the use of spacious premises, and had command of their own time. The Revd Webb was not unique, but one of a type.

When one looks at the first availability of silvered mirror telescopes in the 1860s, in conjunction with such publications as *The Astronomical Register* and *The English Mechanic* magazines after 1863 and 1865 respectively, combined with the rise of serious amateur astronomical societies by the early 1880s, one encounters ordained mechanicians everywhere. Part of this love of optics and mechanics in their various forms included the testing, comparison, and recording of different optical and mechanical configurations.

It was probably in 1864 that Webb acquired his first With mirror, of 8 inches in diameter. He promptly began testing it against his

5½-inch Clark refractor. Soon after, he acquired a superlative With mirror of 9¼ inches in diameter, which was mounted for him upon a stand of ingenious design. This was a Berthon "Equestrian" mount, designed by the clergyman inventor the Revd Edward Lyon Berthon, probably in 1866. Berthon was Rector of Romsey in Hampshire, and his Equestrian mount was designed for amateur astronomers. In the case of Webb's Equestrian with its 9¼-inch mirror, the telescope had a sheet-iron tube, working at the Newtonian focus. The mount was an ingeniously modified form of the German equatorial mount, with the two heavy weights that counterbalance the tube mirror being set down low, like a rider's feet in a pair of stirrups: hence equestrian. The instruments belonging to the British Astronomical Association include an "equestrian" telescope with a mirror by George With, but there is no evidence to show that it was Webb's telescope.[3]

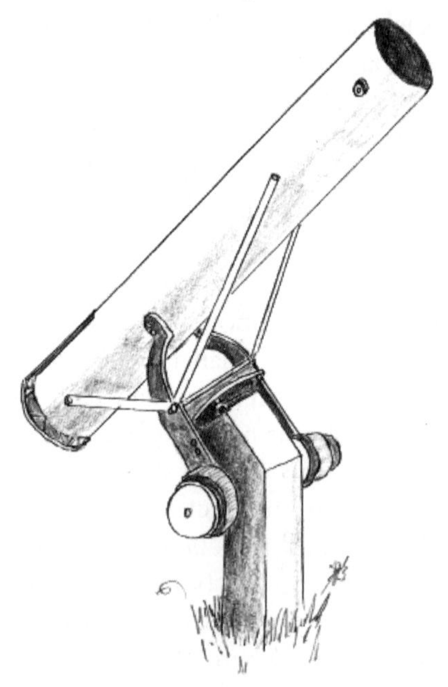

18.2 An "equestrian" telescope with its silvered glass mirror of 9¼ inches aperture. (Drawing by Allan Chapman, based on a photograph of the instrument belonging to the British Astronomical Association.)

Like many technologically minded Victorian clergymen, Edward Lyon Berthon was a pillar of his Romsey community, and very much involved in useful local projects. As the parish was near Southampton, Berthon was interested in maritime safety, and, in the days before the inflatable rubber dinghy, invented the Berthon collapsible lifeboat. With its lightweight wooden frame and stout canvas covering, the lifeboat was light in weight, and could be quickly unfolded and made ready for sea. Berthon lifeboats could carry several dozen people, and their manufacture became a local industry and – it was said – they helped save the lives of thousands of shipwrecked people.

Victorian Herefordshire had several ingenious clergymen and ecclesiastically connected laymen. The Revd Henry Cooper Key, incumbent of the rural parish of Stretton Sugwas about 4 miles from Hereford and not far from Webb's Hardwicke, was an enthusiastic amateur engineer and astronomer. As early as 1859, Cooper Key had built a powered machine for the grinding of telescope mirrors, and was soon figuring high-quality glass specula of up to 18 inches in diameter. One also wonders how much George With's own activities were stimulated and encouraged by the Very Reverend Dr Richard Dawes (no relation to William Rutter Dawes the physician-astronomer), Dean of Hereford Cathedral. Dr Dawes was a pioneer educationalist, concerned with educating and improving the lot of the poor, and had a great interest in science and progress. As headmaster of the Hereford Blue Coats School, and from his duties in Hereford Cathedral, With must have known him well, along with Webb and Cooper Key.

It was a clerical enthusiast for the new silver-on-glass mirrors who first encouraged George Calver to try his hand at making a mirror, in the 1860s. This was the Yarmouth, Norfolk priest the Revd Mr Matthews, who showed his newly acquired With mirror to Calver, who then lived nearby. Calver went on to become a professional, commercial optician, who, down to 1927, made thousands of excellent mirrors, largely for the new amateur astronomy market. With and Calver remained on friendly terms, and the market for fine telescope mirrors became large enough to accommodate both men's creations without any serious competition. In addition to the mirrors and metal tube and mount fittings, the burgeoning popularity of the reflecting telescope generated a market for ancillary fittings, such as 45-degree flat mirrors and eyepieces: a demand met by John Browning and other manufacturing opticians.

Once a serious amateur astronomer became the proud possessor of a fine reflecting telescope, where did he or she keep it? For an 8- or 10-inch-aperture instrument, with a tube of 8 or 10 feet long and set on an engineered equatorial mount, was not portable. The ever-ingenious Berthon provided a model for a "telescope house" or observatory, in the pages of *The English Mechanic* magazine in 1871.[4] He supplied a design and drawings for a wooden telescope house, with a semi-dome rotating on window sash wheels or furniture castors bought from the local hardware shop, which could be built by the local carpenter for £8 to £12, depending on size and complexity. An enthusiastic DIY aficionado might build his own for less, to accommodate his home-made telescope. Simple Berthon observatories began to sprout all over the place: in private and vicarage gardens, and even as part of the prestigious Rugby School Observatory. Some, made in brick or other more robust materials and well-maintained, still survive, and amateur astronomers still utilise Berthon's design. By 1875 it was possible for someone who could afford £50 or so, or who was very good with their hands, to command an optical deep-space-penetrating power that was greater than that which Sir John Herschel had taken to the Cape of Good Hope in 1834.

Astronomical societies and *The English Mechanic* magazine

Buoyed up by their proud industrial city's recent hosting of the British Association for the Advancement of Science jamboree in 1859, a group of Leeds astronomical enthusiasts decided to found an Astronomical Society. The venerable Sir John Herschel agreed to act as its – absentee – President, and even secured them a fine achromatic refracting telescope, which is still owned by the Leeds Astronomical Society. It was erected in a pre-Berthon observatory made of corrugated iron sheets in Love Lane, Leeds. The Society flourished into the early 1860s but enthusiasm began to wane thereafter. Serious amateur astronomy was still doing well, with the new silver-on-glass mirror telescopes, and received two additional boosts in 1863 and 1865 respectively. In January 1863, Sandford Gorton, an Oxford-educated private gentleman of independent means, began publication of the *Astronomical Register* magazine. This was a journal aimed not at Grand Amateur research-driven astronomers, whose needs were catered for by the RAS *Monthly Notices*, but men like Gorton himself: someone who enjoyed astronomy as a serious and informed hobby, but did not see it as an unpaid profession. Primarily *The Astronomical Register* existed to publish amateur observations, of the sun, moon, and planets: topics which then and for later generations formed the essential stuff of amateur astronomy. It also performed the historically invaluable service of conducting a voluntary census of amateur observatories, enabling modern scholars to form some idea of who was observing what, from where, and with what kind of instruments.

Although remaining in print from 1863 to 1886, the *Register* was always close to financial collapse. It was a subscriber-funded magazine, selling for 8d (old pence) per monthly copy: a not insignificant sum when compared to the old *Penny Magazine* and the new cheap mass-circulation newspapers and magazines of the 1870s. In some respects, the *Register* was too specialist in its interest range. It did, however, promote the first national amateur astronomical society: the "Observing Astronomical Society" founded in July 1869. As its name states, it was to be dedicated to serious, eye-at-the-telescope planetary, lunar, stellar, and meteoritic observation. It may have been too specialized in its interests for that date, and its Reports ceased to appear in the *Register* after 1871.

The other boost, in 1865, was from the more wide-ranging, non-specialist, and enduring *The English Mechanic* magazine. In its pages, the technically minded hobbyist was given an identity and focus, and could find articles on how to build a church organ, a clock, or a model railway, set up an electrical generator, make lenses, mirrors, or a camera, and even build a house. It was also a wholly egalitarian publication, cheap, full of drawings, diagrams, and instructions, and inviting letters, comments, and even articles from readers. Correspondents spanned a social range from clever aspiring labourers to Oxbridge dons and great industrialists like Nasmyth. An ingenious village bricklayer might send in a letter requesting advice about optics, and receive a reply from a cathedral canon, a retired Royal Naval officer, or a don. One such working man was Samuel Cooper of Charminster, Dorset, who sometimes wrote for *The English Mechanic* under the pseudonym "The Optical Bricklayer". On a generally cloudy 2 December 1882, he succeeded in securing a probably unique photograph of the Venus transit of that year, using home-made equipment.[5]

Some people saw the magazine as a sort of classless "clever boys' club", as the peer and the paper-hanger came together in its pages through their shared passion for ingenious devices. *The English Mechanic* continued in its original form down to 1925, then thereafter under various titles and formats, such as *Practical Mechanics* (I read this magazine regularly as a boy around 1960, and tried to make the devices it described). Astronomy flourished in its pages, and it became the unofficial journal of the new amateur astronomy movement. One man who wrote regular pieces for it was Captain William Noble, a retired Army officer and astronomical enthusiast. Then, after the 1870s, two further magazines appeared, which, while by no means confined to astronomy in their coverage, were founded and managed by two of the leading astronomers of the decade. One was *Nature*, 1869, the brainchild of Joseph Norman Lockyer and nowadays one of the world's most prestigious scientific magazines, and the other was Richard Anthony Proctor's *Knowledge and Illustrated Scientific News*, of 1881. Their commercial success by the 1880s reveals a burgeoning readership for high quality yet accessible scientific magazines by that date.

Although the time may not have been ripe for national or big astronomical societies in the 1860s, based in the major cities, the decades after 1880 saw a cluster of such societies being founded, most of which still thrive today. Liverpool was the first, in 1881, then came the refounding of Leeds on a permanent footing, along with the Cardiff-based Cambrian Society, Belfast, Newcastle-upon-Tyne, and Manchester: almost all industrial cities. Some societies quickly formed branches, and the Liverpool Astronomical Society soon had one in the Isle of Man, and another, incredibly, in Pernambuco, Brazil, where there was a significant community of expatriate British engineers. Even the astronomically minded Dom Pedro II, Emperor of Brazil, became a member.

In 1890, the dynamic Cirencester resident Miss Elizabeth Brown, Solar Studies Director of the Liverpool Society, became the driving force behind the founding of the national, London-based British Astronomical Association (BAA) (see Chapter 19 for more on her work). The founding of the BAA and the way it would thrive right down to our own day was clearly an indication of a national demand for a society catering for astronomers who wanted to buy and build serious telescopes. These people observed the sky for what might be called – without wishing to sound patronizing – enjoyment and relaxation rather pure research. They might specialize in observing Jupiter or the moon, and submit their excellent observations for publication in the BAA journal, yet they were not expecting to change our fundamental understanding of the cosmos. The much richer Grand Amateurs did that, and by 1890, a growing number of salaried academic astronomers, whose lines of communication were the RAS and Royal Society publications. Secondly, and very significantly, all these amateur societies were open to women members, voted women in to serve on their governing councils, and even appointed them observing Section Directors.

While the serious amateur astronomy movement, with its societies and publications, first got going in a big way in Great Britain, there was no shortage of popular interest in cosmology elsewhere, and here, in particular, an enterprising Frenchman recognized opportunities both to educate and to prosper at the same time.

POPULAR ASTRONOMY IN FRANCE

Camille Flammarion was no less a product of great discoveries, public curiosity, mass markets, and commercial opportunities than were his British contemporaries Richard Anthony Proctor and Sir Robert Stawell Ball. Born in 1825 into a commercial publishing family that would become the Groupe Flammarion, Camille was, like his French contemporary Jules Verne, captivated by the new power of science and technology, and by tracing through its imaginative possibilities. The author of several dozen popular books, he explored ideas such as life on other worlds, cosmology, and spiritualism. Did the souls of the dead "transmigrate", perhaps even to other planets, and could the tails of comets have deleterious effects upon life on earth? While not concerned with fundamental astronomical research, Flammarion owned a private observatory from which he appears to have enjoyed looking at the heavens. In 1887, he founded the Société Astronomique de France, and also helped to set up its published *Bulletin*, establishing astronomy as a serious hobby in the French-speaking world, just as was happening north of the Channel. A similar movement was under way in the USA, bringing about the Astronomical Society of the Pacific in 1889. All of these societies, be they British, French, or American, shared an essentially middle-class membership. But there was also a significant body of working men (so far, I have found no working women, unlike their better-off sisters) who devoted a significant amount of time and energy to astronomy.

JOHN JONES OF BRANGWYN BACH AND OTHER WORKING-MEN ASTRONOMERS

Welshman John (or Ioan) Jones was born in 1818 at Brangwyn Bach, Anglesey, into an impoverished labouring family. Largely unschooled, he taught himself to read Welsh from a Welsh Bible in the house of the local Calvinist Methodist minister, the Revd Cadwalladr Williams. He then taught himself to read English from an English-language Bible. Employing this same "Hamiltonian system" (named after its inventor James Hamilton) of using a text – usually the Bible – in one language to learn another, Jones would acquire biblical Greek, Latin, and other languages as the years went by.

His interest in astronomy came about by an odd route. As a young man, Jones had been crossed in love, and went to the village wise woman to have a curse put on his rival. Finding that the cottage contained maps of the (presumably) astrological sky, he soon forgot the curse, and became captivated by astronomy. Jones then set about teaching himself astronomy (*not* astrology) from a Welsh translation of some of Thomas Dick's popular books (see Chapter 11), such as *The Christian Philosopher* (Welsh edition 1842), and from English works by Sir John Herschel and the American Denison Olmsted. He went to work on Bangor docks, serving the ships which carried Welsh slate to Liverpool and elsewhere for the building trade, and lived in a tiny cottage in Upper Albert Street, Bangor. There, he acquired more languages, studied astronomy, and built a pair of silver-on-glass reflecting telescopes of 6 and 10 inches aperture, along with a spectroscope, the prism of which he ground from a broken piece of glass from a ship's porthole.

Yet in spite of all his erudition, linguistic, scientific, and theological, he remained in his ill-paid labouring job on Bangor docks. His genius and dedication only came to be more widely known after Samuel Smiles included a chapter on his life and achievement in *Men of Invention and Industry* (1884).[6] The Cardiff journalist Arthur Mee featured him in *Young Wales* 4, 48 (1898), as did John Silas Evans in *Seryddiaeth a Seryddwyr* ("Astronomy and Astronomers"), 1923.[7] By the 1890s and up to his death aged eighty in 1898, Jones had become a Welsh cultural icon: bard, poet, *seryddwr*, and the absentee darling of the middle-class members of the Astronomical Society of Wales. As a great admirer of independent ingenuity and achievement, I have always deeply admired John Jones, and some years ago, I received a communication from Dr Gwawr Jones, a descendant of his. She told me the story of his first fascination with astronomy following his visit to the Anglesey wise woman.[8] But when will the University of Wales, or some other Welsh body, name something in this remarkable man's honour?

18.3 John Jones, the Welsh self-educated dock-worker astronomer, linguist, and biblical scholar, with his home-made reflecting telescope in the back yard of his cottage in Upper Albert Street, Bangor, North Wales, c. 1880. (Originally in *Young Wales*, 1898; reprinted in Silas Evans, *Seryddiaeth a Seryddwyr* ["Astronomy and Astronomers"] (Cardiff, 1923), p. 273, pl. 43.)

It would be incorrect, however, to see John Jones as the only working-man astronomer in Victorian Britain. As we have seen, an early one was James Veitch, of Inchbonny, in the Scottish Lowlands, who made Mary Somerville's speculum-metal reflecting telescope somewhere around 1830. Not only was this blacksmith and agricultural tool-maker an astronomer; it seems that his wife Betty was familiar with the science as well, for when once paying a visit to the Veitches, Mary Somerville recorded that Betty pointed out the exact position of Venus in a full daylight sky.[9] It is remarkable how many poorly paid, hard-working people dedicated themselves to astronomical study, and scrimped and saved to buy books and telescopes, and even to build observatories, often of the Berthon type. It has even been suggested that a passion for astronomy might do as much damage to a working-class family budget in 1885 as would a partiality for strong drink – yet with social profit, rather than disgrace.

I have encountered several railway worker astronomers, who enjoyed relatively flexible circumstances. After Roger Langdon had worked his way up the Great Western Railway career ladder, from being a porter to becoming Master of the Devon village station of Silverton, near Exeter, he was, in some ways, his own boss. Having seen the various trains in and out of the station and goods yard, Langdon found the time and resources to build four silver-on-glass reflectors and even to house his 8¼-inch-aperture in a Berthon observatory, built from cast-off railway materials. Langdon turned for advice on making his telescope to *The English Mechanic* magazine, and a friendly and instructive correspondence with a Dr Blacklock. When Langdon had built a superior telescope, he sold his original 6-inch for £10, which helped to pay for the others.[10]

Like John Jones, Roger Langdon acquired an impressive self-education, learning, among other things, how to read the New Testament in the original Greek. In 1871, he travelled to London to present a paper on his Venus observations – through the Revd Webb's kind offices – to the Royal Astronomical Society. He was also an accomplished writer, contributing pieces to various local magazines and writing his autobiography, which was published after his death. Although his wife Anne does not seem to have been especially astronomical, she was nonetheless a strongly intellectual self-educated woman who used their Station Master's house to teach the village children. Roger expressed his indebtedness to her. The Greek-reading astronomical Station Master was also admired by the local squire Sir Thomas Acland, and the Revd F. H. Fox Strangways, both men of high education.

John Jones and Roger Langdon are but two of the well-documented astronomers in humble circumstances of the Victorian age. Others include Langdon's Caledonian Railway porter colleague John Robertson, of Coupar Angus station, near Perth, Scotland.

Born in the early 1830s, he attended lectures delivered to working men by the elderly – and ubiquitous – Dr Thomas Dick. He also read Dick's books, among others, and in later years would even scrape together the gigantic sum of £30 to purchase a fine refracting telescope from the optical manufacturer Thomas Cooke of York. Robertson worked a twelve-hour day, and did his observing either before 6 a.m. or after 6 p.m. He ran up an impressive body of lunar, planetary, cometary, and solar observations, and corresponded with astronomers of the standing of William Christie, who would become Astronomer Royal in 1881, and Dr Ralph Copeland, Director of Lord Crawford's great observatory at Dun Echt, Aberdeenshire.[11]

While Langdon and Robertson may have spent quite lavishly on astronomy, it should be said that railway workers in general were sometimes reckoned the aristocracy of the Victorian working class. Their weekly wages might not have been especially high, but their jobs tended to be secure and sometimes even pensioned on retirement, unlike those of coalminers and factory workers, while in many cases on-site housing was provided as well. Economic booms and slumps came and went, but those vital arteries of communication and distribution, the railways, always rattled on, carrying passengers, coal, food, letters, and parcels – and telescope mirrors and accessories. Steam locomotives were coal-fired and oil-illuminated, and railway workers' houses were generally warmed and lit by the same coal and lamp oil. So perhaps security and an assured access to basic necessities enabled a village station master or porter to splash out on an expensive hobby in a way that a piece-work collier or factory worker would never have dared.

Then there was John Leach, the shoe-maker of Frodsham, Cheshire, who had taught himself French in order to read French higher mathematics, and who even detected an error in one of the mathematical books of the eminent Cambridge mathematician, Isaac Todhunter. Leach also had a correspondence with Sir George Airy, who, as we have seen, preceded Christie as Astronomer Royal.[12]

Thomas William Bush, like Leach, was a modest self-employed craftsman, a baker of Nottingham, who in his religious studies had – after a simple formal schooling – taught himself biblical Greek, Latin, and Hebrew, and astronomy and optics. Unusually for a man of his time, Bush cast and figured an excellent pair of 13-inch-diameter speculum metal mirrors, modelled on that of Warren De La Rue, which he mounted in the equatorial plane, which yielded fine views of objects up to 1,400 times magnification. Bush did very well for himself in business, and almost crept into the modest Grand Amateur class when he bought land and built a fine new house for himself and his wife Martha – she being an accomplished music teacher – and an observatory.[13]

Another astronomical craftsman was Johnson Jex, the "Learned Blacksmith of Letheringsett", Norfolk, who in the first half of the nineteenth century taught himself how to figure and polish fine achromatic telescopic lenses, along with making clocks and watches. When Jex died, on 2 January 1852, the county gentry and clergy, who knew and admired him, commissioned a detailed 3,000-word obituary in *The Norfolk Chronicle*, recounting the many achievements of this truly autodidact figure. Several surviving specimens of his craftsmanship are now preserved in the Bridewell Museum, Norwich.[14]

These working men astronomers are but a selection of those on well-documented historical record. More have come to light since 2002, when the Society for the History of Astronomy was founded, and began to conduct and publish its meticulous regional surveys of astronomers and astronomical sites around Britain. Yet as mentioned above, women were coming to play an increasingly significant role in serious amateur astronomy in mid- and late-Victorian Britain, and would become especially prominent in the British Astronomical Association and in the city astronomical societies. So let us now look at what women achieved in British, European, and American astronomy down into the twentieth century.

Chapter 19

"Ladies of the Night": The Astronomical Women in Great Britain and America

As we have already seen in previous chapters, women were no strangers to astronomy over the centuries, though with recent iconic figures such as Caroline Herschel and Mary Somerville by 1850, female involvement in the science began to advance significantly. Yet while Caroline Herschel and Mary Somerville were both made Honorary Members of the Royal Astronomical Society in 1835, and a handful of later women would receive permission to use the Library or attend particular meetings, it was not until January 1916 that the first five women were admitted to the full Fellowship of the RAS on equal terms with men. Back to the 1880s, women had been admitted not only to membership of the new big amateur societies, but were also elected governing council members and directors of specialist observing sections. So what opportunities existed for formal education and employment for astronomical women in the nineteenth century?

Scientific education for women

Caroline Herschel, Mary Somerville, and many of the serious and Grand Amateur astronomical women who will be discussed in this chapter picked up most of their astronomy and science by private study. Yet by the 1860s, by which time Caroline was dead and Mary was in her still vigorous eighties, the first serious beginnings of female scientific education were being made. This applied not just to science, but to a learned education in general, though some male academics were by no means happy. Some male classicists, for example, foresaw problems in teaching girls and young women Greek and Latin; were not some classical authors so replete with shocking stories and filthy words that the minds of England's innocent maidens would be corrupted? Perhaps science was more suitable, and while biology and medicine contained ideas from which girls should be protected, this did not prevent Dr Elizabeth Blackwell, England's first academically trained physician, from qualifying in the USA, then after 1858 having her name put on the new legal Medical Register. Yet who could deny the moral "purity" of astronomy?

The Revd Webb was at times teaching astronomy and physics to pupils at the Cheltenham Ladies' College by the 1860s, while Bedford College, London, had been founded for female undergraduates in 1849. Then things began to move fast, with the establishment of Girton College and Newnham College, Cambridge (1869 and 1871 respectively), and Lady Margaret Hall (1878), Somerville (1879), St Anne's (1879), St Hugh's (1886), and St Hilda's (1893) at Oxford. The new British metropolitan universities in London (especially the Royal College of Science, now Imperial College, London), Manchester, Leeds, Birmingham, and elsewhere sometimes admitted women students as well, though as a percentage of the small number of British men who enjoyed a university education at that time, female students in higher education were very few indeed. To go to university or a higher college in 1885, a girl not only had to have brains and determination, but also understanding and well-off parents who could imagine a life beyond comfortable domesticity for their daughters.

As has often been the case, things got going earlier in the United States. By the mid-1860s both Wellesley and Vassar Colleges were providing higher education facilities for young women, and by 1865 Vassar, at Poughkeepsie, New York, had a full-scale professional observatory with a fine 12-inch refractor by Henry Fitz, along with a growing battery of other state-of-the-art instruments. Vassar's first Observatory Director was Maria Mitchell: the first woman in history to hold the Professorial Directorship of a major scientific institution. Maria's (pronounced Ma-rye-a) life had been fascinating in its own right. Coming from a family of East Coast, Nantucket Quakers, a religious society where men and women were held equal, she was given a full education. As a twelve-year-old she displayed a flair for astronomy and mathematics, assisting her father to calculate eclipses. As a teenager she became a teacher and also worked as a librarian as well as doing astronomical computations for the United States *Nautical Almanac*. At the age of twenty-nine in 1847, she became world-famous as the discoverer of a telescopic comet which, in spite of a simultaneous discovery by the Jesuit astronomer Francisco de Vico, won her the King Christian VI of Denmark Gold Medal.

In 1865 Maria Mitchell became Astronomy Professor and Observatory Director at Vassar College, with the third largest telescope in the New World at her command. Here she combined her two already well-established talents: as an inspiring teacher of the rising generation, and as an original research astronomer. Her research interests, which were very well suited for the 12-inch Fitz equatorial refractor, were planetary astronomy. At Vassar, she observed planetary surfaces at high magnifications and did drawings, along with making meticulous positional measurements of planetary satellites: the latter playing a major role in the development of our understanding of gravitational mechanics.[1]

Upon her retirement in 1888, Maria was succeeded at Vassar by Professor Mary Watson Whitney, who carried on her dual function of teaching young women and undertaking research into comets, double stars, and other phenomena. Then from the late 1880s, women would find a new professional astronomical role, especially at the Harvard Observatory, as Professor William Pickering began to employ women, both graduate and non-graduate, to meticulously measure and interpret data contained in the fast-growing archive of photographic plates. But we will look at Williamina Fleming, Henrietta Leavitt, Annie Jump Cannon, and their colleagues in "Under New World Skies" in Chapter 21.

Professional astronomy for women in the "Old World"

In some respects, one has to be cautious when using words like "professional" for people doing paid work in astronomy in the nineteenth century. To us, the word "professional" implies something high-powered, whereas most women, and men, who earned their daily bread from astronomy in the Victorian age were employed to undertake routine tasks. The Grand Amateurs may have led the way when it came to technological and cosmological innovation – in conjunction, perhaps, with a handful of not especially well-paid university professors who also held church benefices – but in practice, much astronomy consisted of a hard routine slog.

Sir George Airy, the Astronomer Royal, often warned parents putting forth their bright sons' names for Assistants' jobs at the Royal Observatory, Greenwich, that the aspiring young man would not occupy his nights with pondering the mysteries of the heavens, but in routine observing of familiar objects, followed by equally routine calculation. Airy saw a Greenwich Assistant's job as more closely resembling that of a clerk in an office than that of what we would now think of as a scientist. Germane to pre-digital astronomy was not looking at celestial objects in any kind of philosophical way, but measuring them: measuring endless star transits across the meridian, solar and lunar altitudes, and later spectral lines and other data now harvested on photographic plates. Photography occasioned an explosion in data collection: on a good night an astronomer might successfully expose half-a-dozen good plates of spectra of deep-sky fields. It is true, as Mary Anna Draper and Margaret Lindsay Huggins realized, that these plates could then be measured in the daytime, but it could still take several hours or more to accurately measure and make sense of all the detail recorded by a great 15-inch-diameter lens.

Women's entry into paid astronomy came at the same time as their admission to other, new, technologically created avenues for paid work, around 1880 – as typists, telephone switchboard operators, watch and clock dial painters and assemblers, and paintresses in the new Staffordshire mass-production pottery factories. (My paternal grandmother Sarah worked as such a paintress around 1890.) All of these jobs required intelligence, dexterity, and the acquisition of a skill, were sedentary and did not need physical strength. Unlike textile mill or domestic service work, they were regarded as polite, ladylike jobs or "situations"; ideally suited to the "essentially docile" female mind and temperament.

Most women occupied in paid astronomy in 1900 would have worked in the daytime, and would rarely have been expected to endure the long night watches in a draughty observatory. Instead, they would analyse transit and meridian data, and by the 1890s photographic plates; and one can see Airy's pre-photographic analogy of a male assistant observer to an office clerk extended to female daytime observatory workers. The difference lay in the minds and expectations of the women so employed. When Airy's successor as Astronomer Royal, Sir William M. Christie, began to employ the first "Lady Computers" or calculators in 1890, he was not getting a polite shopkeeper's daughter doing a ladylike job before getting married: he was often getting Cambridge graduates, whose rigorous education and ambitions gave them different and distinctly professional aspirations. The situation of the Greenwich "Lady Computers", however, was demeaning, in spite of Christie, Sir Robert Ball, and others trying to improve their pay and conditions. These high-powered graduates were doing the same work as the normal male Greenwich Computers, who were usually recruited as bright young lads straight from school, on a derisory £4 per month. It is not surprising that few stayed for long, and that the already highly successful astronomical author Agnes Clerke (see below) turned down the offer of a Greenwich "Lady Computership".

Where could serious career aspirations be realized for a professionally trained scientific woman lacking independent means, even in 1900? In many ways, teaching was probably the best option, especially in an academic ladies' college like Cheltenham, or in a good provincial girls' grammar school. With the exception of a Fellowship at an Oxbridge women's college, or at Royal Holloway, there were precious few jobs open to professional astronomical women. Alice Everett was fortunate to be able to resign her Greenwich Computership to take up a better-paid and more interesting job at the Potsdam Astrophysical Observatory, Germany. Alice was commended by Christie as a skilled astronomical photographer. Then Annie Scott Dill Russell had to resign her Computership upon marrying the non-graduate Edward Walter Maunder, of the Greenwich Solar Department, and side-stepped into the semi-

Grand Amateur league, where she, along with her husband, would win a reputation as a solar eclipse photographer.

There were, however, two British women who did come to occupy positions of some standing in astronomy. Annie Walker was a local Cambridge girl who had done well at school, and at fifteen, in 1879, was taken on to the staff of the University Observatory as a general junior assistant. Over the following years, Annie's skill as an observer, calculator, and technical expert saw her promoted to the rank of Second Assistant, her salary rising between 1885 and 1895 from £40 to £100 per annum, close to what a man would have been paid for the same job, making her by far the most senior and best-paid female astronomer in British scientific history to date. The modern astronomical historian Dr Roger Hutchins has found that between 1876 and 1904, the Cambridge Observatory employed about five female Assistants, usually for one or two years at a time.[2]

The other woman astronomer was Elizabeth Isis Pogson (later Mrs Kent), of the Madras Observatory, India. Born in 1852, she was the daughter of Norman Pogson, Assistant at the Radcliffe Observatory, Oxford, who in 1856 discovered an asteroid which won him the Lalande Prize. It was named (like his daughter) after the River Isis, which, along with the Cherwell, flows through Oxford to form the Thames. Her father Norman and family went out to India after 1860 upon his being appointed Government Superintendent of the Madras Observatory, which was primarily a meteorological institution, but also made astronomical observations. In 1873, Elizabeth Isis became her father's salaried assistant, on an official government salary of 150 rupees, which was little more than a servant's wage. Even so, she clearly loved science, and following her father's death was appointed Superintendent of the Meteorological Observatory in 1881, which probably made her the first woman to head – effectively single-handed – an official government scientific research station.

Although proposed by her father for a full Fellowship of the RAS in 1886, she was turned down on a legal quibble hinging on the male pronoun expressed in the Society's founding charter of 1830. But following the change of rules in 1916, Elizabeth Isis was finally elected FRAS in 1920. After various travels, she married a

Captain Henry Kent, an officer in the British Merchant Navy. They settled in Bournemouth, and she received a 250 rupee scientific pension from the Indian government. Elizabeth Isis Pogson Kent, FRAS, died in Croydon, London, in May 1945, aged ninety-two: a true scientific civil servant daughter of the British Empire.[3] Even as late as 1910, paid jobs for men in British astronomy were few in number, and not especially well paid, and for women they were even fewer, and worse paid. Creative astronomical work for women, as for men, still lay in the amateur sector.

Agnes Mary Clerke of Skibbereen, the Irish historian of astronomy

Agnes Clerke belonged to that remarkable and genuinely creative body of Irish astronomers that came to prominence in the nineteenth century. She was unusual, however: not because she was a woman, for there were others, such as Mary Ward, the astronomical cousin of Lord Rosse, but because she was a devout Roman Catholic, as were her sister and brother; the author and historian Ellen, and barrister Aubrey St John. Her colleagues, of both sexes, were invariably Church of Ireland Protestant. Her life and career are also significant insofar as they contradicted a popular stereotype about Victorian and subsequent Ireland. Agnes was not on hostile terms with Protestants; she moved effortlessly among them, despite being a friend of Jesuit priests and patroness of Catholic charities, both in Ireland and in London. Nor was she unique in this respect. She was the child of a happy "mixed marriage". Her father John William Clerke was a Protestant Trinity College Dublin graduate, with a long-standing love of astronomy and science. Her mother, Catherine Mary Deasey, was descended from a family of prosperous Roman Catholic brewers and legal officials of Clonakilty, and her father was a judge's registrar. John William Clerke probably met Catherine through her brother Richard when both men were students in Dublin. In Skibbereen, John W. Clerke ran the local Provincial Bank, and came from a family with an established track record of good relations with the Roman Catholic community. He was involved with several Catholic charities, especially during the terrible Potato Famine which struck Ireland – Skibbereen being badly affected – between 1845 and 1852, when he played an active role in the local famine relief committee.

John W. Clerke had an astronomical observatory on the roof of his bank, containing, among other instruments, a 4-inch refractor, a transit telescope, and a regulator clock, by which the bank manager provided Skibbereen with a public time service. Agnes became fascinated with astronomy as a girl and assisted her father in his observatory. Unusually for the time, the Clerke children were not sent to school, but received an excellent learned education from their cultured parents, and presumably, from private tutors. In particular, they were taught modern and classical languages and history as well as science. The family moved to Dublin in 1861, and when brother Aubrey was sent off to his legal studies, the sisters Ellen and Agnes were taken to Italy, and lived in Florence for a decade. They shared a serious passion for history, and both began to write biographical and scientific articles for *The Edinburgh Review* and other magazines, the ninth edition of *Encyclopaedia Britannica*, the *Dictionary of National Biography*, and the *Catholic Encyclopaedia*, mainly on astronomical and scientific subjects. Broadly speaking, Ellen wrote on Renaissance and modern Italy, and Agnes on astronomical history; and it is on this foundation that Agnes would build her subsequent reputation as the foremost historian of astronomy of her time, whose books and articles are still regarded as authoritative today. In 1877, the sisters returned from Italy, and lived thereafter in London, where they were joined by their brother Aubrey. None of them married.

Agnes did not just write about historical circumstances, however, but came firmly to grips with the hard science of her day, which she communicated with lucidity and elegance. Though not a working observer, she acquired essential observatory skills, primarily while staying with her friends Sir David and Lady Gill at the Cape of Good Hope Observatory, where she learned her way about the southern hemisphere sky. Her knowledge of working, technical astronomy was such that when Christie began to recruit his "Lady Computers" at Greenwich in 1890, he offered her a job. She graciously turned it down, however, as the already distinguished 48-year-old did not need to do routine computational work for a mere £4 per month.

In 1885, she brought out what would become her enduring masterpiece: *A Popular History of Astronomy during the Nineteenth Century*. As has been said before, "popular" in Victorian times did not mean simplistic, so much as rigorous but without equations. Her *Popular History* was, and still is, a formidable treatise on nineteenth-century astronomy, 468 pages long (and growing in subsequent editions). It dealt with all aspects of the rapidly advancing science, from gravitational physics to cometography to spectroscopy, yet in an easy and accessible writing style. It is now a fundamental reference work on astronomy up to 1885. It was followed by five other major astronomical works up to 1905, and a book on the classical Greek poet Homer. Much of the "history" covered in such treatises as *The System of the Stars* (1890), *Problems in Astrophysics* (1903), and *Modern Cosmogonies* (1905) is very modern, and written more in the sense of the Greek word *historia*, or "an account".

While Agnes Clerke's works are magisterial pieces of scientific writing, wrestling with and clearly elucidating the latest discoveries in spectroscopy and planetary, solar, and stellar physics, her cosmology, in many ways, remained Herschelian, as did that of the vast majority of her astronomical friends and colleagues across the world. This cosmology was based upon one great, single Milky Way galaxy, in which all the stars and nebulae were encompassed within one vast system. The earth and solar system, however, were no longer seen as occupying the centre of this broadly circular stellar mass. This view was entirely logical, by the standards of what was known in 1900, for if there were island universes beyond our Milky Way, they still appeared from our terrestrial observing position to be set among and within the Milky Way bodies. All of the nebulae and star clusters were vastly too far away to display parallax angles to an earthly astronomer, meaning, quite simply, that no reliable measuring rod existed to tell us their respective distances and relation to the individual stars of the Milky Way.

In the decades immediately following Agnes Clerke's death, big-aperture telescopic cameras in Californian mountaintop observatories would bring in a host of new and thought-provoking discoveries. By the time of the Great Debate, held in the Smithsonian Institution in Washington DC in 1920, the existence of immensely remote island universes wholly beyond the Milky Way now seemed a real possibility. The whole scale of the universe was changing in the wake of new research technologies, and one really had to wonder how infinite was infinity. Agnes Clerke did not live to see any of this, for she died suddenly of pneumonia on 20 January 1907, at the beginning of her sixty-fifth year.[4]

Women in the new amateur astronomical societies after 1881

Women first found their formal or corporate voice in British astronomy through the amateur societies. As we saw in Chapter 18, these societies, beginning with the Liverpool Astronomical Society in 1881, then the British Astronomical Association (BAA) in 1890, not only admitted female members from their very foundation, but even had them on their governing councils and as directors of specialist observing sections. When one examines the published membership lists and journals of the Liverpool, Leeds, Wales (Cardiff), Newcastle, Belfast, and other regional societies, along with those of the BAA, one encounters a significant – albeit numerically small – female membership. It is often hard to gauge their livelihoods, but they were obviously well educated and, judging by their addresses, living in middle-class districts of their home cities. On the other hand, many of these societies, especially the Liverpool in the 1880s, had members who did not live in or near the home base. The Liverpool Society's Solar Studies Section Director, for example, Miss Elizabeth Brown, lived in Cirencester, Gloucestershire. By that time, though, an efficient and integrated railway system, the next-day delivery Penny Postal Service, and (for the well-off) the electric telegraph made non-local membership for all members, male and female, perfectly feasible.

Many late-nineteenth-century female astronomers, be they single or married, were well-off in their own right, like Elizabeth Brown; others were professional writers like Agnes Clerke and Agnes Giberne, while it is likely that others were schoolteachers. In particular, it would be nice to know more about Miss E. Graham Hagerty, ARCS (Associate of the Royal College of Science, London, now Imperial College, where Joseph Norman Lockyer the solar physicist was professor). In 1900

Miss Hagerty, council member and secretary of the Astronomical Society of Wales, gave her published address as the Higher Grade School, Cardiff: was she the Headmistress? She was clearly a very senior member of the teaching profession, and one of the first scientific women I have encountered to put academic letters after her name. Generally speaking, the big astronomical societies could boast a female membership of a good 10 to 20 per cent, and in 1901, that of Wales had no less than 16.78 per cent women. One wonders how far access to and full participation in the work of learned societies also played a part in the wider women's suffrage and emancipation movement of the late nineteenth century. It certainly did for one young and determined Yorkshire woman.

FLORENCE TAYLOR: FROM LEEDS TO MINNESOTA

On 23 September 1896, then on 28 July 1897, Miss Florence Taylor delivered a pair of lectures to the Leeds Astronomical Society. The first was on "Miss Caroline Herschel, the Great Astronomer", and the second "Mary Somerville, the Great Woman Astronomer and Mathematician". Both lectures were published in the Society's *Journal and Transactions* in their respective years.[5] Caroline Herschel and Mary Somerville were held up as iconic figures and models for emulation to many aspiring young women of the day. In the peroration of her published 1897 lecture, Florence Taylor explicitly associates them with "THE EMANCIPATION OF WOMEN", conspicuously spelled out in capital letters. Responding to her lecture, Charles Whitmell, President of the Leeds Society, was very positive, and reaffirmed the Society's 1895 mission statement to attract lady members. He then went on to affirm the good work that Agnes Clerke, Elizabeth Brown, Annie Maunder, Alice Everett, and other contemporary women were doing both for astronomy and for the wider women's movement. In 1898, Florence married one Dr Hildred, of Nobles County, Minnesota, USA, who appears to have been a relative, and went to live in America. I, and friends living in Minnesota, have been unable to trace any of her activities in the New World after 1898, but the now Mrs Florence Hildred appeared in the published list of members of the Leeds Astronomical Society down to at least 1921. She also made a gift of £100 – a substantial sum in 1921 – to the Society and to Leeds University.

19.1 Florence Taylor, c. 1898. (Leeds Astronomical Society archives.)

It would be fascinating to know more about this remarkable woman, with her active interest in astronomical history and especially in women astronomers. One valuable historical resource for the Leeds Astronomical Society, still in its possession, is its large Album volume, in which letters, pictures, and other documents are preserved. It includes a letter from Florence of 13 June 1898, presenting a "very valuable book" to the Society prior to her marriage and departure for America. Especially interesting is the photograph of Florence in the Album: undated, but probably before or around the time of her marriage. It shows a beautiful young woman, perhaps in her mid-to-late twenties, elegantly attired and confident enough in herself to pose wearing a *pince-nez* pair of spectacles, on a cord: the pose of a self-consciously "thinking woman". It would be wonderful to learn more of her life under the pristine skies of Minnesota, on the other side of the world from industrial Leeds, whether she was happy there, and whether she lived beyond her late forties, the age she would have been when her name disappears from the records of the Leeds Society.

Elizabeth Brown, the sun, and the eclipse-chasers

As we have seen in previous chapters, the sun became an object of increasing study in the nineteenth century, especially when the discovery of sunspot cycles, photography, and spectroscopy transformed the nature of solar understanding after 1840. The sun, like the moon, became an object of intense interest to serious amateur astronomers because of its accessibility to people with relatively modest instruments. As Elizabeth Brown would point out, it was a very suitable subject for female study, as solar observing could be undertaken in the daytime and necessitated "no exposure to night air", which the Victorians regarded as dangerous to the delicate feminine constitution.[6] Besides, she said, many ladies "have ample time at their disposal, and… are often skilled in the use of the pencil".[7] The potential dangers to which Miss Brown and other lady astronomers exposed themselves in their global eclipse travels after *c.* 1880 – such as ocean storms, tropical diseases, and possible hostile attack – would outweigh the risks incurred when observing the moon at night from a chilly English garden.[8]

19.2 Elizabeth Brown, pictured in her own private observatory in Cirencester. (*Journal of the British Astronomical Association* 9, 5 (1899), p. 214.)

In many respects, Elizabeth Brown was a Grand Amateur, being born to wealthy parents in Cirencester, Gloucestershire, in 1830. Her entrance into observational science came via her father, a keen amateur astronomer and meteorologist. By the 1850s, scientific meteorological study was advancing rapidly in Britain, Europe, and the USA, and in observing stations around the British Empire. As we have seen, Elizabeth Isis Pogson and her father in Madras were primarily meteorological observers. In many ways, such study was being driven by James Glaisher FRS, Superintendent of the Magnetic and Meteorological Department of the Royal Observatory, Greenwich, who recognized the importance of recruiting committed, leisured, amateur meteorological observers around Britain who would send him a weekly return of observations, by the Penny Post, in the hopes of building up a wider picture of weather systems. Elizabeth Brown's father did so, and soon found his daughter to be an enthusiastic assistant, and eventually successor. In 1893, the Royal Meteorological Society did Elizabeth the honour of electing her their first female Fellow.

In conjunction with meteorology, she was probably one of the first woman to own an astronomical observatory, and was (after Maria Mitchell, Mary Watson Whitney, and Antonia Maury at Vassar College, USA) the first Englishwoman to be photographed inside one.

As a lady "skilled in the use of the pencil", Elizabeth Brown concentrated on visual solar observing and meticulous drawing. Working primarily by solar image projection, she recorded sunspots, flares, faculae, and other details. Then in 1887, she embarked upon a new line of astronomical activity which had a strong appeal to a growing number of comfortably off, committed amateur astronomers: total eclipse chasing. Professional astronomers had been chasing eclipses, especially those occurring within or near to Europe, since at least the "Turin Eclipse" of 1842, and as we saw in Chapter 15, the Grand Amateur Warren De La Rue joined the largely professional group that went to Spain in 1860. But the new movement of wholly independent and self-funded middle-class eclipse chasers of the 1880s and thereafter was a product of new circumstances. The British middle class, deriving its wealth from the learned professions or inherited private cash, and living in their spacious suburban villas with big gardens (but not great country houses), was getting bigger and richer – a factor which lay at the heart of the new post-1881 regional astronomical societies.

What is more, by 1890 the British Empire had not only reached its maximum global extent, but had created in its wake an extensive infrastructure of efficient organization which was also accessible to private individuals of any nationality. This included fast, comfortable, timetabled and safe global steamship networks. Likewise for railways across India, parts of Africa, and elsewhere, a rapid letter postal service, an international telegraph network, and law and order. Nor were these travel advantages uniquely British, for the Germans, French, Russians, and Americans were doing the same within their own domestic or imperial dominions. So if a person had the time, money, expertise, and inclination, a British, French, American, or any other tourist could venture to Egypt, the Holy Land, India, Siberia, or the Wild West in relative comfort and safety: typhoons, nasty bacteria, or occasional local uprisings notwithstanding. Thus was born the Victorian eclipse chaser, a surprising number of whom were female.

Elizabeth Brown's first eclipse voyage in 1887 was to the safe reaches of Czarist Russia: St Petersburg, Moscow, and then to the comfortable summer country residence of Professor Fedor Bredikhin (or Bredichin) of Moscow. She undertook it with an unnamed lady companion, the congenial English Jesuit astrophysicist priest Father Stephen Perry of the Stonyhurst Observatory, Dr Ralph Copeland, and probably a servant. All went beautifully – except that on the day of the eclipse the sun refused to show its face clearly, and their observations were not as good as they had hoped. It was a common hazard experienced by all eclipse chasers.

Then in 1889, she travelled to see the Trinidad eclipse, again in a party with Father Perry, though the dangers of travel to the tropics in 1889 were brought home by Perry's death from dysentery, after visiting and ministering to the prisons on Devil's Island, as we have seen. After the founding of the British Astronomical Association in 1890 – the Society behind which she had been a major driving force – she travelled to Norway in 1896, and was planning to travel to observe the Portuguese eclipse of 1900 when she died suddenly of pneumonia in 1899: a disease which may have developed from a chill contracted from the night airs when working in the Observatory. She left £1,000 and all her instruments to the BAA, out of a fortune of £63,182 which she shared with her sister Miss Jemima Brown.

After the BAA was founded, eclipse-chasing began to develop an organized, almost corporate, basis, as groups of friends from the Association, including a percentage of both single and married women, travelled to eclipse locations as far afield as India. The Misses Catherine Octavia Stevens, Grace Cook, Alice Everett, K. L. Hart-Davis, Ella Church, and the Revd John Mackenzie Bacon and daughter Miss Gertrude, played their parts in eclipse-chasing. Another stalwart BAA lady eclipse-chaser was Mrs Annie Maunder, whose husband Walter was on the staff of the Greenwich Observatory. Their joint study of historic sunspot cycles led to a recognition of that seventeenth-century sunspot dearth which came to be immortalized as the "Maunder Minimum". Truly daring was Miss Mary Proctor (the late Richard A. Proctor's daughter), lecturer, educator, writer, and astronomical entrepreneur, who at the age of sixty-five hired a biplane to chase the 1927 total eclipse along its path across northern Britain, to become the first aeroplane eclipse observer. (She died aged ninety-five, in 1957.) These were primarily the ladies of the sun. By 1900 or thereabouts, there were also Mary Ackworth Evershed, Irene Elizabeth Toye Warner, Fiametta Wilson, and the lunar cartographer Mary Adela Blagg, all actively involved in serious research in various branches of astronomy. Then, in 1916, with so much collective achievement behind them, the rules of astronomical recognition changed fundamentally.[9]

THE FIRST WOMEN FELLOWS OF THE ROYAL ASTRONOMICAL SOCIETY

From as early as 1828, women had had an ambivalent role in the RAS. In that year the Society had bestowed its Gold Medal upon Caroline Herschel in recognition of her work, and in 1835 she and

Mary Somerville were given the status of Honorary Members; while Anne Sheepshanks had donated her deceased brother's instruments to the Society, and a bequest of £10,000 to Cambridge University astronomy. As we saw above, Elizabeth Isis Pogson was turned down for the Fellowship in 1886, although by the late nineteenth century the Society was coming to an understanding with the growing number of serious and committed women astronomers. The Council agreed that special cards could be given to chosen ladies enabling them to attend the meetings, as regular visitors. Some joked about the Society having to provide potted plants, a piano, and dancing to please the ladies, but the presence of serious female astronomers as a permanent feature of the twentieth-century scientific scene had gone too far to resist by 1916.

Legal modifications to male pronouns in the 1830 charter were made, and at the 14 January 1916 meeting, five women were balloted for the Fellowship, and duly elected. They were Mary Adela Blagg, Ella K. Church, A. Grace Cook, Irene Elizabeth Toye Warner, and Fiametta Wilson. Six more women Fellows were admitted later in 1916, and a growing number thereafter. Since that time, women have come to play an increasingly significant role both in world astronomy and in the RAS (and of course science in general). They have come to hold all the principal offices of the Society, including the presidency, with, to date, Dame Carole Jordan, Professor Kathryn Whaler, and Dame Jocelyn Bell Burnell in the presidential office. Women are now an established and an immensely creative part of the international astronomical scene.[10]

Chapter 20

Astronomy for the Masses in the Victorian Age and Early Twentieth Century

Mass education, on all its levels, was one of the achievements of the nineteenth century. The Protestant reformers of the sixteenth and seventeenth centuries emphasized the need for all people, including the shepherd's boy and the milkmaid, to be able to read the Scriptures in the mother tongue, but things quite often did not work out as smoothly as intended. Schools varied enormously in quality and accessibility, and while both the great public schools and the old, endowed Latin grammar schools had an abundance of free place scholarships for bright youngsters from poor backgrounds, their catchment was immensely variable. If ordinary people saw the inside of a school at all, it was most likely going to be a village dame-school when they were very young, or else classes offered by a well-meaning parish curate.

The age of self-improvement: Sunday Schools, Mechanics' Institutes, and the Victorian "knowledge industry"

The educational opportunities open to most working people in the fast-growing industrial cities, such as Manchester, Sheffield, and Birmingham, were often scanty, as most worked ten or twelve hours a day, six or at best five and a half days a week. This, however, led to a number of ingenious innovations. One was Robert Raikes of Gloucester's Sunday school movement after 1780. Sunday schools were intended not just for little children, as we think of them today, but for unschooled youths and adults who were keen to master literacy and other skills on Sunday afternoons: generally speaking the only afternoon in the week when those older than seven or eight years were not at work. Within a mere nine years, Raikes' movement was educating 300,000 "scholars", and such schools started to acquire libraries of accessible books and magazines.

Between Robert Raikes and the passing of the first statutory compulsory Education Act of 1870, countless millions of folk in Britain picked up what literacy they possessed from an often voluntary church- or chapel-based Sunday school. Once in possession of the priceless gift of literacy, and with access to the printed word wherever they might find it, the intellectually hungry or bored might pursue whatever took their fancy, be it Bible stories, radical political pamphlets, "penny dreadful" murder tales, or books on science. And if a working- or lower-middle-class person had an inclination to science, then astronomy was likely to be a favourite. Many well-placed persons of philanthropic disposition recognized this intellectual hunger among the poor and lower middle classes, and a powerful new educational force emerged: the Mechanics' Institutes (or Institutions) movement, largely the brainchild of Lord Henry Brougham.

The first such Institutes were founded in Edinburgh, then Glasgow and Liverpool, in 1821 and 1823 respectively. Then came Institutes in Manchester, London, Cheltenham, and even in south Devon market towns such as Totnes. By 1840 there were some 600 such Institutes, while it was estimated that those of the Manchester, Bolton, Oldham, and Salford conurbation had 5,000 members by that date. By 1840, they were part of the landscape. Usually founded as an act of local philanthropy, or by a private subscription (like the more up-market county town Literary and Philosophical Societies), the Mechanics' Institutes, Athenaeums, or Lyceums had club rooms, a library, meeting and lecture rooms. Subscriptions were cheap, and while they were principally aimed at mechanics or manual workers, their membership was increasingly swelled by successful artisans, office clerks, and shopkeepers. These were the aspiring classes, not the poor, who had a few shillings to spare, but who could not afford – or might not feel comfortable in – the gentrified "Lit. and Phil." or local subscription library. This invasion of the modestly well-off, it was said, deterred the workman in his "fustian jacket", who felt ill at ease with "too much broadcloth",[1] or office clerks and shopkeepers in their Sunday best.

It would be wrong, however, to see the Mechanics' Institutes as focused primarily upon science, let alone on astronomy. As Henry Solly points out in his *Working Men* (1863), many habitués of the Institutes and organized clubs were in search of a generally broad culture, with lectures on literature, great painters, British heroes and history, as well as science. In many ways, it is hard for a twenty-first-century person to appreciate the grey monotony of life, with its long hours of repetitive work, tiny, cramped houses, and relative poverty, which was the lot of most Victorian working people. If the Mechanics' Institute was going to compete with the gin palace for the working man's leisure time, then it had to offer more than just another form of hard work. This is why lectures on arts subjects, tea and coffee parties, and concerts proved so popular. Intelligent working men and women wanted accessible information, and it was here that the colourful Anglo-Scottish lawyer, and later Lord Chancellor of England, became a force for progress.

Lord Henry Brougham: pioneer of popular education

Henry Brougham was a good friend of Mary and Dr William Somerville, as we saw in Chapter 12 (he first inspired Mary Somerville to write her *Mechanism of the Heavens*, 1831), and was passionate about scientific and astronomical "outreach". He lectured to working men on these and other subjects, in Scotland and around the British Isles. He also cared about social and political reform, and as Lord Chancellor delivered one of the most brilliant speeches ever presented before the House of Lords in the debates leading up to the Reform Bill of 1832. Brougham spoke during a long, all-night Parliamentary sitting, with the thermometer registering 85° F., fortifying himself by drinking *tumblers* of port wine, and ending up on his knees "extremely drunk" at the end – yet still lucid and powerful .[2]

Brougham was a staunch slavery abolitionist, supporter of the Ragged School and other movements, and argued Britain needed a coherent national education system, with a particular bias toward technical and scientific subjects. However, the various Education Bills which he introduced into Parliament were all defeated, being too far in advance of their time. He believed the state itself should take some responsibility for the education of its citizens, especially its poorer ones, as well as advocating self-help and philanthropy. In many ways, the subsequent Working Men's College and Department of Science popular lectures for working men drew inspiration from Brougham's ideas. His inspired ideas, as we have seen, included the Society for the Diffusion of Useful Knowledge (SDUK), the *Penny Magazine* (1832), and *Penny Cyclopaedia* (1833). These and other publications aimed to provide up-to-date information on a range of subjects from cosmology to art history within the financial reach of working people. When you cast your eye over their pages, it is clear that they were *not* mere light pastime reading, on a level with the popular *Varney the Vampyre* tales. The intellectual hunger which they satisfied was evidenced by the tens then hundreds of thousands of copies of these publications sold across the 1830s and 1840s, even if a fair percentage of the readers were office clerks rather than coal miners.

By the 1840s, British society was undergoing rapid change. The population was growing fast (owing, in part, to the control of virulent smallpox after Edward Jenner's introduction of vaccination after 1796); industrial cities, Manchester in particular, were widely coming to be epitomized as phenomena. As we saw in earlier chapters, a rapidly expanding railway network, in conjunction with Rowland Hill's Penny Post, was putting distant places within easy and relatively cheap reach of one another. Steamships put Continental Europe within three or four hours' scheduled sail from England; while the new Cunard Line's SS *Britannia* and Isambard Kingdom Brunel's *Great Western* and *Great Britain* placed the USA within a mere two- or three-week voyage from England by 1845. New wood pulp paper-making machines, picture engraving, and steam-powered great rotary printing presses made books, newspapers, and magazines much cheaper than they had been even in 1800, thereby generating a burgeoning mass knowledge industry to serve the middle, lower-middle, and working classes. Astronomy would soon become part of this knowledge industry.

A major stimulus to the early Victorian knowledge industry was the passing of the Public Libraries Act of 1850, which provided the legal framework for local municipalities to set up public libraries accessible to everyone, male, female, rich, and poor. As a youth I was familiar with the old Peel Park Library and Museum, Salford, Lancashire: a creation of the Act. Its old stacks contained many of the great scientific books of the day, including the works of Sir John Herschel, Richard A. Proctor, Sir Robert Stawell Ball, and Agnes Clerke – tough reading maybe, but through which an intelligent and motivated artisan could explore the wonders of gravitational physics, spectroscopy, and the latest advances in cosmology.

But what was available if an aspiring working- or lower-middle-class astronomer was in search of something less strenuous and more entertaining?

Astronomy shows, demonstrations, and lectures

Irrespective of the perceived imbalance between working-class "fustian" and better-off "broadcloth" seen at the Mechanics' Institutes' events, in the SDUK, and in public library readership, astronomical interest was on the rise in Victorian Britain, especially by the 1880s. This interest, one might say, was primed and loaded through the broader cultural activities fostered by the bodies looked at above, though to get the balance right, astronomy must be seen alongside such popular activities as choral societies, brass bands, model-making, reading, Bible study groups, local history and archaeology, and plant and fossil collecting.

The crossover of documented cultural interests was to be seen in some of the working-men astronomers we met in Chapter 18. John Jones of Brangwyn Bach, for instance, was a Welsh bard and a self-taught biblical scholar and linguist as well as an astronomer; while Devon village Station Master astronomer Roger Langdon made reflecting telescopes and working models, and learned New

Testament Greek from the local curate. The chapel and church hall could be just as active in the propagation of astronomy and wider culture for ordinary Victorians as could the Mechanics' Institute or the urban lecture hall.

So could the street corner: for one of the most accessible places where anyone, of any social class, might encounter hands-on astronomy was at the hands of an engaging street telescope showman, who would demonstrate or exhibit the heavens to passers-by for a "penny a peep". Several such street astronomy exhibitors are on the historical record, one of the earliest being the man who became the subject of William Wordsworth's poem "Star Gazers"[3] of *c.* 1807:

> *"What crowd is this? what have we here! we must not pass it by;*
> *A Telescope upon its frame, and pointed to the sky:"*

This instrument, "Long... as a Barber's Pole", was set up in "Leicester's busy Square" at the heart of fashionable London, beneath the night sky's "resplendent Vault"; "impatient is the Crowd; each is ready with the fee" to behold "The silver Moon with all her Vales, and Hills of mightiest fame".

Yet having set the scene, Wordsworth tells us nothing more about what sort of exhibition this is. Does the "Show-man" delight the audience with a talk about the moon and stars, and the incredible power of his telescope? Does he, as Dr Thomas Dick and others soon would, hold forth about moon-men and a possibly inhabited universe, or perhaps tickle their ears with the sheer *vastness* of the "resplendent Vault", with Dr Herschel's nebulae, or with the possibility of comets crashing into the earth? Well-documented later street telescope "Exhibitors" certainly would. Perhaps the Leicester Square showman just let you have a peep, pocketed your money, then said "Pass on, please" so that the next person could gain access to the eyepiece. According to Wordsworth, the show was not very inspiring, as the casual observers invariably did "slackly go away, as if dissatisfied".

What it is important to realize, however, is that Wordsworth's poem is *not* about popular astronomical education so much as a vehicle for a speculation upon the human condition: that people of all classes, out for a light-hearted evening in the theatres, pubs, or dance halls, should, upon being suddenly brought face to face with divine sublimity, be made to feel "sad" and "dissatisfied" by their innate inadequacy. I suspect that the poem tells us more about William Wordsworth's attitude to life than it does about astronomy. I have been present on countless occasions when local astronomical societies have invited the general public to look through a good telescope at the moon, planets, or nebulae, be they large telescopes in fixed observatories or smaller portable instruments set up in a public place. I have never once seen people leaving the telescope walking "slackly" away as if dissatisfied. Quite to the contrary, in fact; they are more likely to look amazed and delighted, and be inspired to ask all manner of questions about what is up there.

This positive response was something aimed at by one Mr Tregent, a real-life professional telescope "Exhibitor" interviewed at length on 26 October 1856 by the pioneering journalist Henry Mayhew, who published an account in his monumental four-volume *London Labour and the London Poor* (1861).[4] Tregent is included in Mayhew's survey of people who earned their living in the streets of London: a survey which covered a range of honest workers (such as telescope and microscope demonstrators), costermongers, and "criminals" such as prostitutes. Tregent wanted a positive response (as I suspect Wordsworth's exhibitor also did) for one plain and simple reason: delighted customers were more likely to tell their friends and come back for another peep at the heavens, whereas dissatisfied ones were not likely to bother. A good press is good for business.

Tregent's astronomical patter certainly worked, for he told Mayhew that his street activities brought in £125 per year, which in 1856 was the income of a lower middle-class senior bank clerk rather than of a street beggar. Tregent was both intelligent and enterprising: he did not mutely stand by while people peeped through his handsome telescope, but also gave mini lectures about whatever celestial delight he happened to be exhibiting. Tregent owned a fine refractor with a 4½-inch-diameter object glass by an optician named Mull of Clerkenwell, London. Such a telescope, complete with stand, eyepieces, and fittings, would have cost a lot of money in 1856, and Mull's instrument cost £80.[5] It was exactly the type of common telescope that the serious clerical amateur, the Revd Thomas William Webb, wrote about in 1859 (see Chapter 18).

Tregent was a tailor, who had originally tramped to London from his native Yorkshire to find work. As an astute small businessman, he paid Mr Mull, the Clerkenwell optician who made his telescope, not in cash but in suits of clothes. As tailoring tended to be a self-employed trade, Tregent could work flexibly, exhibiting the heavens between making suits and overcoats. It would be interesting to know the source of his astronomical education, for he was clearly knowledgeable about it, and passed on serious scientific information to his pavement audiences as well as letting them look through his great refractor. Was it from working men's lectures, Mechanics' Institutes, or from books? He does not tell us, but he knew a great deal about telescopes, and in outlining his patter to Mayhew, said that he would even tell his crowd something about the really big telescopes in the world, such as the one in Cincinnati, USA, where the "very fine climate" and clear air permitted splendid views of the sky.

Then Tregent would go on to further captivate his listeners. When exhibiting Jupiter, for example, he would draw gasps by mentioning that the planet sped around the sun at an incredible 26,000 miles per hour. This figure could easily be computed in 1856 by knowing the distance of Jupiter from the sun and its twelve-year rotation period, and deriving there from the length in miles of the Jupiterian elliptical orbit. He would also point out Jupiter's orbiting satellites. Tregent (we are not given his Christian name) was not so much a telescope showman as a street educator, who left his students better informed about astronomy, telescopes, and astronomical technology.

20.1 Mr Tregent demonstrating his telescope in the street. (H. Mayhew, *London Labour and the London Poor*, vol. 3 (1851), p. 82.)

In addition to street telescope (and microscope) demonstrators and lecturers, the Victorian era was the golden age of the popular lecture. Public lectures could vary in quality and authority, ranging from those given in Ipswich (1848), South Shields (1854), and Manchester (1861) by George Biddell Airy, the Astronomer Royal, to ill-informed fledgling performances by Mr Robert Children in Bethnal Green, London.[6] In one of his first sallies into astronomical lecturing, Mr Children evoked hoots of laughter and catcalls when he initially introduced his audience to the *consternations* of the sky. Robert Children was a bootmaker by trade, who seems to have been inspired to give astronomy lectures, presumably to a paying audience.[7]

One invaluable source for popular lectures, as I have indicated at length elsewhere, are the volumes of contemporary newspaper cuttings and periodical documents compiled by Dr John Lee and his friends at Hartwell House near Aylesbury, and now preserved in the Museum of the History of Science, Oxford. According to one cutting there was a veritable "Astronomical Mania" in London during the lecture season of 1839. The preserved cuttings contain some fascinating comments about aspiring lecturers such as Mr James Howell, a City clerk and a "regularly dipped Baptist", who possessed a "monkey looking personage", yet who drew paying crowds to the Strand and Adelphi Theatres, London, with his evening astronomical lectures. Like Tregent and Children, James Howell was a "part-time" popular astronomical lecturer, exchanging his daytime clerk's stool and desk for the stage in the evening.[8]

The Hartwell House newspaper cuttings books, along with other preserved magazine advertisement sources, reveal a remarkable world of popular astronomy by the 1860s. There was, for example, the self-educated William Knock, another working shoemaker and part-time Wesleyan Methodist minister, who owned an incredible 3,000 books, and also held forth on the heavens. Likewise, the Revd W. Rees lectured, presumably in his native Welsh, to packed audiences in the Welsh Chapel in Falcon Square, London, in 1850. In the Aylesbury area, an Anglican friend of Dr Lee, the Revd Joseph B. Reade, and the layman philanthropic brewer Thomas Dell, lectured to working men in village halls and to organizations such as the Young Men's Christian Association in the 1850s. Many popular astronomy lecturers had some professional connection – like Thomas Dick in Scotland (Chapter 11) – with Christian ministry or philanthropy.

When reviewing the popular astronomical scene of the mid-century one cannot help but be impressed by the sheer demand and market opportunities that existed for purveyors of astronomical information, often to paying audiences. There was the reliably academic Professor John Pringle Nichol of Glasgow lecturing in Salford in 1839 to 250 men in their "working jackets"; and the gloriously named Jabez Inwards's scaremongering performance in Camden Town Hall on 6 April 1843. Inwards's surviving advertisement handbill reveals a contemporary fear of cometary collisions, with its headline *The Comet. Is the World in Danger?* [9] The comets of 1811 and 1835 (Halley's) are mentioned, but the real focus of alarm is the contemporary Great Comet of 1843. This was "Miller's Comet", named after William Miller in the USA, a self-appointed prophet who had delivered eighty-two lectures on the subject.[10] On the basis of his intensive studies of the book of Daniel in the Old Testament, Miller predicted that the comet which appeared in early 1843 presaged Christ's Second Coming, then Armageddon. We are not told how many people attended Jabez Inwards's lecture, or lectures, but one assumes that his audience would have received a heady mix of astronomy, prophecy, and probably hysteria.

Let us conclude this section by returning to Robert Children. Uninhibited by the catcalls which his early ignorance provoked, Children was a genuine trouper, borrowing money to invest in a sophisticated magic lantern projector, and going on to make an astonishing £6,000, emigrate to America, and become a "Great Farmer".

Richard Anthony Proctor and Sir Robert Stawell Ball: stars of the astronomical lecture circuit

Unlike Children, Tregent, Inwards, and most of the other popular lecturers we met above, both Proctor and Ball were gentlemen born and educated. Born in 1837, Richard A. Proctor went up to St John's College, Cambridge, and soon proved himself a gifted astronomical mathematician. Yet for all his advantages of birth, intellect, and education, he had a serious flaw: his reputation for controversy and tendency to fall out with people.[11] Even when serving as Secretary of the Royal Astronomical Society, he had a serious dispute with Sir George Airy, the formidable Astronomer Royal. Things were made no better in 1866 when £13,000 of his family money vanished in a bank failure, obliging him to make a living with which to support a young family.

In spite of his disputatious disposition, Proctor had two great gifts: he was a brilliant public lecturer and a lucid, captivating writer; and all combined with his first-rate scientific intellect. By 1870, commercial lecturing was already an acknowledged occupation, and there were lecture agents in the same way as there were theatrical agents. If one turned out to be good, then lucrative bookings could be lined up around the country that fitted in nicely with the published railway timetable. Town Halls and Corn Exchanges would be booked in advance, newspaper adverts and handbills printed, and paying audiences attracted. In many big cities, such as Manchester, Leeds, or Cardiff, horse-drawn omnibus and tramcar services could bring in people from the suburbs.

This burgeoning lecture circuit was by no means exclusively astronomical or scientific, and successful lecture agents would bring in celebrity lecturers who specialized in history, travel narratives, missionary work, or the adventures of current British imperial heroes. In this respect, astronomy was on a par with music hall stars and other entertainers, all catering for a growing public demand to be educated, entertained, or both. Proctor did well as a lecturer, and started to rebuild his fortune, embarking on lucrative lecture tours of the USA and a round the world tour incorporating Australia and New Zealand. It was on this tour that, in 1881, the recently widowed Proctor met Sallie Duffield, a Missouri widow, who would become the second Mrs Proctor.

While lecturing could reach great audiences, the best and most profitable form of outreach came from writing. Proctor's *Other Worlds Than Ours* (1870), which we encountered in Chapter 11, went through twenty-nine editions down to 1909, while in less than twenty years he churned out fifty-six other titles, and hundreds of pieces of high-quality scientific journalism. Then in 1881, he established the popular magazine *Knowledge and Illustrated Scientific News*. His accumulated publications, as the modern scholar Michael J. Crowe has rightly said, "introduced thousands" of people to astronomy.[12] Proctor died from yellow fever in New York in 1881 on his way to England to deliver a lecture series. His stalwart widow Sallie immediately stepped in, however, boarded the ship, and gave the lectures herself. Astronomy, one might say, was in the Proctor blood, for Richard's daughter Mary (from his first marriage) forged a major career for herself as a lady lecturer and writer, as we saw in Chapter 19.

Whereas Proctor was known for his disputatious disposition, Sir Robert Stawell Ball was the living epitome of the genial

Irishman. Born in 1840, the son of a senior civil servant, and brought up in a fashionable district of Dublin, young Robert was educated in England, though solidly refusing to give up his gentle Irish accent. A brilliant career at Trinity College Dublin came next, followed by a few years living at Birr Castle, where he was employed as tutor to the Earl and Countess of Rosse's younger sons. As we saw in Chapter 14, Ball's natural sociability soon turned him into a family friend of the Rosses, and a friend for life of his pupils.

20.2 Sir Robert Stawell Ball. (*Reminiscences and Letters of Sir Robert Ball*, ed. Valentine Ball (1915), frontispiece. Author's collection.)

At Birr, the Earl gave Ball the use of the great 72-inch-mirror telescope. On the brilliantly clear night of 13 November 1866, Ball observed – not using the telescope, but standing high upon its gallery – the spectacular meteor stream through which he would contribute to the growing idea among contemporary astronomers that the Leonids and other seasonal meteor streams were caused by swarms of sun-orbiting particles hitting the earth's atmosphere and being burned up at a great altitude. He thus helped to substantiate the explanation for meteor streams and their radiant points which we accept as true today. From Birr, Ball's career took off: in 1867, he became Professor at the new Royal College of Science, Dublin, and in 1874, Andrews Professor of Astronomy at Trinity College, which also gave him Directorship of the University's Dunsink Observatory, and the title Astronomer Royal for Ireland.

Although a passionate and skilful observing astronomer, Ball was a born teacher and inspirer, and a true performer. He tells us that his first general public lecture, entitled "Some Recent Astronomical Discoveries", was delivered on 4 February 1869 in Belfast, for which he received 14 shillings to cover his travel costs. And from there, things took off. He lectured at the prestigious Birmingham Midland Institute, and soon after he was well and truly on the circuit. His academic standing, his affability, and, one suspects, his sense of humour, made him hugely popular, and he was soon filling theatres and town halls.[13] Contemporary astronomical events, such as the Venus transits of 1874 and 1882, gave him topical material, but other lectures dealt with the nebulae, the planets, the moon, and cosmology, with enticing titles such as "A Glimpse through the Corridors of Time". Ball also made full use of new communication technology, and his student and professional lecture careers spanned the development of the gas-powered limelight magic lantern of the 1850s to the electric arc, then on to the Edison light bulb and glass photographic slides by 1890. A born raconteur, Ball tells us how in Dublin an early magic lantern exploded, when the technician got the hydrogen-oxygen mix wrong.

British lecture tours were followed by tours of the USA and Canada, and by the 1890s he was probably making a handsome profit from his performances, for in 1908 the new Manchester Astronomical Society recorded in its minutes that it could not afford Sir Robert's fee for a lecture.[14] He had long since abandoned his expenses only performances. In addition to all of his popular lecturing, academic lecturing in Trinity College, and observing in Dunsink Observatory (and a similar workload upon becoming Lowndean Professor and Observatory Director at Cambridge University in 1892), Ball was a brilliant popular writer. His *The Story of the Heavens* (1885) became a bestseller, providing a wonderful account of astronomy at the time of writing, conveyed in straightforward language with illustrations. His *The Story of the Sun* (1906) did likewise, along with his other publications. By the time of his death at the Cambridge Observatory in 1913, he had probably taken astronomy to many *millions* of people, on two continents, by both the spoken and the written word. He also achieved much as an active Observatory Director and pure mathematician. Ball clearly possessed prodigious energy and a love of his subject, combined with an easy charm which endeared him to people across the social scale. It would not be an exaggeration to say that Sir Robert Stawell Ball FRS was astronomy's first public figure, media celebrity, and mass-educator.

Sir Arthur Stanley Eddington and Sir James Hopwood Jeans: astronomy's first "Knights of the airwaves"

At the time of Robert Ball's death on 25 November 1913, astronomy and physics (and the stability of Europe) were on the verge of fundamental transformation, as we shall see in later chapters. We were also on the verge of a media transformation, as popular education was about to transcend its ancient bounds of the spoken, written, and then published word. The experimental "Hertzian waves" were just being changed into a truly universally accessible technology: radio. For British people, this technology would become the BBC in 1922, as the airwaves began to carry words and music into the gentleman's drawing room, the Pennine farmer's kitchen, and the Welsh pitman's cottage with equal ease. Lord Reith, first Director General of the BBC, saw the British Broadcasting Corporation as having a *mission* of service to the world: to educate and inform, as well as to entertain. This mission statement is still to be seen engraved into the stone frieze of the atrium of Broadcasting House, London. And while commercially produced radio sets could be very expensive, especially in the early days, two factors made radio accessible to working people. One was the cheap crystal set with its simple technology, costing only a few shillings, which had to be listened to through a pair of headphones. The other was the popularity of homemade radio sets, each individual part of which could be saved up for and purchased from a local electrical shop, and the set assembled from simple instructions. (I still have one of my late father's home-made crystal sets of 1927.)

When Sir Robert Ball died, the obvious choice as his successor as Astronomy Professor and Director of the Cambridge University Observatory was the brilliant thirty-one-year-old Arthur S. Eddington, who was knighted in 1930. As we shall see in Chapter 23, Eddington won international fame after 1919 by demonstrating the reality of Einstein's 1916 Theory of Relativity, along with other subsequent discoveries. In 1920 he wrote *Space, Time, and Gravitation*, in which he set out to describe Einstein's theory for the ordinary reader. Then in 1928 came his most influential popular book, *The Nature of the Physical World*, setting out in lucid terms the very latest discoveries in astronomy, cosmology, and physics. This made Professor Eddington a household name, as well as getting him into *Punch* and other popular magazines.

Sir James Hopwood Jeans was five years older than Eddington, and both, in their respective student generations, were members of Trinity College, Cambridge, and High Wranglers in the University's prestigious Mathematical Tripos examinations. While his younger days were marred by bouts of tuberculosis, Jeans held a mathematical professorship at Princeton University while still in his twenties, and the Stokes Lectureship in Cambridge. In 1912, at the age of thirty-five, however, he retired from formal academic work to devote himself to writing and research, becoming, in many respects, a twentieth-century Grand Amateur mathematical scientist of ample private means. It was here that he found his talents and his voice as a popularizer of the new post-Einsteinian astronomy. Friends in Cambridge University Press persuaded Jeans to write a popular book, *The Universe Around Us* (1930), which was immediately followed by *The Mysterious Universe* (1930). They helped to propel Jeans – knighted in 1928 – to popular fame at the same time as his Cambridge rival Eddington. They were rivals in both the academic and the popular sphere; for Jeans had criticized Eddington's model of stellar evolution and radiation stability, as well as other things. Yet Jeans was not only a brilliant writer; his many American friendships also kept him in close touch with the Californian and other observatories which were opening up the "new universe".

Jeans, while an instinctively reserved man, was an excellent public lecturer who, in the grand Victorian tradition, took astronomy to the masses. He also, like Eddington, was popular on the radio: talks, interviews, and programmes such as the *Brains Trust* took both into countless thousands of homes, right across the social spectrum, as did cheap editions of their books and borrowings from public libraries. By the 1930s, astronomy was reaching the masses in a way that the Mechanics' Institutes could never have imagined.

Arthur Eddington never married, dying of cancer in 1944. James Jeans, however, married twice: his first wife being connected with the Tiffany family of New York. His second wife, Susanne or Susi, was Viennese, and like Jeans, an accomplished organist. At their home near Dorking, Surrey, both had their own separate pipe organs upon which they could thunder away. Jeans would play the organ for hours on end, as he evolved his ideas. Did the soaring cascades of glorious sound assist his thinking about the new galactic and extra-galactic stellar wonders being discovered by his American friends at the new Mount Wilson Observatory, then the home of the biggest telescope in the world? Jeans had the honour of being made a Research Associate at Mount Wilson, and it is to the discoveries being made in the great American observatories that we must now turn.

Chapter 21

Under New World Skies: The Great American Observatories

The United States of America possesses several natural advantages when it comes to astronomy. Most of its great continent is much closer to the equator than is much of Europe, thus making the sun, moon, and planets appear higher up in the sky, and hence less prone to atmospheric distortion. The very highest regions of northern Minnesota, for example, are on the same line of latitude as Paris; while New Orleans, at the mouth of the mighty Mississippi, along with California, are further south than Cairo in Egypt. This gives an enormous range of latitude to work within, for in the south in particular one can see deep down below the celestial equator and into the southern hemisphere.

It was their disadvantage of latitude that frustrated British Grand Amateurs in the nineteenth century, as colleagues in Harvard or Cincinnati could routinely record planetary surface detail that might only be available once in a blue moon to someone in Liverpool or Manchester. It was one reason why, on two occasions in the 1850s and early 1860s, William Lassell of Liverpool transported two enormous reflecting telescopes to the Mediterranean island of Malta, which gave him a latitude advantage not possessed by his friendly rival planetary observer William Cranch Bond of the Harvard Observatory. Yet a latitude advantage was only a part of it, for the vast and pristine skies of Columbia had an unpolluted optical purity and transparency that were virtually unobtainable in all but a handful of places accessible to a British or European astronomer. The most gloriously transparent skies I have ever seen are those above the northern Minnesotan forests, and in particular those skies around Lake Bemidji, Minnesota. Here, especially in autumn, one sees with the naked eye the Milky Way spread out like a rich, glowing carpet from horizon to horizon, with the Pleiades and other star clusters sparkling like diamonds and the rising Orion Nebula appearing as a distinct, pearly mist. Take up a pair of binoculars or even a modest astronomical telescope, and be prepared to catch your breath.

With its Appalachian mountain chain in the east, and the Rockies in the west, along with the vast expanse of the Great Plains in between, North America is still today a wonderful place from which to observe the glory of the heavens, and it is easy to appreciate how astronomers assessed its immense research potential in 1860. Even as early as 1881, Dr Henry Draper, the New York physician and Grand Amateur astronomer, was scouting the Rocky Mountains for a prime sky location for a research observatory, just before his sudden death. Yet even by 1881 America had already had a major but perhaps no longer prime sky observatory for nearly forty years.

North America's first big observatories

Astronomy had been taught at Harvard College, near Boston, since the seventeenth century, and the Revd Dr Cotton Mather, colonial America's first FRS, had owned and used telescopes and other London-made instruments there since the late seventeenth century. Then in 1761, John Winthrop, Hollis Professor of Mathematics and Natural Philosophy at Harvard, led an expedition to St John's Newfoundland, to observe the Venus transit of that year, the results of which were passed on to the Royal Society.

The first large telescope in the early United States was built by public subscription in 1845, by the good citizens of Cincinnati, Ohio. It was a truly prestige piece by the standards of the day, containing an 11¼-inch-diameter object glass by Merz and Mahler of Munich. In the 1840s, Cincinnati, on the north bank of the Ohio River, was the last great city westwards from the East Coast: a city of imposing buildings in a then sparsely populated terrain, which had delighted Charles Dickens on his first American tour of 1842. The river steamboats brought the settlers – and Dickens – to Cincinnati, beyond which, and across the Mississippi after St Louis, they joined the waggon trains (Dickens returning to England from Cincinnati), heading for the west. Their world-class astronomical observatory, under the clear and starry skies of the Great Plains, was clearly an object of great pride to the citizens of Cincinnati. It was available to the general public, who could be shown the heavens on prescribed observing nights. I am not aware of anywhere else in the world in 1845 where ordinary folk had the right to view the heavens through an instrument of such power. This public accessibility, however, inevitably limited the instrument's availability for serious astronomical research.

It was at Harvard College that the next great American telescope, with its 15-inch object glass, came into commission in 1847. Made, like the Cincinnati 11¼-inch telescope, by Merz and Mahler of Munich, the Harvard instrument was one of a pair; the other being their great 15-inch for the Russian Pulkowa Observatory, near St Petersburg. It was a matter of great pride to the citizens of the United States that they had a telescope of equal power to that of the Czar's great Observatory. In many respects Harvard was a better location, at 42° north as against Pulkowa's 59° north, thus having a much better command of that ecliptic band within which the planets move.

In many respects, North American astronomy retained many features of that of Britain, such as a Grand Amateur tradition. We saw in Chapters 15 to 17 how Dr John William and Dr Henry Draper of New York pioneered fundamental innovations in big telescope design, photography, and spectroscopy, while the innovative astronomical photographer Lewis Rutherfurd was a private gentleman. The Harvard Observatory, at Cambridge, Boston, grew up in similar circumstances.

In 1839, William Cranch Bond, a Rochester, Massachusetts, manufacturer of clocks and watches, donated his personal collection of astronomical instruments to the university, thereby becoming the first Harvard Observatory Director. Then in 1843, the brilliant comet of that year, the "Millerite Comet" – believed by many to herald the Second Coming of Christ, and causing hysteria in parts of the USA – inspired a public interest in astronomy (see Chapter 20 for English lectures on the "Millerite Comet"). This interest led the entrepreneur David Sears to donate $5,000 to Harvard University. Over the next couple of months, a combination of private individuals, businesses, and charitable bodies subscribed another $10,000, and before long Sears's $5,000 was matched by some $20,000, the whole being roughly equivalent to £4,000 – £5,000 sterling at contemporary exchange rates.[1] All of it was raised privately, in the British fashion. Major observatory foundations in Continental Europe, and also with Czar Nicholas I's great Observatory and telescope at Pulkowa, were funded by royal, government or taxpayers' money. Flush with funds, the Harvard Observatory had placed their order with Merz and Mahler for what would become the famous duplicate of the Pulkowa refractor, as said above, while at the same time constructing the Observatory buildings. The great Harvard refractor came into use in the summer of 1847, becoming America's biggest astronomical telescope. Then, some years later, William Cranch Bond, whose early work at Harvard had been funded by his own commercial activities, received an official yearly university salary of $1,500 for two years, along with $640 for an assistant. The assistant was his son George Phillips Bond, who would succeed to his father's Directorship.

Over the next thirty years, the Bonds would undertake a series of original researches from Harvard, especially into the planets, and into the Andromeda nebula, Messier 31. Over 1850–52, for example, both Lassell – first in England and then in Malta – and William Cranch Bond at Harvard reported the existence of a thin "crepe veil" between the innermost bright ring of Saturn and the body of the planet, later to be styled the "crepe ring" because of its apparent similarity to thin crepe gauze. They were working with instruments of similar power, Bond using the great 15-inch refractor and Lassell his 24-inch speculum metal mirror reflector. Then in 1874, George Phillips Bond believed he had found structural features in the Andromeda nebula. His drawing depicts a bright central galactic nucleus and the smaller, bright satellite nebula, M32, but running across the nebula are a pair of dark bands or arcs, resembling the Whirlpool (M51) nebula's curved arms, yet in the case of Andromeda, tilted at an angle to a terrestrial observer.

What could these arcs be? Even after Isaac Roberts in England obtained the first good photographs of M31, after 1888, the arcs were still a puzzle. Did they indicate that the nebula was rotating? Or could they be part of a distant evolving solar system turning around a condensing central star, the arcs separating off to become planets, after the manner of Laplace's famous nebula hypothesis of 1796? George P. Bond would also produce some superb studies of other nebulae, including that in Orion. But by the 1880s Harvard, like the new Potsdam Astrophysical Observatory, Germany, was moving into the fast-advancing new sciences of astrophysics and spectroscopy, as front-rank astronomical research came increasingly to be the preserve of corporate scientific institutions rather than that of private researchers.

The Harvard astrophysicists

When Dr Henry Draper died of pleurisy in 1882, his wife Mary Anna presented his great archive of photographs, stellar spectra, and other primary astronomical data to the Harvard Observatory. Draper had recently been considering the possibilities of a high-altitude observatory in the Rocky Mountains which, by 1882, were only a few days' train journey from New York, Boston, and the East Coast. This would not become the home of Harvard's high-altitude observatory, however, for having checked over possible sites out West, the Observatory Director, Edward C. Pickering, decided upon a site at Arequipa, Peru, at 8,055 feet elevation, financed by another Harvard benefactor, Uriah A. Boyden. By the 1890s, this Boyden Station was beginning to produce spectral plates of stars visible in both hemispheres, which could be measured and analysed at leisure back in Harvard.[2] This technique, which made for more efficient astronomical time management, was first suggested and used by Mary Anna Draper and Lady Margaret Huggins. What made this possible was not only photography, which recorded the spectral lines in black and white, but also the development of a desk-top plate-measuring machine whereby an observer could measure off the spectral lines through a micrometer microscope eyepiece, at any time of day or night, illuminated by a standard light source.

The ladies of the Harvard Observatory

Williamina Fleming was a twenty-three-year-old Scotswoman from Dundee who had emigrated to America with her husband, who promptly abandoned her and their baby. After earning a living from a variety of jobs, she became a maidservant in the home of Professor Edward Pickering of the Harvard Observatory. Pickering, dissatisfied with some of the men measuring the Harvard plates in the late 1870s, had started to employ women for the task; and in 1881, noticing his "Scotch maid's" intelligence and meticulous skills, he tried her out on measuring plates. She was excellent, and before long Pickering began to recruit other women to what came to be nicknamed his plate-measuring "harem", of whom Williamina became Supervisor.[3]

Some of these women were Vassar College graduates, others were from non-academic backgrounds yet intelligent, painstaking, and accurate in their work. By the early 1890s there were some eight women on the Observatory staff, and there would be more. Many went on to make fundamental discoveries and establish international reputations for themselves as they began to interpret what they were measuring and to draw significant scientific conclusions. Williamina Fleming measured around 10,000 stars as part of the *Draper Catalogue of Stellar Spectra* (1890), and devised a system of star classification designated by the letters A, B, C, and so on. While our knowledge of the physics and chemistry of these stars has advanced greatly since the 1890s, the Pickering–Fleming system is still used today in a modified form. Other Harvard Observatory women, such as Annie Jump Cannon and Antonia Maury, would refine the spectral classes and win credit for themselves as they advanced the science of astrophysics: a science which was still less than forty years old in 1900.

One of the Harvard plate-measuring women would make a discovery that would supply a golden key to facilitating much twentieth-century cosmology. This was Henrietta Swan Leavitt, with her work on the Cepheid variable stars of the Magellanic Clouds of the southern hemisphere, and we will return to her achievement in the following chapter. These meticulous analyses of tens of thousands of photographic plates by the Harvard female astronomers between *c.* 1890 and 1910 would provide an invaluable database upon which many subsequent astronomical discoveries would be built in the twentieth century. But now let us examine the USA's own major contributions to the development of big telescope technology.

ALVAN CLARK AND SONS, OPTICIANS OF BOSTON, MASSACHUSETTS

During the early 1840s, as we have seen, the great German firm of Merz and Mahler of Munich dominated both the European and American markets in building large-aperture and beautifully engineered observatory telescopes. It was to them, as we have seen, that the citizens of Cincinnati, then Boston Massachusetts went when commissioning the first big American refractors. Yet things were all set to change.

Alvan Clark was born in 1804, from a Cape Cod seagoing family of English ancestry, and by the 1880s had become a successful portrait painter, engraver, and artist in Boston. He had an interest in astronomy, and decided to try his hand at grinding and polishing a lens, only to find that he was good at it. He was part of a tradition of self-taught opticians and telescope engineers that included his English contemporary Thomas Cooke of York, Thomas Grubb of Dublin, the mirror-maker William Lassell, Lord Rosse, and William Herschel, perhaps going back to Sir Isaac Newton. Like Cooke, Grubb, and Herschel, Clark not only recognized the commercial potential of fine telescopes, but went on to improve them, mass-manufacturing fine optics at one end of the market and producing increasingly large and powerful big observatory telescopes at the other. (Soon, other American opticians, such as Worcester Reed Warner, Ambrose Swasey, and John Alfred Brashear, would enter the field.) Alvan and his sons Alvan Graham and George Bassett Clark would go on to establish a great American optical dynasty.

One problem faced by early American opticians, however, was finding a good, relatively cheap source of fine optical glass which could be ground into lenses. Clark imported glass blanks made in the glass houses of Chance Brothers of Birmingham, and also from the Parisian Georges Bontemps (who, as a result of French political turmoil, came to work with the Chance brothers), and the French St Gobain glassworks, although before the start of an indigenous American optical glassmaking industry, import taxes on raw materials were a drawback. Clark applied all of his meticulous talents to the fashioning of lenses that soon rivalled, and then overtook, those of the German masters. Deborah J. Warner, a modern biographer of Alvan Clark and Sons and their achievement, rightly styles him an "artist in optics". Clark knew the night sky like the back of his hand, and used close pairs of double stars and binaries as test objects for his new lenses. This led him, on 31 January 1862, to make a discovery that confirmed the physical existence of a mathematically predicted object, a dimmer companion star to Sirius, whose possible existence had been predicted by Friedrich Georg Wilhelm Bessel at the Dorpat Observatory in Estonia, back in 1844. Bessel was the most renowned astrometric observer of the age, specializing in employing the finest German instruments to measure the positions of the stars to within fractions of a single second of arc, using micrometers and a divided object glass heliometer by the great Fraunhofer (see Chapter 9).

By 1844, Bessel had detected a systematic rock or movement in the bright star Sirius, leading him to suggest that Sirius had a "dark companion" star rotating around it; one too small to be seen in its own right, as its light was drowned out by the glare of the brilliant Sirius, yet which still exerted a measurable gravitational pull upon it, thus causing its position to "rock" slightly as the "dark companion" moved around it.

Clark was testing a new 18½-inch-diameter object glass originally destined for the University of Mississippi, but which, owing to the circumstances of the Civil War and finance, was acquired by the Chicago Astronomical Society for the Dearborn Observatory. He had so adjusted things as to have Sirius obscured by a building, intending first to see its glare, then the star itself through the telescope. He was surprised, however, when he saw a dim star shielded from Sirius's glare: it was the pup, or as we now know it, "Sirius B". This is a splendid example of a mathematical prediction drawn from astrometric analysis being demonstrated as an observed reality eighteen years later.

21.1 Merz and Mahler's 15-inch equatorial, Harvard College Observatory. (J. N. Lockyer, *Stargazing* (1878), p. 339. Author's collection.)

AMERICAN LIBERAL ARTS COLLEGES AND ASTRONOMY

One especially American educational institution was the private Liberal Arts College. Modelled partly on the Oxford and Cambridge colleges, and partly on the Scottish universities – Harvard College, in aptly named Cambridge, Boston, Massachusetts, was founded in 1636 by John Harvard, an old Emmanuel College, Cambridge man – these Colleges aspired to give a fine classical and modern education to young American gentlemen. Some would proudly stay as modest-sized colleges, yet others – Harvard, Princeton, and Yale – would grow into world-class universities. By the 1860s Vassar, Wellesley, then others, aspired to give an academic education to young American ladies, as we saw with Maria Mitchell, Mary Watson

Whitney, and other astronomical women in Chapter 19. Science, or natural philosophy, was a natural part of their wider degree courses, along with classical languages, literature, and wider culture, and most Liberal Arts College graduates would leave with at least some knowledge of basic chemistry, physics, and astronomy, in addition to classics – essential skills for people who would build the new nation.

All of these Colleges were independent institutions, being private benefactions like the British public and grammar schools – another example of the Grand Amateur principle. A major aspect of this educational vision was the creation of scholarships, which aimed to make the Liberal Arts Colleges accessible to a considerable degree to the children of the less well-off, depending upon the size of the endowments and their yields. Many of these Liberal Arts Colleges owned, or aspired to own, a fine astronomical observatory. Be an observatory large or modest, it would be used to show the wonders of the heavens to students, and to undertake some kind of research. For the more modestly sized observatories, such researches might involve the regular monitoring of double-star coordinates, asteroid-hunting, or sunspot and faculae recording. With the westward expansion of the United States, great state or public universities were founded, many of which established astronomical observatories, as did the University of Chicago, whose observatory incorporated the 18½-inch Clark lens acquired by the Chicago Astronomical Society. Almost all of these institutions wanted a Clark telescope, be it the Swarthmore College, Pennsylvania's 6-inch, or the US Naval Observatory's 26-inch. By the 1860s Alvan Clark and Sons were telescope makers to the world, with splendid instruments owned and commissioned by British figures and institutions, such as the Grand Amateur double star observer William Rutter Dawes and the Jesuit Stonyhurst College Observatory, for a Clark refractor became a byword for optical excellence.[4]

Percival Lowell, the "canals" of Mars, and Flagstaff, Arizona, in the west

One suspects that Percival Lowell is a more interesting person to read about than he was to work with – a scion of the old Boston aristocracy, very rich, single-minded, and with all the self-assurance and arrogance of a French "aristo" of the *ancien régime*. Graduating from Harvard in 1876, his first love was Oriental studies, but his attention turned to astronomy, following Giovanni Schiaparelli's announcement of the existence of *canali* (channels or pathways, but not necessarily engineered) on Mars at the Red Planet's close encounter with the earth, in 1877 (as we saw in Chapter 11). Becoming convinced that these *canali* were an indication of life on Mars, Lowell resolved to throw his ample resources behind proving that intelligent life existed on the Red Planet.

Studying fine detail on Mars, however, was inevitably a periodic pursuit, for the planet's very elliptical orbit rarely brings the 4,000-mile-diameter Mars close to the earth, although another relatively close approach was due to take place in 1894. Lowell was willing to partly fund a Harvard observing station to see Mars under the pure skies of Arizona for this approach, although his personal management style was far from popular with the Harvard team. So like the millionaire and American Grand Amateur astronomer he had become, Lowell decided to cut the committee work and create his very own Arizona desert observatory. In conjunction with the USA Survey, a site was eventually selected at Flagstaff Bluffs, Arizona, and there, after the Mars opposition of 1894, Lowell built the great observatory that would immortalize his name. It was sumptuously equipped. An order for a superb 24-inch-diameter lens was placed with the Clark telescope company. Lowell had become friends with Alvan Graham Clark, Alvan's son, who, along with his brother George Bassett Clark, now ran the business. Lowell and Alvan Graham had visited European observatories together, and both had inspected the future observatory site at Flagstaff, Arizona.

The way that Americans went in for refractor astronomy may have struck the reader, although in 1909 Lowell did commission a 42-inch-aperture Newtonian reflector, made by Carl Lundin for the Alvan Clark firm. It was the second big-aperture reflector in the USA by that date. The first, George Ellery Hale's 60-inch on Mount Wilson, California, dated from 1908: a giant telescope funded by the industrialist and philanthropist Andrew Carnegie. We will return to the American reflecting telescopes in the next section.

Lowell and his employed staff at Flagstaff were primarily concerned with the astronomy of Mars, and then the other solar system bodies. He did not approve of sceptical comments, as we have seen, that cast doubt on the physical reality of the "canals", and guilty staff members were sacked. Even so, his superbly equipped private autocratic institution did make discoveries, in the fields of planetary atmospheres and asteroid astronomy. Flagstaff also paved the way to two other significant discoveries. One came from the relatively young astronomer, Vesto Slipher, who used the spectroscope to make fundamental discoveries, most notably of the radial motions and velocities of the spiral nebulae, which would help shape twentieth-century cosmology and astrophysics. He also used it to measure the rotation period of Uranus, at 10.8 hours. The second significant discovery came from the post-Lowell Flagstaff employment of the young amateur astronomer Clyde Tombaugh, who would take photographs of a dim, slowly moving distant speck of light which in 1930 would grab the attention of the world: for Tombaugh had discovered the planet Pluto, at the very edge of the solar system.

Percival Lowell died in 1916, leaving few other people convinced of the existence of the Martian canals. Yet he had used his private fortune to create an enduring world-class scientific research institute.

America's two giant refractors: the Lick and Yerkes Observatories

By the 1840s, glassmaking technology and the development of large precision machines to figure and polish lenses were advancing so rapidly that using a term like "giant" with regard to a telescope must inevitably be time-specific. By the early 1870s, the Clark firm had produced yet another large lens that was destined to make history: the great 26-inch-diameter lens for the United States Naval Observatory, Washington DC. It was a relatively rare commission insofar as it derived from a government contract; the great majority of Clark commissions for large research telescopes came from private individuals or foundations.

A true and uncontroversial discovery in planetary astronomy was made at the close opposition of Mars to the earth in 1877 (the same one at which Schiaparelli in Italy had announced his Martian *canali*). Asaph Hall, working with the US Naval Observatory's 26-inch Clark refractor, discovered that Mars possessed a pair of small satellites: providing yet another demonstration of the superlative quality of Clark's optics. The satellites were christened, in accordance with the classical tradition of astronomy, Phobos and Deimos: "Fear" and "Dread", after the sons of Mars (Ares in Greek) who drove their father's war-chariot in Greek mythology.

Then in 1876 a project was begun, using a $700,000 donation from James Lick, a wealthy businessman and philanthropist, to build a true mountaintop observatory containing a capital instrument. When it first came into commission in January 1888, the Lick Observatory boasted the largest refracting telescope in the world, with its splendid 36-inch-aperture object glass, inevitably from the Clark company. Technically speaking, it was the product of two firms that frequently worked together. Alvan Graham Clark figured and perfected the French-cast slabs of glass to form the exquisite 36-inch lens, while the optical engineers Worcester Reed Warner and Ambrose Swasey constructed the tube and precision equatorial mounting structures. The great instrument was assembled and brought into use atop Mount Hamilton, California, under the wider control of the Regents of the University of California. Edward Singleton Holden was its first Director.

Holden however, like Lowell, was not an easy man to work with, and it was his assistants, Sherburne Wesley Burnham, James E. Keeler, and Edward E. Barnard, who did the outstanding work at Lick. Optically, the 36-inch refractor was especially suitable for planetary studies, and its enormous power, combined with the transparent, calm air of Mount Hamilton, led to a number of discoveries, including that of Jupiter's fifth satellite. The 36-inch, along with an array of lesser but specialist telescopes, spectroscopes, and astronomical cameras, led to the Lick Observatory becoming the world's first regularly occupied mountaintop astronomical institution, and discoveries would flow from it in the twentieth century.

21.2 The Lick Observatory 36-inch equatorial, California. (E. Dunkin, *The Midnight Sky* (1891), p. 381. Author's collection.)

In 1897, the world's largest practicable refracting telescope came into commission, at the Williams Bay Observatory, Wisconsin. This was the 40-inch-aperture brainchild of the spectroscopist George Ellery Hale, backed financially by the businessman Charles T. Yerkes, and the handiwork of Alvan Clark and Sons. In it, the refracting telescope reached its apotheosis, for when a 49.2-inch lens was exhibited in Paris in 1900, it did not turn out to be a success. Glass, being a semi-viscous solid, has a tendency to distort under its own weight. So 40 inches would turn out to be as far as one could go with a glass disc supported solely around its edge. In the following section, we will look at the next step forward when it came to making even bigger optical surfaces: very large mirror or reflecting telescopes, where the glass slab could be supported at its back as well as around its edges. George Ellery Hale would play a pioneering role here.

Let us conclude by mentioning another significant innovation at the Yerkes Observatory: the development of a scenario first worked out by the Hugginses and the Drapers at their London and New York private observatories respectively during the 1860s and 1870s. It was the concept of the observatory not simply as a place for telescopes, but also as a research centre incorporating chemical laboratories, photographic darkrooms, and experimental physics-measuring facilities, all essential for the new astrophysical research. This vision was also being perfected at the Potsdam Astrophysical Observatory, Germany, and to some degree at the Vatican and Stonyhurst College, England, Jesuit Observatories. As we saw in Chapter 17, the laboratory preparation, incandescence, and analysis of pure chemical reagents had become an essential component of the spectroscopic analysis of the sun and stars.

AMERICA'S GIANT REFLECTING TELESCOPES

It was shown above that during the eighteenth and nineteenth centuries, big reflecting telescopes were very much the preserve of the Grand Amateurs, with figures such as Sir William Herschel, Lord Rosse, and James Nasmyth building and using their own often idiosyncratic instruments. All were privately owned and used, and the only truly large public reflector was the Great Melbourne, Australia, reflector of 1867, with its 48-inch-diameter mirror and stone and cast-iron engineered equatorial mount, which was not a success. Early big reflectors tended to be quirky in their behaviour, as temperature variations, slight changes in mirror shape and sagging at different elevations altered the delicate optical geometry of their yards long alignments, demanding a user who knew an individual instrument's idiosyncrasies, and exactly how to adjust things to maintain optimum efficiency. For the night shift basis of most public or academic observatories, this was a nuisance, and a *predictable* instrument was a prerequisite, so that it did not matter whether Johannes, Wilhelm, or Otto were the duty observer on any given evening. For this, between1800 and 1890, the equatorial refractor was the undisputed king. These constraints also explain why 95 per cent of the astronomy undertaken at public or academic observatories, such as Greenwich, Pulkowa, Dorpat, Paris, or Harvard, was either of a planetary surface or a mathematical gravitational astrometric kind rather than deep-space nebular astronomy.

By 1900, however, things were changing. The silver-on-glass mirror telescope, optically tried and tested in a range of large Grand Amateur instruments, had rendered temperamental speculum long obsolete, while advances in engineering technology had transformed both the mirror-polishing machines and the big, precision mounts upon which a 36, a 60, or even a 100-inch-aperture instrument might be placed. So by the time that G. E. Hale was planning a large-aperture telescope to set atop a Californian mountain soon after 1900, the big reflector had become much more predictable in its behaviour. Reflectors had an important advantage over refractors for the new and unfolding astrophysical researches of the twentieth century. A 60-inch-diameter mirror could harvest more starlight than even a 40-inch refractor, and it was easier to design out its optical aberrations. As no light passes *through* the mirror – unlike a refractor lens – chromatic aberration (caused by passing through glass) was not a problem: a boon for spectroscopic and astrophysical research.

Improvements in glass-casting technology by 1900 meant that large reflecting optical surfaces could be figured to perfection, and complex structures could be made to evenly support the mirror from the back, thus avoiding distortion.

Cutting-edge technologies can give opportunities to all sorts of people. George Willis Ritchey, for example, was a University of Cincinnati graduate and keen amateur astronomical and stellar photographer with a genius for optics, who had also worked as a furniture maker. Among his many achievements, he had, in conjunction with the Frenchman Henri Chrétien, devised a new configuration for the old Cassegrain reflector, in which improved images were obtained by means of mirrors with exactly complementary hyperbolic curved reflecting surfaces. Originally working at Yerkes, Ritchey began, with G. E. Hale, to plan the optics and mechanics for a 60-inch-aperture reflector to be set up in the clear air of Mount Wilson, California. Ritchey designed all the necessary optical machinery and figured and polished the 60-inch glass blank (from the St Gobain factory in France) into the mirror in the Pasadena workshops. Paid for partly by the Hale family and partly by the Carnegie Foundation, as we have seen, the 60-inch mirror was transported to the top of Mount Wilson to the waiting telescope tube and buildings, seeing first light in December 1908.

The 60-inch, while the biggest telescope in the world in 1908, was not alone on the Mount Wilson site, as first the Snow Telescope (after its original Yerkes donor, Helen Snow of Chicago), and then the 60-foot Solar Tower Telescope, of 1905 and 1908 respectively, had preceded it. Both were specialist reflecting telescopes of 60 feet focal length, and provided the most advanced solar spectroscopic and spectrographic instrumentation in the world at that time. Solar spectroscopy was one of G. E. Hale's favourite fields of astronomical research.

Yet no sooner had the 60-inch come into full commission, and begun to produce a stupendous harvest of new data about the heavens, than he was casting his eyes to an even greater telescope. This was achieved in 1917, when Hale, with Ritchey, the St Gobain glassworks, the Carnegie Foundation, and a new munificent donor, John D. Hooker, who contributed $45,000 to the project, brought the second Mount Wilson giant into use. With its magnificent 100-inch-diameter mirror and massive precision engineered equatorial mount, the Hooker telescope would be the biggest in the world between 1917 and the construction of the 200-inch on Mount Palomar in 1949. This instrument would facilitate some of the most important, perspective-changing ground-based astronomical discoveries of the twentieth century, as we shall see in the next chapter.[5]

Conclusion

It is interesting to note how by 1910, the USA was rapidly moving into the forefront of astronomical discovery. North America's geographical location, as we mentioned at the start of this chapter, was important, combined with its transparent mountaintop and great plains skies. These would become especially attractive, as two World Wars were to do so much damage to the cultural stability and progress of the Old World. The appalling death tolls in battle and the persecution of Jewish and anti-Nazi Christian and other scientists and intellectuals resulted in profound disruption, especially in Germany and countries under Nazi occupation. Many fled to Britain and, in greater numbers, to the USA, most notably Albert Einstein, whom we shall meet in Chapter 23.

As the American economy began to expand, especially up to the Great Crash of 1929, and once again following Roosevelt's New Deal, far more resources were ploughed into all sorts of scientific research than in the shattered and partially bankrupted Old World. As we have seen running through history, astronomy is a high-tech science, depending on abundant spare cash and highly trained specialists for its advancement. Whether these resources came from British Grand Amateurs or American industrial tycoons, the essential dynamic was the same. Yet what is so striking about American astronomy, from William Cranch Bond at Harvard in the 1840s to John D. Hooker at Mount Wilson in 1917, and on to subsequent modern-day benefactors, is its privately funded character: a feature which stood and still stands in contrast to the practices of Russia, Germany, and France. One might suggest that the British "Grand Amateur" tradition never really died out: it simply crossed the Atlantic and re-established itself in a modified form beneath the skies of Massachusetts, Wisconsin, and California.

Chapter 22

On the Eve of the Watershed: Astronomy and Cosmology c. 1890–1920

Astronomy had progressed faster, and perhaps more unexpectedly, during the nineteenth century than through any other since the ancient Chinese and Babylonians first began to systematize the motions of the starry firmament. While key discoveries and realizations had been made by Copernicus, Tycho Brahe, Galileo, Christiaan Huygens, Robert Hooke, and Sir Isaac Newton since 1500, it was the sheer speed and cross-disciplinary character of developments that made the nineteenth century so impressive. As we have seen, these included the fruition of Sir William Herschel's big telescope cosmology, early solar magnetism and physics, spectroscopy, nebular studies, binary star systems, lunar and planetary science, the chemistry of comets – shown by 1890 to contain hydrogen, carbon, and other elements[1] – and the first glimmering of the idea that the universe might be evolving over vast aeons of time.

Yet physical knowledge and astronomical discovery truly escalated between *c.* 1890 and 1920, as we will see in the next few chapters, and while no one knew exactly where things were going, it was clear that radically new cosmological perspectives were unfolding.

Two Victorian technologies lie at the heart of that cascade of discoveries which would lead to the twentieth-century watershed: spectroscopy and photography. After 1859, they would change the way in which we would think about the stars, space, time, and infinity. The recording medium provided by photography would also enable astronomers and physicists to store vast archives of new and often puzzling data, from which profound and novel insights would be subsequently drawn. The Mount Wilson, Lick, and Harvard Observatories in the USA had built up a vast and growing archive of spectra photographs by 1920, while in Europe the Potsdam Astrophysical Observatory, under Hermann Carl Vogel and his successors, would do the same. And the growing archive of photographs of the nebulae – especially of the spirals – from the Californian and South American mountaintop observatories would yield a rich harvest of new cosmological data by the 1920s.

The universe: a steady, stately place?

As we saw in Chapters 6 and 10, Sir William and Sir John Herschel, before and up to about 1850, had developed the idea that stars, nebulae, and flocky matter in space could perhaps be in a state of recycling or remodelling. Did stars form from condensing, compressing flocky stuff, which somehow combusted and began to emit light? Did stars attract each other into clusters which then became so dense that they shattered to smithereens to produce nebulous matter which would set the process going once again to form new stars? For the Herschels, and all those who accepted their overall cosmology, the universe in which all this took place was fixed and "stablished" in space and time – like a city constantly changing and reinventing itself through new buildings, yet still occupying the same site and possessing the same identity. Everything was going on within one great galactic system: the Milky Way.

But what if the universe did *not* possess a uniform time-frame topography? What if seemingly adjacent parts were in reality wholly disparate and scattered, with different laws applying to different parts, with a year in one place being only a week in another? All of these possibilities were beginning to emerge by the first couple of decades of the twentieth century, as space was becoming a truly strange and mind-bending place. For example, why was space so bitterly cold, while at the same time being populated by incalculably large numbers of intensely hot stars? And why, if the sun-like star system possibly stretched out to infinity throughout the breadth and depth of space, did the sky ever go dark at night? Surely, should not the celestial canopy, being composed of infinite myriads of stars, fill every possible spot in the night sky with a blazing light source that made the heavens shine even at sunless midnight? This "paradox" was first suggested by Thomas Digges in the 1570s, then

Edmond Halley,[2] and, most notably in the early nineteenth century, by the German physician-astronomer Heinrich Olbers.

These cosmological possibilities did not result from abstract speculation or imaginative theorizing: on the contrary, they were slowly emerging from observations and new physical discoveries, combined with their possible interpretations. Take light and motion, for example. Philosophers and scientists from classical Greece onwards had known that light must need a "medium" of conveyance, otherwise, how could it travel from the sun to illuminate the world and enter our eyes? Was this medium not the *ether*: a sort of super-subtle, intangible agency that even suffused the assumed vacuum of deep space? The great mid-nineteenth-century advances in electrical and thermodynamic physics seemed to support that way of thinking. James Clerk Maxwell's equations demonstrating the interconnectedness of light, electricity, and magnetism and William Thompson's (Lord Kelvin's) work on the nature of heat and thermal radiation effectively predicated the existence of a medium of conveyance for energy across space.

Did the medium retard the velocity of or deflect the light passing through it, as the Parisian experiments of Augustin-Jean Fresnel and Hippolyte Fizeau in Paris suggested during the 1850s? By the 1870s, most physicists in Europe and the USA accepted the intellectual necessity for a "luminiferous ether", although opinions differed regarding its assumed characteristics, especially as it even appeared to be able to occupy sealed, airless, evacuated glass vessels. Otherwise, how could light pass through an evacuated glass vessel in the laboratory, in a way that sound could not? Was this "luminiferous ether" stationary, just allowing energy waves to ripple through it, or did it move in an "ether stream"? These questions led to one of the classic experiments in late-nineteenth-century physics.

The Michelson–Morley Experiment, 1887

Albert Abraham Michelson had first attempted to detect and measure the ether in 1881. His physical logic was impeccable. If the luminiferous ether moved in a stream, then it would offer more resistance to a light beam moving against it, or moving "upstream",

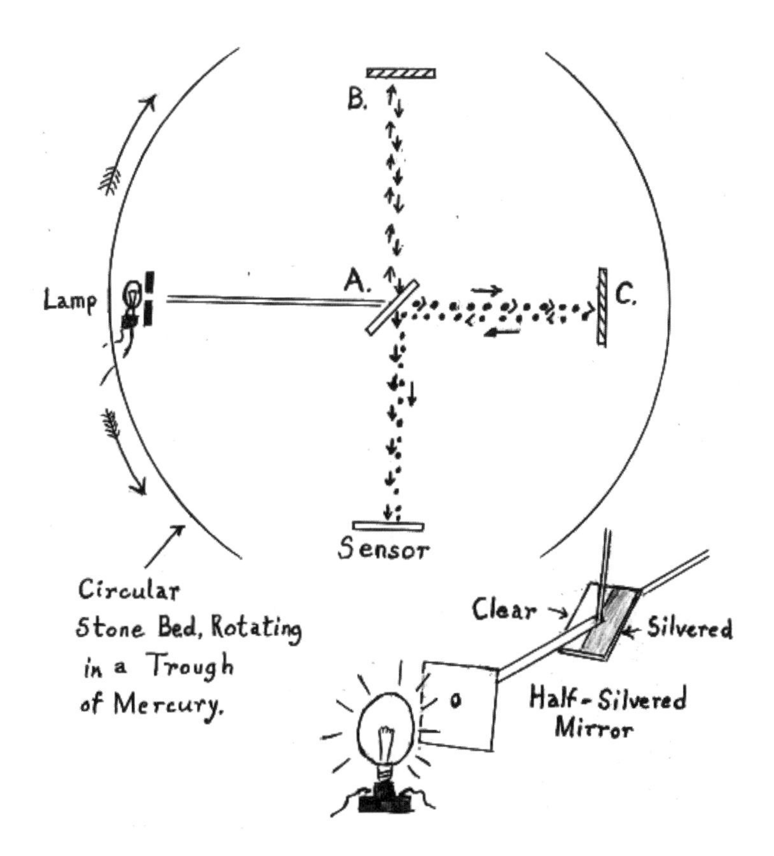

22.1 The Michelson–Morley Experiment, 1887. A single beam of light is split with a half-silvered mirror "A". Half of the beam (its path indicated by arrows) is reflected at 90° to a flat mirror "B", which redirects it back to the transparent half of the half-silvered mirror. The other half of the beam travels in a straight line through the clear half of mirror "A" (its path indicated by dots) to the flat mirror "C", which sends it back to the half-silvered mirror "A". This mirror recombines the divided beam, and brings it to the eyepiece or sensor screen. As the whole apparatus rests upon a circular stone base floating in a bath of mercury, it can be rotated with perfect smoothness. If indeed there was an "ether stream", then the two 90° light beams would have passed through it at precise and predictable intervals apart, sometimes with the stream, sometimes against it (causing a resistance or delay to the light) as the whole was rotated. This would have resulted in asynchronous interference patterns, as the direct and the reflected light rays reached the observing apparatus at slightly different times. Yet no such interferences were found, suggesting that there was no ether "stream". (Drawing by Allan Chapman.)

than it would to something moving "downstream", and even "across-stream" at 90 degrees, as one would find with a stream of air or water. In 1881, Michelson devised an apparatus which would split a beam of light and send it, by means of reflecting mirrors, both up and down the assumed ether stream, and at the same time across it at an angle of 90 degrees. His theory was that when the two sets of light waves were brought into a common eyepiece, the observer would see interference patterns in the light. This would be caused by some of the waves travelling slightly further than others, and arriving at the eyepiece out of synchronicity. His 1881 experiment was not conclusive, however, and in 1887 it was repeated with a larger and more sophisticated apparatus.

Michelson performed this second experiment in collaboration with his colleague Edward Morley in the stone basement of a dormitory at the Case School of Applied Science, Western Reserve University, Cleveland, Ohio. The essential principle of the new apparatus remained the same. But the new apparatus was not only bigger than that of 1881, it was also more stable, being mounted on a great stone slab which floated in a tank of mercury. The purpose of the tank was to enable the whole apparatus to be rotated with perfect smoothness. If there was a luminiferous stream of ether, then the two axes of the apparatus would turn into and out of it alternately, as the whole rotated. Michelson and Morley found no significant phase change in the light as everything rotated.

So was there no ether stream, and more significantly, was there perhaps no ether at all? In its failure to detect an ether stream, the Michelson–Morley experiment of 1887 has been described as the greatest failed experiment in scientific history. Yet that failure was to lead to a momentous rethinking in physics, and, by extension, cosmology, for early-twentieth-century scientists were faced with the prospect of a universe that, from observation, was bursting with energy – light, magnetism, and various forms of radiation – yet there seemed to be no highway along which it could travel. This, in turn, would demand a new and fundamental rethink about the nature and operation of energy, both within the laboratory and between the stars.[3] And in its implication that light had a finite velocity, irrespective of its direction of travel, the experiment would play a significant role in Albert Einstein's Special Theory of Relativity of 1905, which dealt, among other things, with the absolute velocity of light. The old certainties, it seemed, were beginning to crack, as a new physics and a new cosmology began to emerge between 1890 and 1920.

"Twinkle, twinkle, little star; now we know just what you are": The birth, life, and death of stars

In 1806, Jane Taylor wrote "The Star", a poem destined to become one of the most popular and enduring of all children's nursery rhymes: "Twinkle, twinkle, little star, how I wonder what you are." In 1806, not only Jane Taylor, but everyone else wondered what the stars really were, although most astronomers agreed that they were probably very distant suns. In 1806, the astronomical spectroscope still lay half a century ahead, although by 1814 Thomas Young and Josef Fraunhofer had independently discovered that hundreds of black lines crossed the solar spectrum, as we saw in Chapter 9, though no one at that date knew what they indicated. But exactly one century after Jane Taylor composed "The Star", not only were the chemistry and physics of stars being rapidly elucidated, and their remarkable diversities recorded and analysed, but their life cycles were gradually being understood. As we have seen in previous chapters, the Herschels, Lord Rosse, and others had largely surmised that stars might have life cycles, but by 1905–13, a series of physical discoveries and mathematical computations demonstrated these cycles really existed.

Ejnar Hertzsprung was a 30-year-old Dane who had worked at the Potsdam Astrophysical Observatory, and who would be promoted to Director of the Leiden Observatory in 1919. At the same time, Henry Norris Russell was working at Princeton, USA. Both were working with the vast photographic archive of stellar spectra which was coming to be available by *c.* 1910. Along with Hans Rosenberg at Tübingen, Hertzsprung and Russell came to realize between 1910 and 1913 that a new and groundbreaking astronomical discovery was at hand. Since Huggins and Secchi back in the 1860s, the spectroscope had supplied a key to working out the chemistry of individual stars, and through Secchi's recognition of different star types, a spectroscopic scheme of star taxonomy was already well established. Yet this scheme was static, insofar as it still told us nothing about the way in which stars might change, or evolve, to give us a demography and life cycle of the heavens.

In her work on the Henry Draper catalogue of stellar spectra at the Harvard Observatory, Antonia Maury had classified certain types of stars by the width of their spectral lines.[4] Hertzsprung in Europe next noted how stars with narrow spectral lines within a given spectral group tended to have larger proper motions than similar stars with wider spectral lines. As stellar proper motions are the product of a line-of-sight effect as the solar system moves in space, making nearby stars change their apparent position more than distant stars, one might conclude that the stars exhibiting big proper motions were closer to us than those showing small motions. To grasp the idea of proper motions in simple terms, imagine walking through a forest: the trees closest to you seem to shift position more than the distant trees do. So just imagine stars in the Milky Way with big proper motions as similarly placed in space to the trees closest to you in a forest walk. Likewise, when you look out from a railway carriage window, trackside hedges and buildings appear to move past much faster than does a distant hill.

By 1910, the possibility was emerging of creating a linear scale of star distances in the galaxy so as to place stars with relation to the sun. Some star distances had been known back to Bessel, Struve, and Henderson in the late 1830s, their distances established by means of meticulous trigonometric parallax measurements. But the geometry of parallax measurements was inevitably limited to the relatively local stars of only a few light years distance, as beyond that the parallax angles simply became too small to detect by conventional micrometric instruments. The parallax method was rather like a land surveyor measuring the exact distances of objects in one's local neighbourhood in England with reference to one's front door. Such a local survey would never tell you how far away Sydney, Australia, might be. On the other hand, if the spectral type of individual stars could be established, it might be possible to compute "spectroscopic parallaxes". All that one needed to do, in theory, was use a star's proper motion to place it on a purely relative distance scale, such as A is nearer than B, and C is further away than B. If the correct distance of a nearby spectral star could then be measured, and a yardstick established, it could be used to derive the purely relative distances of the other stars by means of their respective proportions.

There was still another snag: what was a star's *absolute* brightness? A simple pocket torch can produce blinding light if shone into the eye at the distance of a foot, while the beam of a powerful searchlight seen at 30 miles from a high mountain top will be no more than a glimmer, although there is a profound difference of absolute luminosity between them. If, on the other hand, astronomers could devise a way of measuring a star's absolute or intrinsic brightness, as opposed to its mere apparent brightness, as it looks to us in the sky, then we might be able to create a scale by which we could arrange the stars in their correct order with relationship to the earth or solar system. Sir John Herschel had tried to do this for local stars with his experimental astrometer in 1836, as mentioned in Chapter 10.

In the early twentieth century, astrophysical discoveries, in particular those relating to the spectral types, or classes, of stars made this possible. By the 1910s, therefore, it would become feasible to calibrate stars by their respective absolute or intrinsic brightness, or magnitudes. This would enable us, analogically speaking, to place both the pocket torch and the searchlight at the same distance from the observer's eye – let us say at half a mile – in a totally dark environment. Now we would see the lights in their true brightness: the torch a mere glimmer, the searchlight a powerful beam. Once we could do this with the stars, then we could begin to put yardsticks out into the Milky Way to establish respective stellar distances. As astrophysical techniques became increasingly refined and values established, a star's absolute magnitude came to be defined as the brightness which it would display when at a distance of 10 *parsecs*, or 32.6 light years, from the earth. And once astronomers could arrange the stars by their true, absolute distances from the sun, in three-dimensional space, we could next begin to fathom what made them twinkle.

Ever since Huggins, Secchi, and Draper, the new astronomical science of astrophysics had involved as much work in the laboratory as in the observatory, for in the lab it was possible to discover how the chemical elements of the newly discovered Periodic Table behaved under different controlled conditions. While we could never go to a star and test its chemistry and physics on the spot, we could, by using the rich package of data exhibited in the star's spectrum, tie up the laboratory behaviour of a specific chemical with its stellar spectral behaviour, to unravel an individual star's physics and chemistry.

Lying at the heart of all this thinking was our ever-increasing knowledge of the uniformity of nature. For just as Newton's laws of gravitation apply throughout the known universe, so do the laws by which hydrogen, calcium, and all the other chemical elements behave. The spectroscope enabled astronomers to take Sirius, Vega, or any other star and analyse it in the laboratory. The tens of thousands of star spectra plates available by 1910, being analysed and elucidated by Williamina Fleming, Antonia Maury, Annie Jump Cannon, and their colleagues in the USA and Europe, provided a rich foundation of physical stellar data with which to grapple with the ancient mystery of starlight.

The vital key to our understanding of the star population distribution – or stellar demography – of the Milky Way would be supplied by those stars present in star clusters and star clouds, such as the Pleiades, the Hyades, and the Magellanic Clouds. In clusters and clouds, the stellar inhabitants were all, in cosmological terms, at more or less the same distance from the earth. This meant that a bright star, or one of high magnitude, was no nearer to our solar system than a dimmer, low magnitude companion. Now if one could discover and tabulate the spectral differences of these high and low magnitude stars all within the same cluster, then one might be able to use this information to elucidate the chemistry, physics, and relative distances of individual stars that did not appear in clusters.

In 1910, Hans Rosenberg constructed a diagram of the stars in the Pleiades cluster, in which he placed their Hydrogen "Balmer" and Calcium "K" spectral lines against the stars' magnitudes. As these lines were known to bear a clear relationship to temperature,

Rosenberg had found a way of calibrating a star's luminosity against its temperature. By 1910, therefore, astronomers were gradually moving toward devising a technique by which a star's apparent, or visual, magnitude could be used to establish its intrinsic or absolute magnitude, to give us some idea of its relative distance, and then to establish its basic chemistry and even its temperature range.

Yet what was the source of that prodigious energy that made the stars shine, or twinkle, in the first place? What sort of life cycle could a star go through during the millions of years of its existence? We will return to the source of the energy of the sun and stars in later chapters, but we must now turn to how Hertzsprung's and Russell's work enabled astronomers to understand the different types of stars which existed, and how a star might pass from birth, through life, and on to extinction. This brings us to Hertzsprung's and Russell's famous "Diagram".

The Hertzsprung–Russell Diagram, 1910–13

Since its first devising, the "H–R" Diagram has become one of the most famous diagrams in physics.

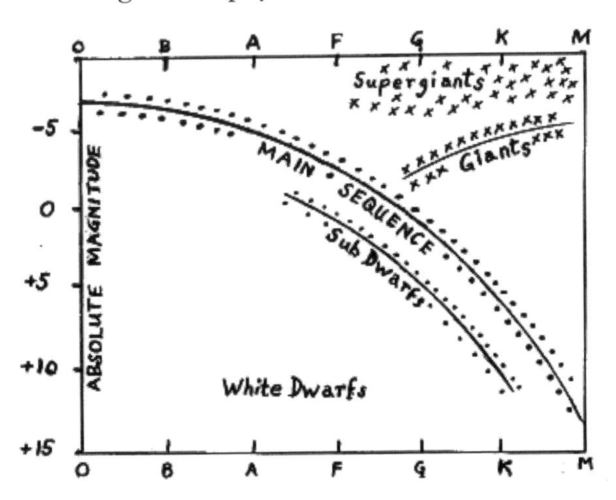

22.2 The Hertzsprung–Russell Diagram, showing the "Main Sequence", giant, and dwarf stars. (Drawing by Allan Chapman.)

It made it possible to define and illustrate the characteristics of all known star types, classified within a set of sequences and groups. The horizontal axis of the graph represents the spectral class of the stars, while the vertical axis shows the stars' absolute magnitudes.[5] From left to right along the horizontal axis, the star types are arranged into seven classes, O, B, A, F, G, K, M, with "O" stars being the hottest and "M" stars the coolest of the spectral classes. There is an old mnemonic which has been used for decades to remember the correct order, although it has been frowned upon in our more politically correct time: "O Be A Fine Girl, Kiss Me."

Running from the upper left and down to the lower right axes of the Diagram are the absolute magnitude and spectral classes of the great majority of stars in the night sky. This is called the Main Sequence, with surface temperatures descending, left to right, from hot to cool. Our own sun, which is classed as a middle-aged currently stable star, belongs to this Main Sequence category. On the other hand, some stars pass into a giant phase, being large but relatively cool, including well-known stars such as Aldebaran, Capella, Pollux, and Arcturus. Then down in the bottom left-hand corner are stars which are the opposite of the giants: these are the white dwarfs. These stars are intensely hot and gravitationally powerful, such as the tiny star rotating around Sirius (Sirius B, first seen by Alvan Clark in 1862, as we have seen). Many white dwarfs are smaller than the earth, yet have masses so dense that one cubic centimetre of their material weighs a metric tonne – hence their intense gravitational power.[6] White dwarfs are generally considered to have started life as hot white stars, then swelled, and aged in the "Main Sequence" of the Hertzsprung–Russell Diagram. Further aging, and nuclear fuel exhaustion, next led to their collapse into intensely dense white dwarfs. Like humans, stars can die in a variety of ways: some can just peacefully burn out, others swell, inflate, or explode as supernovae.

The ingenious thing about the Hertzsprung–Russell Diagram was that it provided an interpretative scheme within which to monitor and trace star types and star life cycles. Once it became possible to establish absolute magnitudes, in the hands of Harlow Shapley and others, and see these in relationship to spectral type and temperature type, then whole new and wonderful vistas for cosmology began to open up; based upon a vast and growing international archive of stellar spectral plates, interpreted in conjunction with new discoveries in laboratory physics, as ancient speculations became physical realities.

One crucial new discovery, between 1908 and 1912, provided yet another way by which an astronomer might extend a reliable measuring rod into space to understand not only the population of the Milky Way, but, after 1924, the nebulae as well. This involved a type of variable star known as a Cepheid, which was discovered by another of the ladies on the Harvard Observatory team.

Henrietta Swan Leavitt and the "Cepheid" stars

The first astronomers to observe and draw attention to that class of stars known as Cepheid variables were John Goodricke and Edward Pigott in York, between 1781 and 1786. As part of their wider study of stars that varied in brightness over time, they noticed that the star designated by the Greek letter δ (delta) in the constellation of Cepheus displayed a very regular cycle of visual luminosity over an exact period. There things stood for a good 120 years.

Henrietta Swan Leavitt was born in 1868, the daughter of a Massachusetts Congregationalist minister. Educated at Oberlin and Radcliffe Colleges, she first distinguished herself as a fine classical scholar, philosopher, and geometrician before taking a course in astronomy, which would change the future, and make her world-famous as an astronomer. Having obtained her BA, she lost her hearing and was deaf for the rest of her life.

Joining the Harvard team of lady spectrum plate measurers, first as a volunteer and then as an ill-paid professional, she began to study the stars of the smaller Magellanic Cloud, and she made a profound realization on plates taken at Harvard's Arequipa, Boyden Station, Observatory, Peru. From a painstaking study of the spectra of some 1,777 variable stars in the Cloud,[7] by 1908 she recognized a crucial fact: there was a direct relationship between an individual star's light variation cycle over time and the star's absolute brightness. As all these stars were at the same approximate distance, cosmologically speaking, from the earth, being in the same star cloud, it was clear that this period-luminosity relationship had nothing to do with a star's distance from the earth.

After adding more data, she published her discovery in a major paper in 1912, and this period-luminosity relationship would come to be referred to as Leavitt's Law. The crucial significance of her work, as she realized, lay in that if all the Cepheid variables were in a single region of the sky (the smaller Magellanic Cloud), then it would be possible to use their spectral and apparent and absolute magnitude data to calibrate them as potential yardsticks into space. This had a parallel to Hans Rosenberg's above-mentioned discovery of the Calcium "K" and Hydrogen "Balmer" lines of the Pleiades cluster stars, as a means of correlating stellar temperatures to spectral type through the laboratory of their known space location. From Henrietta Leavitt's work, Cepheid variable stars would become "standard candles", the known and defined characteristics of which would, within a few years, enable twentieth-century astronomers to plumb the deepest depths of space, to develop distance indicators not only for zones within our own Milky Way but even for nebulae beyond.

Henrietta Leavitt became head of the Photometric Department at the Harvard Observatory. Her entire life was blighted by bouts of illness, including the one which deprived her of her hearing while still a young woman, and she succumbed to cancer at the age of fifty-three. She was buried in the family plot in the Cambridge, Massachusetts cemetery. She never married, lived simply, and devoted the rest of her energies to her wider family, and to the Congregationalist Church of which she remained an active and devoted member throughout her life. Within only a few years of her death, however, Leavitt's Law would play a major interpretative role in those discoveries made in the Andromeda Nebula in and after 1924, which would lead in turn to a fundamental rescaling of the entire cosmos, and to a realization that the spiral nebulae were galaxies remote from our own Milky Way.[8] We shall return to these discoveries in Chapter 24.

Harlow Shapley, the spiral galaxies, and the Milky Way

One of the fascinating things about so many of the great astronomical discoverers is the unusual career paths which brought them to the work for which they became famous. William Herschel was a musician, Henry Draper a physician, Alvan Clark an artist, Henrietta Leavitt a classical scholar, and Williamina Fleming a domestic servant. As we shall see, Edwin Hubble began as an academic lawyer and a star athlete, while Harlow Shapley started life as a Missouri farm boy, before becoming a crime reporter for the Kansas *Daily Sun*, and then a world-famous astronomer.

By the 1910s, astronomers were accumulating photographic data at a remarkable rate, and puzzles and paradoxes were pouring out. Not only were there the spectral data discussed above, but there was that growing archive of photographs of the deep-sky nebulae, the most puzzling of which were the spiral and elliptical nebulae. Since Lord Rosse's first visual identification of the spiral structure of the Messier 51 nebula in 1845, increasingly powerful telescopes and photographic plates, most notably from the big Californian telescopes, had yielded a harvest of hundreds more.

At the Lick Observatory, James Edward Keeler had begun to study variable stars in the spiral nebulae early in the twentieth century, whereas Vesto Melvin Slipher undertook a different approach. In his spectroscopic study of spirals, Slipher noted that many displayed spectral lines that tended to be shifted toward the red end of the overall spectrum. In other words, they displayed "red shifts".[9] A smaller number displayed "blue shifts", or some shifting toward the spectrum's blue end.

As a physical phenomenon, red-shifting was explicable in terms of the Doppler Shift, named after Christiaan Doppler, who, as we have seen, noticed in 1842 that the pitch of an approaching sound increased as its source approached and dropped as it receded. This effect was produced by the approaching sound waves bunching up and the receding ones stretching out. As both sound and light were well-established wave phenomena, it was possible for physicists to apply the Doppler effect to explain spectral red shifts, as the light waves arriving from a receding galaxy were drawn out in a similar way to that of the sound waves of a receding train whistle.

By 1917, Slipher had most definitely discovered that the spirals had their own independent linear motions through space, receding or approaching, fast or slow, yet each one with its own individual movement. Yet exactly where were they moving? Were their motions contained within what was then seen as the Galaxy, or the Milky Way, or were they extra-galactic island universes in their own right, travelling at great speed in open space beyond the Milky Way? Whether there were whole galactic systems of stars *outside* our Milky Way would become the great question of the pre-"watershed" decades before 1924.

We saw above how Henrietta Leavitt's work on the Cepheids and her period-luminosity law opened up the possibility of calibrating the distances of individual Cepheids and the clusters or nebulae in which they might occur. But as things stood when she published her definitive paper in 1912, Leavitt's Law could only give the proportionate distances of the Cepheids, one to another, but could not supply absolute distances. As yet, no individual Cepheid distance was known precisely in light years by which an astronomer could calibrate the rest. A solution would come via Harlow Shapley, working primarily with the world's largest telescopes: first with the Mount Wilson 60-inch, and then, after its completion in 1917, with the great 100-inch Hooker reflector.

What physical mechanism was it that made Cepheids display their characteristic period-luminosity feature? Since Goodricke and Pigott in the 1780s, it had been generally assumed that they were eclipsing binary star systems: a dimmer companion to a brighter star passing in front and partially diminishing its brightness. Yet from gravitational physical and other reasons, Shapley abandoned the eclipsing binary model for Cepheids. Instead, he proposed a pulsating model, the individual star swelling and collapsing due to its own internal energy generation processes, and systematically varying its light output accordingly – a model largely substantiated by Sir Arthur Eddington in Cambridge, England, in his own work on the physics of stars.

As a part of his study relating to Cepheids, Shapley was struck by the apparent asymmetry whereby the globular clusters of stars (often rich in Cepheids) were not evenly distributed across the Milky Way, but tended to be bunched into particular zones: most notably in the region of Sagittarius, which is in the direction of the Milky Way galactic centre. This led Shapley to conclude that our solar system was not at or even near to the Milky Way galactic centre, but well out toward the edge. He was quite correct in this conclusion, which has been confirmed by a century of subsequent observations.

To take things further, however, Shapley needed a technique by which he could accurately calibrate Henrietta Leavitt's proportionate law, and express galactic stellar distances in terms of relatively exact light years. To this end, he recognized the interpretative significance of the new Hertzsprung–Russell Diagram. Among the new technologies being brought to bear on astronomy between 1890 and 1920 – the "pre-watershed" era – was statistical analysis. This was being used by Henrietta Leavitt and the Harvard ladies in their analyses of spectra plates, and by Hertzsprung, Russell, and others as they attempted to interpret the structure of the Milky Way galaxy and whatever might be beyond from the growing mountain of photographic data.

Using data recorded on the photographic plates first of the Mount Wilson 60-inch telescope, and then after 1917 the 100-inch reflecting telescope, Shapley devised a statistical technique by which Leavitt's Law of the period-luminosity of Cepheids could be used in conjunction with the Hertzsprung–Russell Diagram. From the spectral class and absolute magnitude axes of the H–R Diagram, Shapley worked out a way of using the known and measured period or graphically represented light curve of a Cepheid to reveal its true luminosity, or absolute magnitude, and from there, calculate the star's distance in light years. Once the technique could be made to work – hopefully – for some local Cepheids, then it could be extended to any Cepheid, irrespective of its distance, once its precise period-luminosity characteristics had been established. So all the cosmological proportions might, at last, be given some actual distance weightings, by which one could ascribe light-year dimensions to the Milky Way.

By 1920, Shapley had come to argue that all nebulae, clusters, Cepheids, and other phenomena were contained within our one single vast galaxy. He had even scaled it to an incredible 300,000 light years across, with our own sun and solar system some 50,000 light years from the centre. But he did later amend this figure to some 100,000 light years across, with our solar system some 30,000 out from the centre.[10]

A very important contributory factor to Shapley's scaling of this one, gigantic galaxy, however, had been the researches of his Mount Wilson friend and colleague, Adriaan Van Maanen, who, between 1916 and 1923, was using the 60- and 100-inch telescopes to study the potential distance relationships of the spiral nebulae. This brings us back to yet another motion which it was believed the spiral nebulae might display in addition to their motions along our line of sight. Since Lord Rosse's discovery of the spiral, or Catherine-wheel-like structure of M51, the Whirlpool nebula, astronomers had been studying the spirals in the hope of detecting a rotary motion, and by 1916, several hundred had been photographed. The logic behind their study was simple: if, within a given period of time, the stellar features or axial orientation of the nebula had changed, even by a precisely measured fraction of a single second of arc, then it must be relatively close to us, in cosmological terms (otherwise the angle would have been too tiny to detect). As with stellar parallaxes, a large angle of rotation would suggest that the spiral was closer to us than would a small angle, as the line-of-sight displacement would appear bigger close up: just as in the case of the positions of the trees in a forest walk, as we mentioned above with reference to the Cepheids.

In the 1920s, however, it was realized that Van Maanen had made a systematic error in his spiral measures – and that no rotations were detectable. His errors were so tiny, as to amount to a mere 0.002 of a millimetre on a large glass plate photograph of a nebula. So the obvious conclusion to be drawn from the now non-rotating spirals was that they were at such vast distances as to be beyond our own galaxy (an idea which would dovetail elegantly into Edwin Hubble's discoveries in the Andromeda M31 spiral after 1924, as we shall see in Chapter 24). The status and behaviour of the spirals and the Cepheids, along with other cosmological components, would all come together in the so-called "Great Debate" about what Sir William Herschel had styled the "Construction of the Heavens".

The Great Debate: Smithsonian Museum, Washington DC, 26 April 1920

Like Henrietta Leavitt, Heber Doust Curtis was a Greek and Latin linguist before he became fascinated by astronomy, obtained a PhD from the University of Virginia in 1902, and joined the staff of the Lick Observatory. He later became Director of the University of Pittsburgh's Allegheny Observatory. Curtis's own work on the radial motions and their red-shift velocities, along with studies of novae, or new stars, in the spirals, had led him to conclude that the spirals were extra-galactic, or outside the Milky Way star system. The meeting of the National Academy of Sciences at Washington's Smithsonian Institution in the spring of 1920 featured a discussion between Curtis and Shapley, each presenting his case for the "Construction of the Heavens": Shapley arguing for a "one big galaxy" cosmos, with Curtis presenting the case for island universes beyond our own galaxy.[11]

Of central significance to both men were features observed in the biggest of the spirals, the ever-fascinating nebula M31 in Andromeda. If M31 was not a part of our one big galaxy, Shapley argued, but well beyond in extra-galactic space, then it must be stupendously large. The nebula displays a total angular diameter of almost four degrees as seen from the earth, so if this angle were extended into deep space, it must be bigger than anything that the mind could conceive. If these supposed extra-galactic spiral nebulae were rotating, as Van Maanen's measurements then suggested was the case, their rotational velocities must exceed the speed of light, which would have violated the laws of physics. Similarly, a recently observed nova in M31 was so bright that, were the nebula in the depths of extra-galactic space, its sheer energy output would again challenge the basic laws of physics.

Curtis responded by pointing out the paradox of why M31 appeared to contain such a large number of recorded novae if it were a mere part of our own galaxy. Likewise, why did the dark dust bands in M31 (which had even been observed visually by William Cranch Bond and his son George Phillips with the Harvard 15-inch refractor) have parallels to the dark bands within our own galaxy? And to top it all, why did so many of the spirals display such enormous spectral red shifts, which strongly suggested that they were receding from us at stupendous speeds? Did not these evidences argue for the spirals, including M31, being independent, extra-galactic island universes flying into the very depths of visible space?

As things stood in 1920, it was largely a matter of push against shove between the two cosmological models. Both Shapley and Curtis presented convincing arguments based upon the evidence as it then stood. It is important to remember, though, that the legendary character of the Great Debate was in reality a *post factum* creation, built up in many ways on the subsequently published scientific and newspaper literature. At the Debate, Shapley was being cautious in how he presented his case, knowing that there were Harvard men in the audience who were scouting for a candidate for the university's prestigious Observatory Directorship: a post which Shapley coveted, and which he was offered in 1921.

Things were set to change radically as the post-1920 decade unfolded. By1920, that floodtide of new astronomical evidences, discoveries, puzzles, and paradoxes which had been welling up across the last thirty or so years had reached its high-water mark, and following further discoveries made by Edwin Hubble with the 100-inch telescope at Mount Wilson after 1924, the watershed would be crossed. Thereafter, a new cosmological schema would emerge which, in its essentials, would become the one within which we still work today.

Chapter 23

It's All Relative. The "Alice in Wonderland" World of Early Twentieth-Century Physics

In some respects, one can see Lewis Carroll's *Alice in Wonderland* (1865), *Alice Through the Looking Glass* (1871), and his subsequent *Alice* stories as prefiguring the shape of physics and cosmology to come. For in Carroll's brilliantly imaginative narratives, the normal laws of physical reality are suspended. A little girl shrinks, then extends, enters strange worlds, and encounters a mysterious Cheshire cat and a watch-carrying white rabbit. Whole time-frames seem out of synchronicity with normality, while space, reflections, and reality do weird things.

Writing mind-bending children's stories was not Carroll's day job, nor was he called Carroll. The Revd Charles Lutwidge Dodgson was an ordained Deacon, Oxford mathematics don and an official Student (Fellow, or Tutor) at Christ Church, Oxford. His more official publications were devoted to the higher branches of mathematical geometry, which very few of the fans of his *Alice* books would have made head or tail of, and were written under his real name. Even the pen-name that made him a globally famous author in a short period of time was a linguistic play upon his two Christian names: Lewis being an English variant of the German Lutwidge, and Carroll a Latinization (Carolus) of Charles. Oxford dons have always loved playing word-games, and they still do.

Ironically, it was from the 1860s onwards that new discoveries in the nature of energy, heat, light, and chemistry began to advance rapidly, especially as astronomers started to wrestle with how new thinking about physics, matter, and energy might be taken from the laboratory and applied to stars and to deep space. So in some ways, Charles Ludwidge Dodgson's *Alice* stories might be seen as heralding the shape of the physics to come. We saw in Chapter 22 how the Michelson–Morley experiment in the 1880s undercut and destroyed the plausibility of the cosmic ether, though in doing so it presented a genuine puzzle: how could light and energy cross the vastness of space without some kind of highway or conveyance? And by extension, what were light and energy in the first place?

The "physics quake" of the 1890s: X-rays, atoms, and radiation

Just as Henry Cavendish took the earth into the laboratory, using his sealed torsion balance to measure its density in 1797–98, so the building blocks of cosmology were similarly taken into the laboratory, especially in the 1890s. If, that is, one assumed cosmology's most fundamental building blocks to be atoms and the forces that held them together, upon which all things, from planets to galaxies, were built. The discoveries of the 1890s had an ancestry stretching back to Sir Humphry Davy's and Michael Faraday's electro-chemical researches of the early nineteenth century. What occurred when an electric current was passed through a chemical compound – such as water – thus breaking it down into yet smaller units?

Faraday (at the suggestion of William Whewell in Cambridge) had proposed the term ion or ions for those components of chemical compounds released during the electrolysis process. Then in the mid-century, William Thomson, later Lord Kelvin, was further researching the internal structure of chemical substances, as well as the physics of heat – to develop the science of thermodynamics. By the 1860s, the term molecule was coming to be used, especially by chemists, to denote the smallest possible component of an element capable of possessing a free existence and reacting chemically. These molecules were exceedingly tiny, for Lord Kelvin suggested the following scale of comparison. Imagine that a single drop of water were magnified to the size of the earth; then the size of a molecule would fall somewhere between the size of a small lead pellet (of the type fired from shotguns) and a cricket ball.[1] So what were the sizes of the atoms themselves?

The famous electrical equation of James Clerk Maxwell in the 1860s had opened up new lines of investigation into the nature of electric, magnetic, and electrically related energy, giving rise to a discovery in physics that would, in its practical dimension, exert a profound effect upon the twentieth century. These were the waves of electrically generated energy discovered in the 1880s by Heinrich Rudolf Hertz, which would make possible radio and television, and in the 1930s would open up a new radio window into the universe. But how did these Hertzian waves relate to the new, fast-developing realms of matter, energy, and space?

Physics was entering upon an Alice in Wonderland scenario, in which big, small, and invisible energies entered strange new realms of being, far removed from normal experience. What were atoms? How did energy come in a proliferation of forms, while at the same time remaining preserved in balance, in accordance with Lord Kelvin's thermodynamic laws? And what was *solid* in a world where matter had become a collection of forces of various kinds, and old-fashioned hard reality was an illusion?

All of this was only the prelude to even stranger things to come after 1890. The next wonder arrived in 1895, when Wilhelm Conrad Röntgen was experimenting with fluorescence and electrical discharges in evacuated and cathode ray tubes. The discharges emitted a new form of energy that would pass through the human body and fog photographic plates wrapped in sealed packages placed within closed drawers. The new radiation – which was only generated so long as electricity passed through Röntgen's apparatus – would not pass through coins, metals – or bones. Because of their unknown nature, they came to be styled X-rays. They would go on to transform medical diagnostic, and would later be found to be coming to earth from deep space, along with alpha, beta, gamma and other forms of radiation, thereby opening up new possibilities in our understanding of stars and galaxies.

X-rays, light, Hertzian radio and other forces from space all appeared to travel in wave forms. Could the wave form, therefore, be a universal characteristic of all types of energy, and what decided the form it would take – radio, X-ray or light – was the mathematical geometry of the wave, with radio, for example, having a much longer wavelength than light?

Captivated, like many other physicists, by Röntgen's discovery, the Frenchman Antoine Henri Becquerel, who had long been interested in fluorescence, began early in 1896 to work with the uranium salt potassium uranyl sulphate. He found it could cause sealed photographic plates to darken – just like X-rays – and realized something was being emitted by the uranium. He even discovered a mysterious burn appearing on his stomach, after carrying a vial of uranium salts around Paris in his waistcoat pocket: the first realization that radiation could alter natural human tissue growth. It was also discovered that uranium radiation possessed some kind of physical charge.

Also working in Paris in 1897 was a young Polish scientist, Marie Sklodowska Curie, and her husband and fellow-scientist Pierre Curie, who were investigating the curious properties of uranium salts. It was found that when uranium salts and the mineral earth pitchblende were placed near a new and highly sensitive electrometer which Pierre and his brother had built, the instrument responded to something being emitted by the substances. This was especially puzzling, for no one in the laboratory was doing anything to activate the uranium and pitchblende, such as bombarding them with X-rays. Yet with no external energy being applied, the electrometer responded to a constant stream of something being emitted by minerals. On the other hand, the reactive materials did not at that time appear to diminish as they radiated, although we now know that they do diminish, very slowly. Unlike other physical energy systems known in the 1890s, such as electric batteries, engines, or even hungry human bodies, the uranium salts and pitchblende never went flat and ceased to radiate. What first Becquerel, then the Curies, had discovered was radioactivity. Marie would go on to isolate the radioactive elements polonium (named in honour of her native Poland) and radium, both in 1898, and in 1903 would win one of the first Nobel Prizes for her work, which she shared with Pierre and Becquerel. Pierre Curie died in a road accident in 1906.[2]

Radiation and radioactivity were taking physics into a wholly new realm by 1900. If an element could radiate without apparently losing mass or substance, this posed big questions about the nature of atoms. Astronomically speaking, the new radioactive elements furnished a puzzle parallel to that of the sun and stars; what enabled them to radiate continuously through vast aeons of time without burning out? It would be Ernest Rutherford and Frederick Soddy who would develop an atomic model for radioactivity in 1908.

The mighty atom

The new physics of the atom was advancing rapidly by 1900, as physicists began to further experiment with cathode ray electrical discharge tubes and the strange fluorescent effects which they produced when fired against their metal foils. One of them was Joseph John Thomson. The son of a Manchester bookseller, he won scholarships first to Owen's College, Manchester, where he studied engineering, then to Trinity College, Cambridge, which provided the foundation for his subsequent achievements in physics. Working in Cambridge University's Cavendish Laboratory, Thomson discharged cathode rays at thin metal foils, and was puzzled by what seemed to come out behind the foils. This led him to conclude that cathode rays contained something smaller than what was then understood to be an atom – an electrical charge that was no more than a component of an atom, and which, after 1897, came to be christened an *electron*.

The next major development took place in Manchester during the following decade, largely at the hands of Ernest Rutherford and his colleagues. Born in 1871, Rutherford came from a rural New Zealand background, and after showing his youthful brilliance at home, won a scholarship to Trinity College, Cambridge. From Cambridge, he went on to McGill University, Montreal, Canada, which possessed excellent physics research facilities. In 1907, his truly international career brought him to the Chair at Manchester: a university which even at that date already had a world-class scientific reputation. Working with his colleagues Ernest Marsden and Hans Geiger (who in 1908 invented the Geiger Counter radiation measuring instrument), Rutherford was experimenting with radiation-generated alpha particles. Using thin metal foils coated with zinc sulphide as his targets, he fired alpha particles at them, noticing that when the high-energy particles hit the treated foil, the screen flashed or scintillated, emitting light.

But then, he noted, strange things began to happen. Some particles travelled straight, others were slightly deflected; yet approximately one in 8,000 was widely deflected, or even bounced back. Why? How could something as thin as metal foil deflect high-energy alpha particles? Rutherford expressed amazement at this phenomenon, saying "It was as incredible as if you fired a 15-inch shell at a piece of tissue paper and it came back at you."[3] Not only were atoms made up of tiny constituent parts, but some of these sub-atomic particles possessed truly bizarre physical properties.

By 1911–12, Rutherford had developed a new model for the atom, which was a universe removed from the old one of chemical-element-specific billiard balls upon which post-Lavoisier chemistry and physics had been built after *c.* 1780. The new Rutherford atom consisted essentially of empty space, in which a proton mass existed at the centre, with electron electrical charges hurtling around it. Over the following decades, more elementary particles would be discovered: most significantly, James Chadwick's discovery of the neutron in 1932. Either way, the atom still remained an empty space in which charged particles moved at incredible velocities.

In 1919, after moving to the Cavendish Laboratory, Cambridge, Rutherford, along with Chadwick, was able to put this model to a sort of *experimentum crucis* test. Nitrogen atoms were now bombarded with alpha particles, resulting in protons being thrown off to produce hydrogen. Quite literally, Rutherford and his colleagues had split the atom, re-engineering its particle structure using an alpha particle stream, to change one element into another. Rutherford would style this and his other work with Frederick Soddy on atomic structures *The Newer Alchemy* (1937). In a way, they had achieved the goal of the medieval alchemists, and performed a transmutation, albeit not quite lead into gold.

All of these discoveries from 1895 onwards, however, had opened up a myriad of new questions about the nature of mass, energy, radiation, and even substance. They would give birth to the new science of quantum physics, as Niels Bohr, Max Planck, and Erwin Schrödinger in particular began new theoretical investigations into the way energy behaved. Did it act in a stream, or in waves, pockets, or quanta? This became especially significant after Rutherford's orbiting or planetary model of the atom after 1911–12, and Niels Bohr began to compute precisely how an electron might change its orbit, and how radiation energy might be generated.[4]

All of the above experimental work, from Becquerel and the Curies up to Rutherford, Chadwick, and beyond, was conducted in the laboratory, not the astronomical observatory. Yet its astronomical implications were obvious, for once we had coherent models for how energy behaved under the controlled conditions of the physics lab, we might extend them toward answering the time-honoured questions of how the sun and stars burned, how energy crossed space without an ether highway, and how stars passed through the energy life cycles described in the Hertzsprung–Russell Diagram. All of this was happening incredibly quickly, with a mere 24 years between Röntgen's discovery of X-rays and Rutherford's splitting of the atom. Physics was moving into a realm of increasing strangeness. Yet while all this had been going on in the physics lab, equally strange things were going on in space, all within the same quarter-century after 1895.

Mercury, Vulcan, and the Problems of Gravity

Newton's theory of universal gravitation first presented to the world in 1687 stood, and still stands, as one of the great monuments to the human intellect. It could be used to account for most aspects of planetary motion and to unify various branches of physics into one mathematical synthesis. Yet one particular observed phenomenon continued to nag at astronomers: the advance of the perihelion of Mercury. The *perihelion* is the point at which a planet, in its elliptical orbit, makes its closest approach to the sun: *peri helion* in Greek meaning "around the sun". (A planet's orbital opposite to perihelion is *aphelion*, when it reaches its greatest distance from the sun.) Instead of this point moving in conformity with the path computed for it by gravitation theory, Mercury appeared to have an independent motion of its own, and this baffled the high-precision-measuring astronomers of the nineteenth century. The obvious solution, surely, must reside in the as yet undiscovered presence of a small planet rotating rapidly in an orbit between that of Mercury and the sun. Because of the intensely hot region within which it must be rotating, this theoretical planet came to be given the generic name Vulcan, from the Roman blacksmith of the gods.

In December 1859, a French provincial doctor and amateur astronomer, Edmond Modeste Lescarbault, believed that he had detected this tiny world crossing the solar disc. Urbain J. J. Le Verrier, Director of the Paris Observatory (who along with but entirely independently of John Couch Adams in Cambridge had discovered Neptune in 1845–46) wanted to learn more about what Lescarbault had seen. Le Verrier wanted to be convinced that Vulcan really existed and was the cause of Mercury's unpredictable orbital behaviour. But conclusive sightings of Vulcan, even by astronomers with much better telescopes than that of Dr Lescarbault, were not forthcoming, and as the Vulcan theory lost credibility, the puzzle remained.[5]

The patent clerk of Bern: Albert Einstein and relativity

Albert Einstein was born into a modest Jewish family of Ulm, Germany, in 1879, though later became a Swiss citizen. He had not been especially gifted at school, in Germany or in Italy, and after gaining his PhD from Zurich and marrying, he was unable to obtain a regular academic job. As a result, he took a modest clerical post working in the Bern, Switzerland, Patent Office. With only evenings and Sundays to devote to theoretical and mathematical physics, he embarked upon a series of studies and calculations which, first after 1905 and then after 1915–16, would entirely change the twentieth-century understanding of space, time, light, velocity, astronomy, and cosmology. This was his celebrated Theory of Relativity: a theory which respected, yet advanced far beyond, Newtonian gravitation. It would, among other things, account for the perihelion advance of Mercury without the need to invoke the non-existent Vulcan.

A wholesale cosmological rethink was beginning by 1900, with Newton's original laws of gravitation looking less universal than had been assumed since 1687. Could other, as yet undiscovered, forces be at work, such as the ones influencing the orbit of Mercury and the behaviour of light and energy across those vast tracts of space that were being opened up at the Lick, Mount Wilson, and other observatories, in the new physics? It was not that Newton's laws were wrong, so much as that they might well be limited, or local, in their application.

Einstein addressed these questions, and his solutions would pitch him into international celebrity before he was forty. Yet what was his Theory of Relativity? Leaving aside the sophisticated differential geometry and field equations that supplied demonstrations and proofs, Einstein's Theory was developed over two great publications, spanning ten years.

In 1905, while still in Bern, he published his Special Theory of Relativity, dealing primarily with light and velocity. The previous year, the Dutchman Hendrik Lorentz (the 1902 Nobel Physics Laureate) had published a paper showing that bodies moving near to the speed of light (186,000 miles per second) would shrink. As they shrank, they would correspondingly increase in mass. Back in 1889, the Irish physicist George Francis Fitzgerald had also concluded that an object would shrink as it approached the speed of light, although Fitzgerald's and Lorentz's work had been performed quite independently of each other. In deference to both men, however, this contraction would be named the Fitzgerald–Lorentz Transformation or Contraction.

This Transformation would constitute a major component of Einstein's Special Theory, as Einstein began to elucidate, from a mathematical perspective, the behaviour of bodies as they approached the speed of light. Einstein concluded that time was not an absolute in the Special Theory, as it was in classical physics. Not only did an object contract in the direction of its motion as it approached light velocity, but its time slowed down as well, and the mass of the body would increase correspondingly. Velocity, time, and mass, therefore, were shown to respond in relationship to each other, or to be *relative*. The one fixed and unchanging point in this new mathematical cosmology, however, was the speed of light, which always travelled at 186,000 miles per second irrespective of circumstances: the speed of light was an *absolute*.

Einstein's Special Theory drew together several anomalies emerging in classical physics, things which could not be adequately explained by post-Newtonian physics, and gave them coherence and context. For example, while in 1905 Einstein had been unaware of the results of the Michelson–Morley experiment, his Special Theory made possible a mathematical explanation of the behaviour of light in an ether-free universe. This was just one of the problems for which the theory could provide an explanation. The significance of his 1905 paper did not go unnoticed, and before long he had migrated from his Patent Office stool to a Professorship in the University of Bern. He would go on to occupy senior Chairs in Zurich, Prague, and Berlin, as he developed the physical implications of light, velocity, contraction, and mass. In 1911, for example, he predicted, mathematically, that light, while absolute in its velocity, was nonetheless susceptible to gravitational attraction. A ray of starlight just grazing the edge of the sun would be deflected by 0.83 arc seconds (later amended to 1.7 seconds). (We will return to starlight deflection in the next section.)

In 1915–16, Einstein published his General Theory of Relativity, which built upon the postulates of the Special Theory. Essentially, the great paper of 1915 in which he set out the General Theory dealt with the interrelationship of gravity, acceleration, space, and time, and in it he outlined a "spacetime continuum". This was a four-dimensional universe, where the usual coordinates of length, breadth, and height only made sense when seen in the context of time: hence "spacetime". At the heart of the Special Theory was the realization that mass, energy, and gravity possess the power to distort the space around them, thus giving rise to the Einsteinian concept of warped space and time. By way of analogy, think of how a powerful magnet might affect the path of a metal ball suspended from a string. Swing the ball near to the magnet, and the magnetic field warps the natural geometry of the swinging ball's orbit. Although gravity and mass are not the same as magnets, they both have the capacity to control the geometry of the space around them.

One consequence of this gravity-mass-space relationship is that Newton's Laws no longer apply in the presence of very strong gravitational fields. What is more, a strong gravity field (like near-speed-of-light acceleration) slows down time, making time relative rather than a constant. This gives rise to one of the paradoxes implicit within the Special Theory: if two people in two separate spaceships were flying through space at significantly different velocities, their time zones relative to each other would be different (so they would age at different rates, relative to each other), although each person would feel perfectly normal and their individual spaceship's on-board clocks would be running normally. Implicit within Einstein's Special Theory was his 1911 realization that a strong gravitational field would alter or deflect a ray of starlight, which would be dramatically confirmed during the total eclipse of the sun in 1919.

What Einstein's Special Theory also did was provide an explanation for the old problem of Mercury's orbit, based essentially upon strong gravity fields, and discounting any need for a Vulcan-like planetoid between Mercury and the sun. Although the Special Theory, dealing largely with properties of light and motion, won rapid acceptance, the General Theory was more problematic. For one thing, its expression required a new mathematical geometry, while many found the new Field Equations, treating as they did the curvature of spacetime owing to a gravitational source, hard to follow and to integrate into existing frames of reference.

Einstein's General Theory was essentially one which saw cosmological forces as operating in a *continuum*, or stream of energies, yet this was in itself hard to reconcile with new ideas of quantum physics, which saw energy in all its forms as acting through *quanta*, or separate packages or units, rather than through continuous streams. This problem has not gone away, in spite of the proposal of several models for a "Grand Unified Theory".

Over the century since Einstein's General Theory was published, physicists and cosmologists have discovered physical confirmations for many key components, including in laboratory experiments with cyclotrons. Einstein's two Theories have also facilitated insights into some of the main cosmological puzzles of *c.* 1910, such as how the stars burn. How do they remain stable, become unstable, burn out, or blow up? The super intense gravitational fields treated in Einstein's two Theories, and especially in the General Theory, provided the conceptual and mathematical tools whereby we might understand the physics of what would come to be called the Big Bang, along with the formation of the chemical elements in the early universe. They also enabled us to make physical sense of black holes, gravitational collapse, and neutron stars, and to predict the gravitational waves emanating from the Big Bang: gravitational waves since discovered and confirmed by observation. Without Einstein's Theories, the great richness of modern cosmology, including ideas of dark matter and dark energy, would not really have been possible.

But astronomy is an observational, and, wherever possible, an experimental science. Mathematical theories and clever conceptual models are all well and good, but to make them migrate from the realms of metaphysics to hard science, physical *proofs* are demanded. Einstein's brilliant yet still conjectural General Theory would score its first hit here, on the island of Principe, just off the coast of West Africa, on 29 May 1919.

SIR ARTHUR STANLEY EDDINGTON, EINSTEIN, AND THE SOLAR ECLIPSE OF 1919

Back in 1911, Einstein had calculated that gravity fields should have an effect upon beams of light, and the total eclipse of the sun of 1919 would provide the acid test. If, at the moment of totality, when the sun's light was entirely blotted out by the moon, the positions of stars close to the sun were photographed, their positions could be compared with the same stars' normal or uneclipsed positions. In the simplest terms, all that one needed to do was photograph at night, some weeks or months before, that region of the sky where the eclipse would take place, then photograph it again during the eclipse. The star positions on the two plates would then be carefully measured with micrometer microscopes and the results compared. Crucially, would the stars appearing very close to the darkened solar and lunar limbs appear slightly *closer* to the sun during eclipse, thus indicating that their light was being bent by the sun's powerful gravitational attraction?

Arthur Eddington (knighted 1930) had been convinced of the significance of Einstein's General Theory since soon after its first publication in 1915–16. But in 1916, Britain, Germany, and their respective allies were gridlocked into the bloodiest war in history, with horrendous casualties on both sides: World War I. Almost overnight, from August 1914 when war broke out, Germany suddenly flipped from being thought of as one of Britain's closest friends on the international and cultural fronts, to being its hated enemy. Long-standing English shops, whose owners had their German names above their doors, were vandalized, and even the once-popular German "Sausage" Dachshund pet dogs were kicked in the street.

Things were made no better in the public mind by the fact that Einstein was a pacifist and a Swiss national, and Eddington a Quaker pacifist or non-combatant, so that his advocacy of Einstein's discoveries almost savoured of treachery to many. Even at meetings of the Royal Astronomical Society it was reported that men got up and walked out when Eddington took the stand to talk about the General Theory of Relativity. Perhaps these men, it should be remembered, had lost sons and grandsons in the battles raging across northern France and Belgium. In 1916, moreover, Eddington was still of military age, being in his thirties, and although Plumian Professor and successor to Sir Robert Ball (died 1913) as Director of the Cambridge University Observatory, he still needed the protection and influence of Sir Frank Dyson, Astronomer Royal, to save him from being arrested and probably imprisoned as a pacifist.

When the war ended in November 1918, however, things began to return to a vague semblance of normality, though a shaky sort of normality with the political, demographic, economic, and social disruption of the war, plus the slaughter of the Russian Imperial family and the rise of Soviet communism. Even so, the eclipse expeditions went ahead. Two separate British expeditions were sent out, one to Sobral in Brazil, where eclipse totality would first become visible, and the other to Príncipe, off the west coast of Africa, over which the eclipse shadow would swing as it crossed the Atlantic. Eddington, assisted by Edwin Turner Cottingham, directed the Principe expedition, where the eclipse was photographed using "astrographic" cameras, telescopes, a portable darkroom, and a plate-measurer.

By 1919 Eddington had become so committed to Einstein's Theory that he felt he *must* detect the deflection of starlight as a clinching proof. So much so, that Dyson had joked with Cottingham beforehand that he should take along a straightjacket to Principe, just in case no deflection was seen, for then "Eddington will go mad and you will have to come home alone". Eddington was clearly in on the joke, however, for upon measuring the star positions on the freshly exposed and developed eclipse plates in Principe, he said "Cottingham, you won't have to go home alone."[6]

So a deflection of the star positions had been photographed and measured. The expedition's success and Eddington's claimed physical proof for a key part of Einstein's General Theory were immediately telegraphed home, and within hours the news was spinning around the world, to be published in newspapers and magazines. Some scientists, however, questioned the neat accuracy of Eddington's measurements for the light deflection, wondering how far his enthusiasm for the General Theory might, perhaps, have coloured his interpretation of the data. Yet this was on the level of detail and prediction: most astronomers accepted that General Relativity had passed its first observational test.[7] (We shall be meeting Sir Arthur Eddington again in Chapter 25).

ALBERT EINSTEIN THE AFFABLE CELEBRITY

In the wake of the 1919 eclipse observations, and with the Great War mercifully over at last, Einstein's name and Theory were soon familiar to people the world over wherever there was Western science. His relativity arguments utterly baffled all but a handful of specialists worldwide. This was a major part of the Theory of Relativity's fascination, and it, and Einstein himself, soon took on an almost mystical aura in the public imagination. One story goes that a journalist interviewing Eddington asked him who the mere handful of people who understood Relativity were, to which he is said to have responded that he, Eddington, understood Relativity, Einstein understood Relativity, but who was the *third* person who understood it?[8]

The Swiss–German physicist soon enchanted the world, not only because of his profound and far-seeing genius, but also by his personality. Whereas his great cosmological genius predecessor Sir Isaac Newton had often been reclusive and sharp-tongued, Einstein was approachable. In an age of visual humour, as was the early twentieth century, the eccentric tail-coated professor with his great mane of hair and often unkempt moustache was a gift to the cartoonists. By 1925, there was probably not a person within the reach of Western civilization who had not heard of Professor Einstein or seen a photograph, newsreel, or cartoon of him. No small part of his mystique was his status as an all-seeing super-genius who had taken the conventional certainties of life, such as time, space, light, and motion, and turned them topsy-turvy through a species of mathematics that only one person in a million could even begin to understand.

On 30 January 1933, Adolf Hitler became Chancellor of Germany, and a law was soon passed to exclude and remove Jews from all public offices, including university professorships. This was the thin edge of the wedge which within a few years would lead to the Holocaust. Germany, Austria, and the lands to be occupied by the Third Reich were especially well-blessed with great scientists, many of them Jewish, and those who could began to look for sanctuary in England and the USA. Many of the world's leading theoretical physicists of the age were of Jewish descent, which led the ideologues of the Reich to style physics "corrupt Jewish science", contrasting it with pure experimental disciplines such as chemistry, which it deemed to be pure "Aryan" science.

While legally speaking Einstein was a Swiss citizen, he still held a German university chair, and it soon became clear that, in spite of his fame, the doors were closing around him – and goodness knows what might happen next. Before Hitler came to power, however, Einstein had spent parts of the years 1931 to 1933 in Oxford, where he became a Student (don, or professor, as Lewis Carroll had been) of Christ Church. A beautiful profile drawing of his head still proudly hangs on the wall of the Senior Common Room, of which he had been a member – a picture always pointed out to Common Room guests.

23.1 Albert Einstein. Pastel portrait by F. Rizzi, 1933. (By kind permission of Christ Church, Oxford, Senior Common Room.)

Although but a child myself when Einstein died in the USA in 1955, I have had the honour in Oxford to know two people who had been in his company. The first was Dame Kathleen Ollerenshaw (née Timpson) of Manchester (1912–2014) who, as an undergraduate mathematician at Somerville College, had attended Einstein's Oxford lectures, "squeezed into a back row", in Rhodes House.[9] The other was Dr Paul Kent (1923–2017), Doctor Lee's Reader in Chemistry at Christ Church between 1956 and 1972. Paul Kent told me that in 1949 he was working at Princeton as a young visiting scientist, and often saw Einstein walking around the campus and chatting informally with students. It was Einstein's seventieth year, and bumping into the great man one day he was invited to his momentus birthday party – a social occasion which Paul never forgot.

One of the most prized, gazed-upon, and photographed treasures in Oxford's Museum of the History of Science is one of the blackboards, covered with equations, which Einstein used in his lectures. The story is that after his Oxford lectures, the blackboards were being routinely wiped down for reuse, when some perceptive soul realized their historic significance. This particular board was spared, and its chalked equations stabilized and preserved for posterity. I have often wondered whether it was one of the boards upon which the young Dame Kathleen saw him write.

Postscript

It is perhaps not without irony that Albert Einstein, whose Theories of Relativity took twentieth-century physics and cosmology into new realms of time and space, should have spent time, intermittently over three years, between Germany and Princeton, in the same Oxford Senior Common Room, Christ Church, in which the real-life Alice's multi-dimensional *Wonderland* had been born back in 1865. As he sipped his afternoon tea, and looked through the window, did Einstein ever glimpse the White Rabbit, the Mad Hatter, and the grinning Cheshire Cat frolicking among the flowers and foliage in the Common Room garden outside?

Chapter 24

Crossing the Watershed: Edwin Hubble, the Celebrity Astronomer of the Galaxies

As we saw over the previous three chapters, the whole shape of astronomy, physics, and cosmology was changing rapidly between 1890 and the 1920s. Yet one great problem remained: did the "universe" consist of one big Milky Way galaxy, or were the thousands of nebulae, now photographed and filed away in the archives of Harvard, Lick, Mount Wilson, Potsdam and elsewhere, "extra-galactic"? Were they what Immanuel Kant in his *Universal Natural History and Theory of the Heavens* (1755) had speculated might be island universes scattered throughout space and possessing no physical connection with our own Milky Way?

This was the watershed which would be crossed after 1924, and in the wake of which the rest of twentieth- and twenty-first-century cosmology would take its form.

From small-town Missouri to self-created English gentleman

Edwin Powell Hubble was born in Marshfield, Missouri, in 1889, though he, with parents John and Virginia, would later move to Wheaton, Illinois. The Hubbles were a successful middle-class family, and John Hubble was an insurance executive. As a boy, Edwin's most conspicuous talents were athletic rather than academic, and he excelled in American football, basketball, baseball, and track athletics. He had, however, an early general interest in astronomy, though when he went to the University of Chicago, he read law, in deference to his father, who wished him to become a top-flight lawyer. Even so, he got on to the University of Chicago basketball team, as well as distinguishing himself in other sports.

These sporting accomplishments, no doubt, played a significant role in his winning one of the new and highly prestigious Rhodes Scholarships to Oxford. These scholarships, created under the trust of the late Cecil Rhodes, aimed not just to educate purely academic young men, but men who combined academic with physical prowess. Young Hubble fulfilled both criteria.

In Oxford, he became a member of The Queen's College, on the High Street, and settled down once more to take an Oxford law degree. Then, in addition to his beloved sports, he began to attend science lectures, read science books, and talk with scientists. At Queen's, he also fell in love with the ideal of the English gentleman, dressing in tweeds and developing something of an English accent and manner of behaving, of which I shall say more below. Back in America, and feeling honourably released from his beloved recently deceased father's desire that he should pursue a legal career (he already had law degrees from Chicago and Oxford), he turned to what was becoming his true passion: astronomy. Working at the Yerkes Observatory, Hubble took his PhD at Chicago in 1917 and was all set for a scientific career.

Then in 1917, the United States declared war upon Kaiser Wilhelm's Germany, and as an American patriot, an Anglophile, and a vigorous sportsman, Edwin Hubble joined the US Army. Posted to Europe in July 1918, his infantry regiment never saw battle on the Western Front: which was probably good news for the future of astronomy, because Hubble the energetic risk-taker could well have ended up in an early grave. After the Armistice in November 1918, he was posted home, eventually leaving the army with the official rank of Major, although he may have served as a brevet Lieutenant Colonel. With his PhD and military title, Hubble sailed back to Europe, completing his already impressive education with a year of astronomy in Cambridge, and receiving his final polish as a true American–English gentleman.

In 1919, George Ellery Hale, Director of the Mount Wilson Observatory, head-hunted the thirty-year-old Hubble for the Observatory staff. With access to the world's largest telescope – the newly commissioned "Hooker" 100-inch-aperture reflector – he was to remain at Mount Wilson for the rest of his career, and make his great discoveries there, doing more, perhaps, than any other individual to carry astronomy "over the watershed".[1]

24.1 The 100-inch-aperture "Hooker" reflecting telescope, Mount Wilson, with which Hubble and his colleagues made their key discoveries. (Drawing by Allan Chapman.)

HUBBLE, RED SHIFTS, AND THE "EXTRA-GALACTIC" UNIVERSE

As things stood after the Shapley–Curtis Great Debate in Washington on 26 April 1920, the evidences for a single big galaxy universe and, contrariwise, for one of island universes were roughly even: with a tilt, perhaps, in favour of the big galaxy. The key to fathoming the true nature of the deep-space universe, however, was seen to lie in the spiral galaxies, the largest of which, to an earthly observer, was the M31 nebula in Andromeda, along with their spectroscopic red shifts. Much hinged, as we saw in Chapter 22, on Vesto Slipher's discovery of the apparent red-shift recession of the spirals. The next advance came in the wake of the completion and full operation of the Mount Wilson 100-inch, as the high resolving power and superior optical technology of that telescope revealed rich fields of individual stars within the Andromeda nebula. What, in lesser instruments, had appeared as areas of glowing, misty light in Andromeda now resolved into dense carpets of separate stars, each one eligible for spectroscopic analysis.

Many of the men and women who addressed themselves to atoms, energy, and the structure of the cosmos were theoreticians and mathematicians. While Hubble had an academic mathematical training behind him, his natural inclination lay in observation, experimentation, and meticulous physical techniques rather than theoretical modelling. Having the great 100-inch telescope at his – shared – disposal, he began a detailed study of these Andromeda stars. Even then, however, obtaining spectra for individual Andromeda stars was difficult, for they were so faint, and exposures had to be long, even under the skies of a Californian mountaintop. Hubble's mastery of painstaking observational techniques was crucial.

Between 1922 and 1923, primarily in Andromeda but also in some other high-definition nebulae, he detected Cepheid variable stars. Back in 1911, as we saw in Chapter 22, Henrietta Leavitt at Harvard, studying the Cepheids present in photographs of the lesser Magellanic Cloud, had worked out a period-luminosity relationship for these variable stars, and following Harlow Shapley's calibration of Leavitt's Law, a practical deep-space yardstick had become available. Using the distance-calibrated Andromeda Cepheids in particular, Hubble was able to establish that the nebula was not a part of the Milky Way, but a quite separate island universe. Andromeda was a galaxy in its own right, and a crucial cosmological watershed had been crossed in the final resolution of a problem with which Sir William Herschel and Immanuel Kant had wrestled. By 1924, it was understood on firm physical grounds that our own galaxy was but one of countless large, complex star systems scattered throughout space, extending to the very edges of optical visibility. Nor had we any reason to suppose that our own galaxy was in any way special, let alone at the centre of the universe; for spiral and other galactic systems receded in all directions, though some regions of space were more densely populated with galaxies than others.

Hubble, working with the Mount Wilson 100-inch telescope, continued to collect photographic data about the nebulae, including the detection of novae and supernovae, especially in the Andromeda nebula. The old term nebulae was replaced by the designation extra-galactic nebulae. It would be incorrect, however, to assume that Hubble's discoveries, and their coverage in *The New York Times* of 23 November 1924, represented a sweeping victory for the "extra-galactics". It took time for some of the Harvard astronomers around Harlow Shapley to come to terms with the new universe. Much of their objection lost weight after 1935, when Adriaan van Maanen revised his 1912 values for the assumed rotation of the spiral nebulae, making it possible to regard them as situated much further away in space than his original values had implied.

On the basis of the Cepheid calculations in particular, Hubble now began to compute some actual distances, and announced that the Andromeda galaxy was some 750,000 light years away from our own galaxy, and in the region of 25,000 light years across.[2] These figures were later modified as more and more new observational data became available, especially after 1948, when the 200-inch-aperture Mount Palomar telescope came into commission Yet the distances still remained vast.

Following the realization that the Andromeda Cepheids made feasible a potential scaling of the stellar universe, the next move was to establish a relationship between the Cepheid proportions and the galactic red shifts. This was based upon the original 1912 and subsequent work of Vesto Slipher, expanded by Milton Humason at Mount Wilson. After these two sets of data were effectively calibrated by 1929, astronomers had at their disposal a technique whereby the distances of new, faint galaxies could be measured. Such measurements had all manner of potential when it came to extending measuring rods into space: could the universe be expanding, and were some parts receding faster than others? From these values, could one even calculate how old the universe was? The possibilities seemed enormous, and would set much of the agenda for cosmological research thereafter.

HUBBLE'S LAW AND CONSTANT

Hubble's Law and Constant have been styled the most *inconstant* law and constant in physics. This was not to cast aspersions upon Hubble's brilliance and originality as an astronomical researcher, but a consequence of the speed at which astronomy, astrophysics, and cosmology were advancing over the thirty years or so after he announced his Law in 1929. The figures first proposed by Hubble were rapidly superseded in the light of fresh observational data – although the basic principle behind the Law remained unaltered.[3]

In its basic first expression, Hubble's Law laid down a scale of proportions relating to the radial velocities of the nebulae: a radial velocity being the nebula's computed rate of recession along a straight line from earth into distant space. The increasing velocity displayed by presumably remote nebulae was significant, in contrast with the slower velocities of nearer ones, such as M31 in Andromeda. This seemed to suggest that the universe was *expanding*, with the most remote objects having the greatest acceleration. Hubble arrived at his conclusions through a combination of Cepheid star period-luminosity light curves and increasing red shifts displayed in the spectra of the nebulae. Galaxies are so vastly remote that it is impossible to measure their distances by conventional trigonometrical techniques of the kind used by Bessel and others to measure star distances, such as those of 61 Cygni or Vega. Consequently, Hubble and the extra-galactic astronomers of the early twentieth century used the degree of red shift exhibited by a nebula spectrum as an alternative deep-space distance indicator.

By 1931, Hubble, his expert astronomical photographer Milton Humason, and their colleagues had amassed an archive of nebula radial velocities for galaxies estimated to be as remote as 150 *million* light years. From these, Hubble worked out his distance scale, proposing that at 1 million light years a nebula would recede at 100 miles per second. At 100 million light years, it would recede at 10,000 million light years per second, and so on. This regular increase in red shift and velocity came to be styled "Hubble's Constant".

From this data and corresponding computations, Hubble was able to calculate the distance and proportions for the Andromeda nebula outlined above. Even so, on the basis of these proportions, it still appeared to be significantly smaller than our own star system: the Galaxy. Likewise, the globular clusters of stars near to Andromeda seem scaled-down when compared in size to the clusters around our own galaxy. Could this mean, therefore, that our own star system had some kind of primacy as far as dimensions were concerned?

Dimensions notwithstanding, one conclusion did appear to emerge from Hubble's work: the physical expansion of the universe. Otherwise, how could one explain the radial recession and increased red shifts of dim and distant nebulae? Yet Hubble, who was a practical observer and laid no claim to being a theoretical cosmologist, was wary of drawing too many conclusions in advance of further evidence. No one could be quite sure what as yet undiscovered physical laws might be acting upon the red shifts at such prodigious cosmological distances.

In less than a decade, however, between 1923 and 1932, Hubble and the photographic plates of Milton Humason at Mount Wilson had given pretty conclusive answers to two questions about the cosmos and the nebulae which extended back through Lord Rosse, the Herschels, and on to Edmond Halley. Firstly, Hubble demonstrated that the long-puzzling nebulae were vast star systems remote from our own galaxy. Secondly, he showed that the cosmos was not static but dynamic, and in a state of expansion. While not himself a religious man – Hubble had been brought up in a Christian household but later became an agnostic – the radial recession of the nebulae (or what would later be recognized as galaxies in their own right) strongly suggested that they had once been much closer together in space than they were now. This realization would feed into a related body of theoretical cosmological ideas during the 1930s onwards, that the entire universe had originated from a single, intensely dense primal atom, and something had caused it to fly apart, as we shall see in the next chapter.

From known data regarding the red shifts of the receding nebulae, it appeared by the early 1930s that one might date the age of the universe to around 2,000 million years. This date seemed too young for the geologists, who were already arguing that terrestrial rock strata were older than that. So while Hubble's Law and Constant held as a series of proportions, it became obvious in the wake of yet more recent data that their originally computed values needed to be extended back further in time, and amplified with yet more data. As we have seen on many occasions already, the development of new technologies would herald fresh thinking, by presenting new primary observational data. Where Hubble's Law and its implications were concerned, this new technology would come in the form of a giant new telescope: the 200-inch-aperture reflector on California's Mount Palomar. Coming into use in 1948, the Palomar 200-inch was the crowning glory of American big telescope building, and as the Mount Wilson 60-inch of 1905–08 had been almost doubled in size in the great 100-inch of 1917, so the Mount Palomar doubled the optical aperture yet again. The Palomar would be the last great terrestrial single primary mirror telescope, and future instruments would either be multi-mirror arrays, or telescopes orbiting the earth in space, as we shall see in later chapters.

Before looking at how Hubble's original findings were subsequently developed, in many cases at the hands of his own assistants and colleagues, notice should be taken of his highly influential Hubble Sequence of Galaxies. Here, he produced a visual classification system for nebulae which, after the inevitable adjustments, is the one still generally used today. What is more, his *The Realm of the Nebulae* (1936) and *An Observational Approach to Cosmology* (1937) did much to take his techniques, approach, and discoveries in astronomy and cosmology to a wider audience than those who read academic papers.

THE SUBSEQUENT DEVELOPMENT OF HUBBLE'S COSMOS: MILTON HUMASON, WALTER BAADE, AND ALLAN SANDAGE

Creativity and ingenuity are fascinating things, and can never really be planned in advance or systematically organized, which makes them so interesting to the historian. Who in 1917 would have imagined that a twenty-six-year-old dropout from formal education, who worked on ranches, loved the mountains, and earned a living as a mule-driver on Mount Wilson would play a major part in the development of twentieth-century cosmology? Yet Milton Humason did.

Milton Humason

In 1917, Humason began to work at the new Mount Wilson Observatory: not as an astronomer, but as a night watchman and general handyman. Although a school failure, Humason was clearly highly intelligent, good with his hands, and had a friendly and obliging manner. This led George Ellery Hale, the Observatory Director, to take him on as a member of the technical and scientific staff: an unprecedented career change for a man without one school examination certificate, let alone the otherwise de rigueur PhD degree. Humason soon showed himself to be an expert photographer, and to have great patience and manipulative dexterity, as was evident in his masterly photographs of deep-space objects taken with the great 100-inch telescope. In spite of the great power of what was then the world's largest and most powerful telescope, photographic plates were still relatively slow (or of low sensitivity), and the objects being focused were extremely dim. To record fine detail in a distant nebula, or obtain a spectrogram of a distant Cepheid star, with all of its Fraunhofer lines sharp and in place, demanded great skill. It was not a matter of locating the object, then simply snapping away with the great telescopic camera. Instead, the exposure needed to be guided, largely by eye, in conjunction with the telescope's equatorial drive motor, which helped the astronomer to track an object's path across the sky. Faint spectra in particular might require an exposure of several hours, perhaps even over a couple of nights, to register all the essential data which could be used for subsequent analysis. Humason was a master of such painstaking work. These skills brought him into contact with Hubble, and the ex-mule-driver was to play a significant part in Hubble's achievements, through the quality of his nebular photographic plates, and in helping to form aspects of Hubble's "Law", and the calibration of the recessional velocities of nebulae.[4]

At the age of fifty-nine, in 1950, Milton Humason was formally honoured when the University of Lund, Sweden, bestowed a DSc degree upon him.

Walter Baade

Like others before them, the Mount Wilson 100-inch and the photographic plates of the post-1917 era hit a technological ceiling. As was mentioned above, the next leap forward would come in 1948, with the great Mount Palomar 200-inch. With this instrument, Walter Baade would observe details invisible in the 100-inch after 1952, and go on to revise Hubble's earlier values for the size of space. So much of the early distance debate hinged upon the globular clusters and their Cepheids around the Andromeda nebula. Even with the mighty 200-inch, details of individual stars within these hazy clusters of light were hard to resolve. Did this mean, therefore, that they were much further away than the distances computed by Hubble, based upon observations with the less powerful 100-inch telescope, the primary mirror of which could resolve less detail than that of Mount Palomar?

Another problem lay in the apparent absence of a certain type of star, known as an "RR Lyrae", in the Andromeda clusters. RR Lyrae stars are named after the first star of this type discovered in the classical constellation of Lyra, the Harp. The prototype RR Lyrae star was a very short-period variable star in a tight cluster within Lyra, and displayed special spectroscopic characteristics. Yet in spite of their high absolute brightness in comparison with our own sun, no RR Lyrae stars were detected. Once again, this absence suggested that Andromeda and its star clusters were much more remote from us than should have been the case according to Hubble's original Law and distance computations. And another problem emerging from the newly exposed 200-inch telescope photographic plates was the relative dimness of ordinary nova – "new stars" – observed in the Andromeda system compared with Milky Way galactic novae. Such stars seemed at least two or three times fainter than Milky Way novae when distance-calibrated in accordance with the original Law.

In short, the findings with the 200-inch Mount Palomar telescope suggested a need to recalibrate Hubble's Law and period-luminosity weightings for Cepheid variable stars. None of this meant that either Henrietta Leavitt or Edwin Hubble were wrong in the way in which they elucidated the cosmos. It simply meant that, in the light of the new telescopic evidence, their original values needed revision, while the basic principles upon which the values had been derived still held good. Several conclusions were drawn from the work of Baade and others in the 1950s. For instance, the Andromeda nebula was a full-scale galaxy just like our own, and probably of comparable size, its clusters also being of a similar size to those in our Milky Way. The Andromeda galaxy and system was also considerably further away than Hubble's originally computed 750,000 light years. This recalibration of the cosmos was further necessitated by terrestrial geological discoveries relating to radioactive rocks, which strongly suggested that the earth itself was over 2,000 million years old. So clearly, the universe had to be much older. By the 1950s, it was being reckoned at round 5,000 million years, which also chimed in better with independent geological estimates of the ages of the earth's oldest rocks.

None of these elegant calculations of distances and ages were without caveats, for no one knew what sort of obscuring matter might exist in space, and what effect it could have upon all of the proportions. Likewise – as Hubble himself had been aware – no one could be sure how the light from a rapidly receding galaxy might affect its red shift, and hence its distance and age estimate. Such were the questions and revisions emerging in light of the findings of the latest big telescope in the world after 1948.

ALLAN SANDAGE

Born in 1926, when Hubble was beginning to astound the world with his discoveries, Allan Rex Sandage stood very much in the Hubble tradition of being an observational cosmologist and a masterly user of great telescopes, in his case the Mount Palomar 200-inch. Walter Baade had been his doctoral supervisor at the California Institute of Technology, and Sandage also fell under the intellectual influence of Hubble, before the latter's sudden death in 1953. Just as Baade had modified Hubble's Law (largely from data gathered from new studies of Cepheid stars within the Andromeda galaxy), so Sandage refined it yet further. In 1958, he published a modified value for Hubble's Constant concerning the recession of the galaxies. Using the new data, Sandage computed a new Hubble Constant, one-seventh of the size of Hubble's original value of 1929. By the early 1970s, he computed a fresh Constant that was only one-ninth of Hubble's original.[5]

In numerical terms, Hubble's extra-galactic nebulae expansion rate was some 500 kilometres per mega parsec, but Baade and Sandage would reduce this to 180, 75, then 55 kilometres per mega parsec. To express these distance and velocities in conventional terms such as miles or kilometres is about as practicable as giving the earth–moon distance in millimetres: one soon runs out of zeros. Instead, astronomers use the parsec and mega parsec. To get a sense of the sheer vastness of the distances involved, and of the numbers required to express them, a parsec is equal to 3.259 light years, and a single mega parsec to 1 million parsecs, or 19 trillion kilometres, or about 12 trillion miles.

By reducing the size of the Hubble Constant from 500 to 55 mega parsecs, however, Sandage made the universe both bigger and older. *Older*, as one now had more time for galactic matter to recede to immense distances, while allowing the geologists space to account for the development of our terrestrial rocks, using measured long-term radioactive decay rates. *Bigger*, as the long-time recession of the galaxies enabled us to confidently place them much further away, making it possible to explain how the Andromeda galaxy and its globular clusters could be just as large as our own galaxy and its clusters, rather than simply a relatively local miniature version of our own Milky Way galaxy. From this new set of proportions, all the other galaxies could be increased in size pro rata.

By the early 1990s, a cosmology was taking shape within the astronomical community which, while possessing a clear intellectual lineage back to Hubble in the 1920s, had developed into something vaster and more diverse than anything that would have been envisaged in 1929. This included a universe which might have expanded and collapsed upon itself, and the collapse of a gas cloud to form our own Milky Way, and a universe populated by radio sources, pulsars, and dark matter. All of this had its roots in an advancing telescope and image-processing technology, and even when the resulting data formed the basis for theoretical computations, the observed data were always potentially there to test them by. Allan Sandage was a living link between the cosmology of Hubble and that of the new millennium, with a fund of stories and insights into the character of his old boss, Edwin Hubble.

Edwin Hubble and the stars of Hollywood

Between 8 and 10 April 1992, the National Astronomy Meeting of the Royal Astronomical Society was held in Durham University, with Allan Sandage as the special guest speaker. On the evening of 9 April, Sandage delivered his lecture, on the current state of cosmological thinking, including his personal reminiscences of Hubble. I felt honoured over the three days of the conference to have several conversations with Sandage, where once again he was keen to talk of the astronomical history he had witnessed. Hubble figured prominently. We saw above how Edwin Hubble began to reinvent himself as an English gentleman after his Rhodes Scholarship to Oxford, 1910–13, and during his time at Cambridge after 1918. One sees clear traits of the natural actor in Hubble, who became accomplished in projecting his image and playing the leading role at Mount Wilson, then at Mount Palomar.

Hubble was a handsome man, standing over 6 feet tall, athletic, and obviously charismatic, always impeccably dressed, with well-polished shoes. This persona received its final touch from the pipe which he used to carry, and with which he might gesture, even when it was not lit: playing the perfect English gentleman, always measured and precise, with a carefully cultivated accent and a whiff of pipe tobacco about his tweeds.

Sandage said that in the Mount Wilson and Mount Palomar dining rooms, Hubble always sat – as if by right – at the head of the table, and such was his style that his fellow-diners obediently fell in with him. Was he cultivating the persona of an Oxford or

Cambridge head of college, presiding at high table? I suspect so. It is hardly surprising, therefore, that the superstar of the world of astronomy also enjoyed the friendship of the stars of the silver screen, in Hollywood. When not studying the heavens, Hubble could sometimes be found at Hollywood parties, where celebrities of the stature of Charlie Chaplin, Harpo Marx, and Lillian Gish ranked among his friends. Like the Hollywood stars, Hubble knew how to strike the right pose for the photographers, facial expression and position of pipe perfectly poised, to immortalize the image of the man who had fathomed the "length, breadth, depth… and profundity" of the cosmos.[6] Like the musician-astronomer Sir William Herschel, who had coined the above biblically resonant phrase in 1817, Hubble was an instinctive performer.

Another Hubble trait to which I warm was his lifelong love of cats: he owned a beautiful black creature whom he christened Copernicus. A very apt name, for whereas Nicholas Copernicus had rearranged the solar system in the sixteenth century, so the master of his feline namesake would rearrange extra-galactic space in the twentieth. Despite the profound scientific and wider intellectual impact of his work, Hubble never received the Nobel Prize for Physics: for in his day, the Prize did not cover astronomy. He suffered a heart attack in 1949, and then, while a movement to modify the Nobel rules to enable the Prize to include astronomy was under way, he died from a blood clot on the brain on 28 September 1953, two months short of his sixty-fourth birthday. Sadly, the Nobel Prize cannot be awarded posthumously.

Chapter 25

The Belgian Priest–Cosmologist and the "Cosmic Egg"

Scientific research has always attracted those who approach fundamental problems in nature from two different perspectives. Some are fascinated by the pure intellectual possibilities of mathematical and philosophical ideas, while others are essentially motivated by the power of practical experimentation and technologically related observation. In the seventeenth century, one might think of the contrasting approaches of the deductive and mathematical René Descartes and the essentially experimental Robert Hooke: then in the eighteenth century, the philosophical cosmology of Immanuel Kant as opposed to the big-telescope studies of William Herschel. In the nineteenth century, there were the contrasting approaches to reality exhibited in the metaphysics of Hegel and his disciples, as against the deep-space quest of the big telescope engineer-astronomers such as William Lassell and Lord Rosse. The twentieth century saw an intensification of these two approaches to fathoming cosmological truth, exhibited on the one hand in Albert Einstein, and on the other in the Harvard spectrum plate analysts. Yet they were, perhaps, most clearly exemplified in the researches and conclusions of Edwin Hubble and the Reverend Monsignor Georges Lemaître.

Father Georges Lemaître of Leuven

Although later a working diocesan Roman Catholic priest rather than a member of any religious order, Georges Lemaître, who was born in 1894, was given a rigorous classical education by the Jesuits of Charleroi, Belgium. During World War I, he joined the Belgian Army (on the Allied side), serving as an artillery officer. After the war he returned to his studies, now concentrating on theology, in preparation for his priestly ordination in 1923, and on astronomy and astrophysics. Lemaître studied in Cambridge, where Arthur Eddington exerted considerable influence on his thinking, and did research into stellar astronomy and contemporary cosmology. Lemaître next worked at the Harvard College Observatory, and was further trained by Harlow Shapley, in the wake of the Great Debate of 1920.

Returning home to Belgium in 1925, Lemaître at first took up a junior post at the Catholic University of Leuven (Louvain). There he began to develop ideas originally explored in his doctorate, on the mathematical characteristics of variable stars, and then moved on to other issues in the cosmological physics of the early 1920s, such as Einsteinian relativity, and the possible expansion or stability of the universe. Lemaître was very much of an analytical mathematician and theoretician in his approach, rather than an observer or user of instruments. However, he fully appreciated the significance of instrument-based discoveries that provided the physical facts alongside which one could evaluate theories.

During the 1920s, Lemaître's work was largely confined to the Belgian academic world, and while fully aware of what was going on in America, Cambridge, the rest of Europe and elsewhere, he tended to circulate his ideas via the Belgian academic press. Hence his monumental paper of 1927, published in *Annales de la Société Scientifique de Bruxelles*, largely escaped notice until after 1930, when he sent a copy to Eddington, who had it translated into English and published in 1931 in the international *Monthly Notices of the Royal Astronomical Society*.[1] Entitled "A homogeneous universe of constant mass and increasing radius as accounting for the radial velocity of extra-galactic nebulae", it gave an early mathematically coherent model for an *expanding* universe. It also accounted for the rotation and recessional radial velocities of the spiral nebulae, first recorded by Vesto Slipher after 1912 and later Edwin Hubble. In 1927, however, when Lemaître's paper went to press in Brussels, Hubble's work was still in its developmental stages.[2]

So where did cosmological ideas stand by the end of the 1920s? Was the universe stable, expanding, or collapsing? What was the true significance of the red shifts, and how did all this relate to what, in 1930, was Einstein's *static* model of the curved relativistic cosmos? Eddington and others on the governing council of the Royal Astronomical Society decided to devote its January 1930 meeting to an exploration of the evidences and arguments for and against the various current cosmological theories.

Making sense of modern cosmology: the Royal Astronomical Society discussion meeting, Burlington House, Piccadilly, London, 10 January 1930

Exactly eighty years later, at the RAS meeting of 8 January 2010, the Society revisited the circumstances of 1930, and I was honoured to be invited to deliver the historical lecture examining where things had stood, eighty years ago.[3] This meeting was not as well attended as usual, however, though not because of the historical nature of the meeting. The RAS enjoys historical lectures, but snow blizzards and ice were currently gripping England and making travel difficult. It is, for example, the only time that I recall going to Burlington House, on Piccadilly, wearing heavy wellington boots, and carrying my shoes in a bag.

In 1929, as we have seen, Hubble had recently published his major paper and graph showing the distance–velocity ratio of the extra-galactic nebulae. Not claiming to be a theoretician, however, he provided no theoretical structure to account for his observed data, and was still holding his fire regarding the true cause of the galactic spirals' red shifts. Hubble and observational astronomy apart, though, let us review how theoretical, mathematical cosmology had evolved in the fourteen years since Einstein's General Theory of Relativity in 1916.

In 1917, Einstein had proposed a relativistic cosmology that was homogeneous in its structure. It was also static, insofar as it was not expanding, and was contained within its own spacetime continuum. Theoretically, however, such a universe could have contracted in upon itself, so as to maintain its stability. Einstein posited a Cosmological Constant, counterbalancing both internal implosion and outward expansion. Then in 1922, the Russian mathematical astronomer Alexander Alexandrovich Friedmann examined Einstein's relativity equations and proposed an alternative interpretation. Friedmann calculated that the universe – theoretically speaking – could have a constant mass and density, yet could also be expanding. This expansion could also incorporate Einstein's space-time curvature. In short, Friedmann was discussing a relativistic, expanding universe. It was not, in 1922, based upon observed or experimental criteria, being essentially a mathematical model derived from relativity physics. Nor was it without the potential for physical testing, for it was compatible with relativity predictions about the way in which gravity bent light, as well as explaining the advance of the perihelion of Mercury's orbit. By 1930, Alexander Friedmann was no longer alive. The 37-year-old had recently returned from a holiday in the Crimea, where it is thought that he had contracted typhus fever.

Also in 1917, Willem de Sitter, who would become Director of the Leiden Observatory, Holland, was inspired to undertake the computation of *theoretical* universes, based upon relativistic and other modern physics postulates. It is fascinating to see how theories in physics developed over the previous decade or so were so inspiring to mathematical cosmologists at this time. Willem de Sitter, unlike Einstein, had come up through an observational astronomy route. Born in 1872 into a distinguished Dutch legal family, Willem (rather like Hubble) had been encouraged to study law at university. But at the University of Gröningen, he had fallen under the influence of the astronomy professor Jacobus Kapteyn. Kapteyn was working on star densities in the Milky Way, especially those stars visible around the southern hemisphere pole, as recorded in an archive of glass plates, as were those of the Harvard and other astronomers when he published his list of 454,000 southern stars.

These stellar density studies enabled Kapteyn to calculate that the galaxy, or Milky Way, was a vast lenticular structure, round in shape and thicker in its middle than at its edges: a calculation which was broadly congruent with Sir William Herschel's Milky Way star density counts of the 1780s. Modern observational and mathematical techniques, however, had enabled Kapteyn to ascribe dimensions to the galaxy, which he suggested was 55,000 light years across and 11,000 light years thick in its centre.

Under Kapteyn's inspiration, the young de Sitter had worked in observational astronomy, both in Holland and at the Cape of Good Hope, but it is his mathematical cosmology in the wake of Einstein's General Theory of Relativity in 1916 that concerns us in this chapter, and which involved him in the RAS discussions of January 1930.

Deriving from earlier mathematical models that de Sitter began to develop after 1917, his paper to the RAS meeting of January 1930 argued for an expanding universe, with parallels to that of Friedmann. However, this theoretical universe contained *no matter*, being, rather, a mathematical study of forces in space and time. This was in contrast with Einstein's relativistic cosmology which saw a universe of real matter and energy in curved space and time, but where the Cosmological Constant kept everything stable, maintaining the laws of physics in all places and preventing fundamental change. Consequently, one could sum up the situation in January 1930 as a choice between a theoretical universe which contained matter and was motionless, and an equally theoretical relativistic universe which had motion and expansion, yet which contained no matter![4]

Father Lemaître and Sir Arthur Eddington

As we saw in Chapter 23, Arthur Eddington was probably Britain's most famous astronomer by the 1920s, and he played a leading role in the RAS discussions of 1930. In the wake of the publication of the proceedings of the January meeting – de Sitter's paper having been printed in *The Observatory* journal in February 1930 – and their receipt in Belgium, Georges Lemaître wrote to Eddington and included a copy of his own paper of 1927. Lemaître explained to Eddington how he had developed an Einsteinian relativity universe model which not only contained matter, but was also expansive. In many respects, this was a revelatory moment for Eddington, as Lemaître's model, by combining both relativistic theories, showed a clear way ahead. It was also able to incorporate Hubble's observational findings at Mount Wilson, thereby bringing theory and observation elegantly together.

We must not jump to the conclusion that his Cepheid and extra-galactic studies placed Hubble on the side of the expanding universe – at least in 1932. He had his reservations about the cause of the red shifts, considering that as yet unknown forces in "inter-galactic" space might be causing them instead of a seemingly universal expansion. The pragmatic celebrity astronomer of Mount Wilson remained cautious.

Once Eddington and others had grasped the implications of Lemaître's paper, however, various things began to fall into place, especially by 1933, when Lemaître and Einstein had met and discussed these ideas on two or three occasions. An implicit (if not mathematically explicit) feature in Lemaître's expanding cosmology was the idea that the universe had come into being at a particular time, and that from its present size, and measured rate of expansion, it might be possible to calculate its age. On the most simplistic level, one had only to back-calculate from what was currently known of the rate of expansion, to ascertain when it had all begun. This was variously described as everything deriving from one single primal atom or "cosmic egg", or latterly by the dismissive Sir Fred Hoyle as a "big bang".

"IT'S ALL A 'BIG BANG'": SIR FRED HOYLE AND HIS STEADY STATE COSMOLOGY OF 1948

Sir Fred Hoyle was the son of a wool merchant of Bingley in Yorkshire, and a grammar school scholarship boy, who distinguished himself as a physicist and astronomer at Cambridge. Fred was not renowned for his retiring disposition, and always enjoyed being provocative, sometimes too much so. He was direct in his manner, had no time for fancy airs and graces, and drank tea all day long. Fred was thirty-three in 1948, when, along with Hermann Bondi and Thomas ("Tommy") Gold, he announced a new approach to cosmology at the Royal Astronomical Society out of Town meeting, held in Edinburgh. This was the "steady state" theory, embodying a new interpretation of relativity cosmology incorporating an expansion principle, yet with no primal atoms or big bangs, but with matter thinning out endlessly from its hypothetical point of origin.

The steady state model, however, was of an eternal universe, a feature that it shared with the cosmology of Aristotle of *c.* 350 BC: a universe with no indication of a beginning, and no sign of an end. So how could such a stable, steady-state universe expand in a way that was congruent not only with contemporary Einsteinian theory, but also with the numerous observations and measurements made by Hubble and his colleagues on Mount Wilson, which would soon be dramatically confirmed and amplified with the new 200-inch telescope on Mount Palomar?

Hoyle and his colleagues proposed that the Cosmological Constant, upon which the coherence of the laws of physics depended, could indeed be maintained if new matter came into being. So instead of all matter coming into being in a single big bang, atoms might come into being on a slow, regular basis – atoms not just of hydrogen but also of deuterium, lithium, and other elements essential to stellar physics and chemistry. But were Hoyle and his colleagues not contradicting one of the fundamental laws of physics: that while matter might be changed, it could not be continuously created from nothing? Yet this "creation" of new matter was essential if the universe were going to expand and become bigger, while at the same time maintaining a stable, or cosmologically constant, uniform density and distribution of matter.

Although never having as many adherents across the astronomical community as the big bang, the steady state theory had no shortage of well-wishers in the 1950s. Some may have been attracted by its aesthetic elegance, balance, and unity: always significant factors in the minds of cosmologists. Others perhaps favoured it because they were uneasy with the implicit creation moment in the big bang: it was a bit too biblical, especially bearing in mind that its leading proponent was a Roman Catholic priest. But as we shall see below, Father Lemaître, as a rigorous and thorough-going scientist, was careful to distance himself from those who read overtly religious implications into his theory.

Observed physical evidences began to mount up against the steady state theory in the 1960s, and these, like most other evidences in astronomy and science, resulted from new discoveries made possible by new research technologies. One was the new technology of space probes and radio telescopes (to which we shall return in Chapters 26 and 27), and the detection of "radio galaxies". These were distant, red-shifted galaxies which not only gave off light, but also radio wave energy. Similarly, the discovery of "quasars" posed a problem for the steady state theory. These "quasi-stellar objects" (hence quasars) were truly strange beasts in the cosmological zoo, resembling small stars, even in the new post-1948 Mount Palomar 200-inch telescope, yet being sources of prodigious radio emissions. (The same was found to apply to the newly discovered radio galaxies.)

When quasar distances were first computed from their optical red shifts, it was discovered that they were extremely remote, and sometimes receding from us at as much as one-half the speed of light. Their distances were largely confirmed following the development of radio telescope interferometry in the 1960s and thereafter. We will return to interferometry as a distance-measuring technique in Chapter 27. Quasars upset the credibility of the steady state theory by demonstrating that matter was not homogeneously spread across the cosmos. All the quasars and radio galaxies were immensely remote from us, none being found in our more "local" galactic universe. This was not only good news for the big bang theory, but even confirmed a prediction implicit within that theory. For if quasars were so very remote and travelling so rapidly, this indicated that the matter and energy they contained had probably left the formative explosion before the stuff making up our own local part of the universe. Was energy behaving differently in the quasars than it was in our own cosmic backyard? The universe was not a place where matter and energy were uniformly distributed and being continuously created *ex nihilo*.

The true death knell for steady state cosmologists came in 1964, following the discovery of the cosmic microwave background. This is a uniformly spread body of energy permeating space, of a type known as "black body" radiation. The existence of such an atmosphere of radiation suffusing the cosmos was even predicted within the big bang theory, the microwave energy being a product of the bang. It was not easy to explain this microwave background in steady state terms, though attempts were made, generally along the lines of the energy deriving from scattered starlight. If there were a form of continuous creation of matter and energy – for ever pushing the universe outwards to make it expand – as the steady state theory postulated, then there should have been pockets of microwave energy. The observed reality was that this microwave energy was evenly and uniformly spread throughout the universe, suggesting an ancient big bang origin. This cosmic microwave background was made detectable in the first place by advances in radio, and in X-ray, microwave, and other short-wave energy particle branches of astronomical technology.

By the 1970s the steady state theory was being faced with increasing challenges as one new discovery after another fitted in with the big bang theory, rather than with the steady state. Yet Fred Hoyle, the doughty Yorkshireman, with probably a fundamental distaste for anything resembling a one-off, divine-sounding ex nihilo creation, tweaked his theory whenever possible. He even branched into other fields, such as geophysics, new gravitation theories, and even his panspermia theory of the existence of life-related molecules distributed throughout the cosmos, thereby undermining the potential uniqueness of planet earth, and hopefully, any implied *singular* act of creation.[5]

RETURN TO THE STARS

As we saw in earlier chapters, the spectroscope and photography revolutionized our knowledge of the stars after 1860. Yet how the stars continued to burn, and the elusive nature of the fuel that powered them, remained largely unknown even by 1920. To make sense of an expanding, galactic universe, it was also necessary to make sense of the individual components that made up the galaxies: the stars themselves.

Einsteinian relativity physics offered a way forward, and just as Henrietta Leavitt had played a fundamental role in cosmology by recognizing the importance of the period-luminosity of Cepheid stars, so the Cepheids provided a key to stellar physics. What mechanism lay at the heart of the periodicity of Cepheids, which resulted in their regular cycles of light and energy output? Karl

Schwarzschild had worked on the problem, but it was Sir Arthur Eddington after 1916 who began to show a way ahead. Much hinged upon what kept a star physically stable, especially if it was a variable star. The generally accepted model at the time was that a star was a ball of gas, held in equilibrium by thermal energy pushing outwards – providing the light and energy of the star – counterbalanced by a gravitation pressure, pulling the star mass to a centre. Significantly, Eddington recognized the outward, driving force from the star's centre was not straightforward thermal pressure but atomic radiation, meaning stars could be seen as powered by an atomic process.

This derived from the Special Theory of Relativity (1905) recognition that mass could be converted into energy in the stars, one aspect of which was the presence of helium in stars, which opened up all sorts of atomic astrophysical possibilities. In his classic works *The Internal Constitution of the Stars* (1926) and *Stars and Atoms* (1927), Eddington readdressed the physics of the stars not from a classical thermodynamic but from an atomic physics perspective. Working mathematically – and sometimes speculatively – he came to see an electron-proton reaction resulting in a nuclear fusion energy release as lying at the heart of how the stars shine without rapidly burning out. In consequence, he calculated that stellar interiors must be many millions of degrees hot, and creating prodigious radiation pressures within themselves. Stars, in accordance with Eddington's model, became generators of atomic energy, actuated by the energy exchange between hydrogen and helium.

This essential stellar mechanism would provide a model by which to explain the behaviour of different types of stars, such as Cepheids and variable stars, red giants, and white dwarfs. Specific parameters might be modified to account for different star types (giants, dwarfs, and such), but the basic underlying atomic physical process remained the same. (In biological terms, it was similar to the way in which the heartbeat, blood chemistry, oxygen, water, and nutrition are common to all mammals, be they giraffes, cats, or humans.)

Eddington's atomic physics model for star energy sources played an incalculable role in framing the way in which stellar physics developed after 1926. Eddington was, however, by no means the only figure in the new atomic physics of the star field by the 1920s, although at that time he was the giant.

Subrahmanyan Chandrasekhar and the White Dwarfs

One particular type of star in the increasingly strange and extreme place of deep space was attracting particular attention: the white dwarfs. One of the first to be discovered had been the so-called "dark companion" around Sirius, known as Sirius B. As we saw in earlier chapters, its existence had first been predicted mathematically by F. W. Bessel in 1844, on the basis of the slight rocking motion displayed by Sirius, suggesting an encircling companion which exerted a prodigious gravitational force upon the parent star. Sirius B was finally seen by Alvan Clark in Boston when testing a new large-aperture refracting telescope in 1862 (Chapter 21). White dwarfs are stars possessing a relatively small physical size, yet consisting of a gravitational mass that is so dense that Eddington said one might put a ton weight of white dwarf matter in a match box.[6] That was subsequently found to be a low estimate. What were such stars, and how did they evolve?

In 1924, Sir Ralph Howard Fowler in Cambridge had invoked the new science of quantum physics to explain white dwarfs. Such stars were wholly congruent with relativity and quantum theory, being old stars that had used up so much of their nuclear fuel, they had collapsed in upon themselves. Their surviving atomic structures had become much more compressed than was the case with normal atoms, making them immensely dense, and by extension, gravitationally very powerful. So much so, tiny Sirius B could exert a gravitational drag effect upon the giant Sirius. By way of comparison, imagine an object the size of a car rotating around the earth, yet possessing such an enormous density as to cause a gravitational rocking effect upon the earth. White dwarfs were the first in a menagerie of other strange and extreme cosmic beasts emerging in the 1930s.

Their existence also came to be explained in light of new discoveries taking place in modern physics, especially after James Chadwick's elucidation of the properties of the neutron in 1932:

another sub-atomic particle but possessing different properties from those of already-known protons and electrons. Then in 1938, Otto Hahn and Lise Meitner discovered nuclear fission, by atomically degrading uranium – a discovery that would soon have frightful unexpected human consequences in the atomic bomb. The rapidly advancing atomic physics of the laboratory, and their relativistic and quantum mathematical theoretical origins, would also play a vital role in explaining what went on in the stars – which brings us to Subrahmanyan Chandrasekhar.

"Chandra", as he was generally known, came from a distinguished Tamil, Indian, family. His father worked as a senior official in the Indian railway system, though his uncle Sir Chandrasekhara Venkata Raman was the Nobel Laureate in Physics in 1930, and his mother was a prominent literary scholar and translator. As a boy, Chandra had been captivated by mathematics and astronomy – disciplines running deep into ancient Indian civilization – and in 1930 he won a scholarship to Trinity College, Cambridge. While still a student, and using the latest relativistic techniques of the day, the brilliant young Chandra began to calculate what might happen to old stars as their fuel supplies approached exhaustion. If a star shrank as part of this process (as was the case with white dwarfs), what were the parameters of contraction?

At the 11 January 1935 meeting of the Royal Astronomical Society, the twenty-five-year-old Chandra proposed what would come to immortalize his name as the "Chandrasekhar Limit". This was a stellar physical mechanism which he had been exploring since 1930–31. Now, a star generates its atomic energy, or fuel, by converting hydrogen to helium deep within its interior, the resulting atomic reaction releasing radiation energy. This helium thus forms a shell around the central hydrogen "reactor", but this can only continue until the hydrogen has diminished to a certain level. Now if an old star shrinking into a white dwarf does not exceed a mass of 1.44 (or 1.4) times that of our own sun, then it can remain stable. If, on the other hand, it exceeds this Chandrasekhar Limit of mass, then it will become unstable, and while in the early 1930s no one could be fully sure what this would entail, Chandra's Limit would provide a physical mechanism to explain later stellar discoveries.[7]

For example, such a star, exceeding 1.4 solar masses, might explode as a supernova as its gravitational and radiation pressures became unstable. Afterwards, when much of the matter was blasted away into space, the remaining mass might collapse further. It might also become what a later generation of astronomers would define as a "neutron star": an intensely dense mass of neutrons.

Of all the great astronomers of the 1930s, especially in Cambridge, Chandra most respected Arthur Eddington. Yet Eddington had failed to accept Chandra's "Limit", and even ridiculed it as "absurd", although this disagreement did not destroy their friendship and mutual respect. It was even reported that they went together to watch tennis at Wimbledon. In 1983, via the Cambridge University Press, Chandra, now residing in the USA, wrote his *Eddington: The Most Distinguished Astrophysicist of His Time.*

Chandrasekhar's Limit would provide twentieth- and twenty-first-century cosmologists and astronomers with one of their most valuable mathematical tools when it came to understanding stellar evolution, including that of neutron stars, pulsars, and black holes. The dramatic proof for the existence of neutron stars came in 1967, when Dame Jocelyn Bell Burnell – then a doctoral student in Cambridge – and Professor Anthony Hewish were investigating radio stars. Jocelyn Bell (as she then was) detected a rapidly pulsating (one pulse every 1.33 seconds) radio signal of great regularity coming from the constellation of Vulpecula. Not only was the scientific world agog, but so too were the media, and some even speculated that the pulses might be signals from an extra-terrestrial intelligence: nicknamed "Little Green Men". The pulses, it was later discovered, were generated by a rapidly rotating magnetic collapsed neutron star. Jocelyn Bell's pulsars supplied an observational substantiation for a hitherto theoretically predicted type of star: one that had shrunk to little more than a small, intensely dense, rotating body of neutrons.

Nine years before Jocelyn Bell's discovery, in 1958, in the United States, David Finkelstein was building upon the earlier work of Karl Schwarzschild, and had computed the possibility of a star that had contracted to such a density that no energy could escape from it. This was the first clear articulation of what would come to be called a "black hole". Emitting no light or other forms of energy, they cannot be directly observed, photographed, or detected with radio telescopes like pulsars, although there is a way around the detection problem. As a black hole must, by its very nature, bend and deform the spacetime surrounding it owing to its intense gravitational attraction, astronomers believe that the presence of a black hole can be detected from the behaviour of the light and energy reaching the earth from behind it. It is even considered that a black hole might be present at the centre of our own Milky Way galaxy.

25.1 Dame Jocelyn Bell Burnell. (By kind permission of the Royal Astronomical Society.)

All of these discoveries only emphasize the fundamental significance of Chandrasekhar's and Eddington's researches, and of their work on the physics of individual stars. In his particular concern with white dwarfs, Chandra opened up a whole range of possibilities for cosmology and astrophysics, such as collapsed stars, neutron stars, and black holes. By our own time, all of these initially mathematically modelled deep-space objects have either been directly observed or else – such as black holes – confidently defined from the behaviour of their surrounding spacetime.

Lemaître, Pope Pius XII, and the Big Bang

In the early 1950s, Monsignor Georges Lemaître became uneasy when His Holiness Pope Pius XII claimed that the model of a universe expanding from a single source provided evidence for a divine creation. While he did not doubt the universe was divine in its origin, Lemaître was too much of a thoroughgoing scientist to read theology directly into a physical theory. This was an especially sensitive area in the early days of the steady state theory, when, as we have seen, some people backed steady state because it did not explicitly involve a perhaps biblically related concept of an origin for everything. Pope Pius heeded the advice offered by scientists within the Vatican and did not push the analogy further. Even so, Lemaître was one of the leading Catholic scientists of the twentieth century, in a tradition extending back to medieval times, and was not only a Member of the distinguished Pontifical Academy, but even served as President of that august body.[8]

Between the publication of his original expanding universe theory in 1927 and his death in 1966, Lemaître had lived through a period of intense development in astronomy and physics. Within that time our knowledge of the universe, from the macro level of a general expansion to the micro level of the possibility of neutron stars (confirmed after 1967) had been revolutionized, and space, with its relativistic laws, collapsed white dwarfs, supernovae, radio galaxies, and theoretical potential for black holes, had become a very strange place indeed. Truly, a watershed in human understanding had been crossed.

Stephen Hawking and the Black Hole

Perhaps the most iconic scientist of our own day is the late Professor Stephen Hawking, who passed away on 14 March 2018, aged seventy-six: an astonishing age for a man who had fought a courageous lifelong battle against a progressive, paralysing neurological disease that left him wheelchair-bound for decades, capable of speaking only through an electronic voice. Icons apart, Hawking's scientific reputation derives from his fundamental work on the theory of black holes: those bizarre creatures of post-Einsteinian cosmology which relate, as we have seen, to the big bang itself.

Since his postgraduate student days, Stephen Hawking had been fascinated by the idea of a "singularity" or an infinitely small, infinitely mass-dense entity from which nothing could escape, and in which the concept of time itself was irrelevant. Could such a singularity lie not only at the heart of galactic black holes, but perhaps also at the heart of the big bang, before it rapidly inflated to form the expanding universe? Yet any pre-big-bang singularity must, by definition, precede the creation of time and space themselves. These theories require us to go beyond even relativity physics, into the strange domain of quantum, with the possibility of particles appearing, seemingly, from nothing. Such ideas might also posit multiverses, in their own space–time environments, each expanding from spontaneous quantum particles, yet wholly unobservable by us.

The science of Stephen Hawking and his fellow mathematical cosmologists belongs to a world where words fail, and mathematical concepts must take over. It poses all manner of profound philosophical questions about beginnings and endings, including Hawking's own quest to discover why our universe is so beautifully fine-tuned in all its operations. Hawking hoped to find some kind of purely physical, non-religious explanation for all this coherence.

Conversely, other modern cosmologists find no problem in seeing a yet greater Mind lying behind the Grand Design.

Chapter 26

Sir Bernard Lovell and the "Radio Universe"

It was the great solar flare and magnetic storm of early September 1859 that first alerted astronomers to the sun's hidden depths of energy: forms that went beyond the known ones of heat and light, in its various wavelengths. As Michael Faraday had first shown, and as was demonstrated theoretically in the equations of James Clerk Maxwell in the 1860s, magnetism and electricity were both aspects of the same physical phenomenon. But if the sun had shown itself to be a magnetism- and electricity-inducing body, what other strange powers might it possess, and what hitherto unknown wavelengths within the electromagnetic spectrum might it transmit? Following Heinrich Hertz's demonstration of a new type of electromagnetic energy phenomenon in 1888, radio waves, several scientists began to ask whether the sun might be a radio transmitter. Julius Scheiner, Johannes Wilsing, Sir Oliver Lodge and others had built experimental equipment in the hope of detecting such solar waves, though at that early date all had failed.

Then in 1902, a discovery was made which seemed to foreclose on the possibility of ever detecting radio waves from space: the discovery of the ionosphere around the earth. In December 1901, Marchese Guglielmo Marconi succeeded in using radio waves to transmit a Morse Code message some 2,000 miles across the north Atlantic. But how could this happen when the radio waves appeared to travel in straight lines? How could they follow the curvature of the earth to cross the Atlantic without flying off into space? By 1902, it had come to be realized that the earth's atmosphere possessed a high-altitude ionized layer, and that this layer, enfolding the entire planet, was capable of bouncing, or reflecting, radio waves inside itself, thus making it possible to send radio signals around the globe. Yet if this ionosphere could reflect radio waves within itself, surely it most probably prevented *external* radio signals, such as those from the sun, from ever penetrating down to the earth's surface. Would they not be bounced back into space? So the discovery of the ionosphere, while unleashing an enormous developmental potential for terrestrial radio, seemed to close the shutters down on any kind of radio astronomy. There things remained for the next thirty years.

Karl Jansky's "merry-go-round" and the birth of radio astronomy

Telephonic communications had come a long way since Alexander Graham Bell built his original speaking apparatus in 1876. Telephone wires criss-crossed the planet by 1930, and even more significant was the growth of wire-*less*, or radio, telephony that developed after Marconi's early radio sets, after 1900. Yet radio telephony posed serious problems. Why, for example, was there so much interference in the form of hissing and crackling, especially when long-distance calls were being made, including calls made on conventional trans-continental telephone wires? Why was this interference so unpredictable? In 1932, the Bell Telephone Laboratories in the USA were employing a twenty-seven-year-old engineer named Karl Jansky to investigate it. He had constructed a large moveable aerial at Holmdel, New Jersey, consisting of a large wooden frame about 100 feet long on the wheels of an old scrapped Model T Ford car, which could be rotated to face different directions. Its purpose was to seek out and identify potential sources of static and other terrestrial electrical phenomena that might cause problems on long-distance radio communications.

A motor would slowly rotate the "merry-go-round" once every 20 minutes to scan the entire sky. It could pick up various forms of static, such as that produced by distant storms, and he learned to identify the different crashes, crackles and hisses with various natural or man-made sources. But one source remained baffling. This was "Very weak... very steady, [and] causing a hiss in the [head] phones that can hardly be distinguished from the hiss caused by set noise".[1] It was a sound so faint and persistent that most researchers would have dismissed it as electrical circuit-related, but it captured Jansky's curious intellect, and he resolved to take the matter further.

Unlike most other sources of static, however, this hiss seemed to rise in the east and set in the west. Was it in any way connected with the sun? After a careful analysis of his results, though, Jansky realized that it could not be solar, for instead of correlating with the sun, the hiss started 4 minutes earlier each day. Checking with astronomer friends, Jansky realized that his period coincided with the sidereal or star day, of 23 hours and 56 minutes, rather than with the sun's day of 24 hours.[2] Further painstaking analyses of his data revealed the noise was somehow associated with the constellation of Sagittarius. As it was already well known that the centre of our Milky Way galaxy lay in the direction of Sagittarius, it dawned on Jansky that he was picking up a radio signal from within the depths of our own galaxy, or some 26,000 light years away. Significantly, that signal was penetrating the ionosphere and getting down to the earth's surface.

Jansky's discovery caused a sensation, and *The New York Times* of 5 May 1933 carried a front-page story entitled "Radio Waves Traced to the Center of [the] Milky Way". The discovery soon became global news, and the deep-space "hiss" was played on public radio, producing a sound resembling "steam escaping from a radiator".[3]

Surprisingly, Jansky never took things further, and this has caused much speculation. His own statements were contradictory. At one stage, he supposedly claimed that, in spite of his discovery, he was really interested in electronic technology, not astronomical research. On another occasion, it was claimed that he wanted to build a more sophisticated apparatus than the "merry-go-round" but his employers, the Bell Telephone Laboratory, were not encouraging, and as he liked his telecommunications work at Holmdel, he had no wish to change jobs. His health may also have played a part, for he suffered from kidney disease and dangerously high blood pressure, and excitement was to be avoided. He died of a stroke in 1949, at the age of forty-four. The celestial "hiss" that he discovered, and had the curiosity and persistence to pursue, opened up an entirely new branch of astronomy which, in conjunction with optical telescopic work, would transform humanity's understanding of the universe.

The "radio window" and how the radio telescope works

In some respects, the origins of radio astronomy might be dated back to 1800 and 1801, when Sir William Herschel and then Johann Ritter detected "black light" infrared and ultraviolet rays. Michael Faraday, James Clerk Maxwell, Heinrich Hertz, Wilhelm Röntgen, and even Marie and Pierre Curie added their vital pieces to the growing jigsaw puzzle of strange new energies undreamed of before 1800. Most of these energies would be identified as belonging to what twentieth-century physicists would style the "electromagnetic spectrum". With the exception of atomic radiation, they were all manifestations of wave frequencies, with long radio waves at one end of the spectrum and short, penetrating X-rays and then gamma rays at the other, with IR and UV flanking the narrow wave band of visible light.

The optical or visible light window of everyday perception is the narrowest part of the entire electromagnetic spectrum, and what radio and electronic techniques made access possible to vast swathes of invisible energies bombarding us from space, and to realms of study that no one could have imagined in 1930. When one adds the new research potential of radio to the post-1916 universe of general relativity, galactic recession, the big bang, white dwarfs, and possible black holes, one appreciates how far things had travelled since the "pre-watershed" world of *c.* 1915. But how was all this new radio-related energy harvested, analysed, and made sense of on a practical level?

Radio telescopes are reflecting telescopes, though they collect radio waves rather than the light waves of a normal optical mirror telescope. A radio telescope needs to have at least two key components. Firstly, it needs a curved solid metallic or metallic mesh "mirror", to collect the radio waves and bring them to a focus, in much the same way as a parabolic glass mirror focuses light in a Newtonian reflecting telescope. Secondly, there must be an electronically sensitive receiver that picks up the focused waves, placed at the point of radio focus: the radio equivalent of the observer's eye or photographic plate. Once harvested, the radio waves can be sent – as in the days of Jansky – to a pair of headphones, or to a pen recorder mechanism to trace variations in frequency onto a moving strip of paper. In our own time, it is more likely to be fed into a sophisticated computer, which can analyse the data in a variety of ways.

Some radio telescopes were long, lateral structures, sometimes moveable, like Jansky's "merry-go-round", while others were set up permanently on the ground. The post-1945 Cambridge University array of long, connected aerials or wires on wooden poles and covering some four acres of land was such a fixed radio telescope. Switched on twenty-four hours a day, and unaffected by clouds and rain (but not electrical storms), the Cambridge and similar fixed arrays simply let the sky drift past overhead and recorded the continuous stream of electronic energy descending from it. This enabled the astronomers to build up a continuous map of the radio universe "visible" at a given latitude.

In addition to the big, fixed field-array type of radio telescope was the steerable dish, capable of being turned to any part of the sky to collect data from particular objects, and track them across the sky. The most famous steerable "big dish" telescope was Sir Bernard Lovell's great 250-foot-diameter dish at Jodrell Bank, Cheshire, to which we will return presently

26.1 The radio telescope. The metallic dish functions in the same way as a silvered mirror in an optical reflecting telescope. Radio waves from space hit the curved surface of the dish and are reflected to a focal point at the top of the central mast. Instead of an eyepiece or photographic plate, an electronic receiver picks up the focused rays, and relays them to an electronic laboratory where the data they contain are analysed. (Drawing by Allan Chapman.)

Unlike optical telescopes, which can be focused to angles a single arc second or less, single or individual radio telescopes were blunt instruments with a crude focus and covering such a large area of sky that it was often impossible in the early days to associate a radio source with a specific visual source. That was until the development of an ingenious radio technique known as interferometry.

Pioneered at Cambridge, and then used globally, interferometry is based upon the use of a pair or more radio telescopes collecting data from the same celestial radio source simultaneously. As radio energy travels in waves with peaks and troughs, it means that two different radio telescopes some distance apart will receive slightly different parts of the wave at any one time, being slightly out of phase with each other. Should the same radio source, at the same time, be observed and recorded, let us say in Cambridge and at Jodrell Bank, Cheshire, which were about 2½ degrees of longitude apart, then the signal would appear to be slightly out of phase. By measuring the phase, or interference, discrepancy, and knowing the distance between the two radio telescopes, one could pin down the signal to a much smaller area of sky than could be done with any single telescope. And if radio telescopes in England, Europe, Australia, and the United States, with their thousands of miles base lines between them, worked in unison, then a radio source could be pinpointed in the sky with even greater accuracy. Even by the mid-1950s, interferometry was making it possible to match up visual with radio sources in the sky with great accuracy.

But we are jumping ahead, and must now return to what happened in radio astronomy in the wake of Jansky's discovery, and how it was turned from being an electromagnetic curiosity into a world-changing science.

Grote Reber of Wheaton, Illinois: an amateur leads the way – yet again!

As we have seen on many occasions in earlier chapters, most branches of astronomy were identified, researched, and advanced by independent amateurs, who first showed the way which professionals would later tread. Radio astronomy was no different, for while the first detection of radio waves was by a professional radio engineer – Karl Jansky – he was not able, for various reasons, to pursue the *astronomical* aspects his work. This would be left to Grote Reber.

Born in 1912, Reber fell in love with radio as a teenager, and by the age of fifteen had already built and was transmitting with a radio set. At first, he was interested in long-distance terrestrial radio communication, but news of Jansky's discovery changed the direction of his thinking. By the time Jansky presented his main paper to the radio engineering world in 1937, Reber was experimenting with various long-wave configurations. He was even hoping that he might be able to bounce radio signals off the moon – a decade ahead of the first earth–moon "bounces" achieved by World War II Signals Corps engineers. By August 1937, and having spent about $2,000 (or *c.* £400) of his own money, Reber had completed the world's first custom-built radio telescope. It consisted of a great parabolic dish, 31 feet in diameter, covered with contoured iron plates and supported by a great wooden frame which enabled it to be tilted to different parts of the sky. It was built in the back garden of his house, and attracted attention from curious passers-by.

Having tried, without much success, to pick up short-wave signals from space, Reber next moved to the longer radio wavelengths. In October 1938, his home-made "pie plate" iron dish picked up a signal in the 6-foot-wave part of the radio spectrum, which Reber realized was coming from the Milky Way. His discoveries led him to reorganize his day, for he worked for a Chicago company that manufactured domestic radio sets.[4]

Living in an urban environment near Chicago, Reber appreciated that cars, trains, electrical devices, and many other factors could cause static interference with his 31-foot radio telescope. Consequently, on coming home from work, he would dine early, go to bed, and be up again by midnight. He would then spend the next six hours in the basement of his house listening to, measuring, and trying to identify the sounds picked up by his dish in the back garden. This was the quietest time, as far as human-generated static was concerned. His dedication paid handsome results, for he not only confirmed Jansky's findings regarding the radio activity of the

Sagittarius region of the Milky Way, but identified several other sources, including the constellations of Cygnus, the Plough (or "Big Dipper"), Cassiopeia, and Aquila. Yet the more Reber looked into the radio sky, the more peculiar it became. The big, bright stars were invariably silent in the 31-foot dish, whereas apparently starless regions of the sky seemed to blaze with radio noise.

One truly puzzling phenomenon emerging from Reber's work concerned the nature and source of the radio noise. What generated such powerful signals in objects at such incredible distances from our planet? Was the radio noise part of the internal nuclear processes by which stars were known to burn by 1940? Astronomy and astrophysics needed to think in terms of energy bands that went beyond the ancestral visual optical regions of the spectrum, if we were to develop a more complete view of the universe.

SIR ALFRED CHARLES BERNARD LOVELL AND JODRELL BANK, CHESHIRE

Without doubt, Reber first demonstrated the radio universe merited serious scientific attention to the world, in publications such as the *Astrophysical Journal.* But by the early 1940s, both Britain and the USA had more urgent uses for radio innovation and research than scouring the heavens for celestial hisses and crackles, for World War II was in full swing. This provide a powerful impetus for all manner of radio and telecommunications research, as Sir Bernard Lovell showed in detail in a major historical paper in 1977.[5] The development of radar, for example, came about primarily in Britain, following Robert Watson-Watt's 1935 experiments, as a way of detecting the possible menace of German aircraft on the way to bomb London. Radar worked by transmitting a signal into the sky, so that if it encountered a hard, reflecting surface, such as an approaching aircraft, it would be reflected back. Then in 1942, the British radar defence apparatus suddenly ceased to operate. Had the Nazis discovered a way to "jam" it? Deep relief was felt soon after when the Astronomer Royal and his staff pointed out that a solar storm had been beaming a burst of intense electromagnetic energy toward the earth, and the sun was the cause of the radar blackout.

Somewhat reminiscent of the great magnetic storm of 1859 which first showed that the sun could emit a magnetic flux that could induce dangerous electric currents in telegraph wires, in 1942 it was shown that the sun could also obliterate terrestrial radio signals.

The radio and telecommunications developments during the war were so intensive, that when peace eventually returned, radio knowhow stood in a very different league from 1938. This new knowledge and technology would make large-scale professional radio astronomy possible. Besides the technological innovation, post-1945 Britain and America had two further advantages: a large number of men, and a smaller number of women, who had become radio and radar experts and whose knowledge and skills could be adapted for radio astronomy work; and an abundance of army surplus radio equipment being sold off cheap. This included radar dish receivers, electro-mechanical equipment such as oscilloscopes and automatic pen recorders, and an abundant supply of radio valves, rectifiers, tuners, and other equipment.

Some of these factors enabled Martin Ryle to develop the Cambridge University array of early radio telescopes, using wooden poles, chicken wire, and even old military hospital beds. When this and other apparatus was switched on and the sky circled overhead, a great harvest of data about the radio universe flooded in, to be interpreted as best it could. Former War Department equipment lay at the heart of the world's first large, steerable radio telescope observatory, as we have seen, at Jodrell Bank, Cheshire, some 21 miles from Manchester, the brainchild of Sir Alfred Charles Bernard Lovell – always known as Bernard.

Sir Bernard Lovell, born in 1913, had obtained his PhD in physics from Bristol University in 1936, when he was appointed lecturer in physics at Manchester University – which, along with Cambridge, was Britain's premier university for physics research. After the war broke out in 1939, Lovell did important work on radar and became an acknowledged expert on the subject. He had been struck by how cosmic rays and meteoritic matter left electronic traces on the apparatus. After 1945, he obtained trailers of surplus military radar equipment and had it set up at the Manchester University research field station at Jodrell Bank, and began to study cosmic rays there.

In 1951, Lovell was elected to Manchester University's new Chair of Radio Astronomy and to the Directorship of Jodrell Bank. He developed and built the world's first large-scale fully steerable radio telescope there, with a vast dish of 250 feet in diameter.[6]

Taking almost a decade to design and build, Lovell's Jodrell Bank telescope became a flagship achievement for post-war Britain, and made it possible for this rainy island to once more enter the big league of observational astronomy: a position it had lost when big telescope optical astronomy migrated to the clear, pure air of California and Illinois after 1900. The radio universe could be studied even in the pouring rain, for the electronic energy came through regardless. Jodrell Bank became a symbol of intense national pride at a time when Britain was exhausted after giving its all to beat Nazi Germany (a pride shared by the patriotic, cricket-loving Sir Bernard – who described being away from England in spring as a form of masochism).

Lovell realized, however, that the design and construction of a fully steerable 250-foot dish represented a voyage into uncharted territory, engineering-wise. For this, he brought in the engineer Charles Husband. Construction finally got under way in the late summer of 1952, with the laying of the massive foundations and circular track on which the great dish and its supporting pillars and ancillary structures would rotate. Gradually, the great lattice of steel girders, the height of a cathedral spire, began to ascend above rural Jodrell Bank. This was still semi-austerity post-war Britain, with useful second-hand commodities being bought cheap to fulfil specific functions. For example, the two great steel bearings upon which the 250-foot dish would rotate, nearly 150 feet in the air, were former gun turrets from the battleships HMS *Royal Sovereign* and HMS *Revenge*, purchased cheaply from the breaker's yard. Originally designed to turn the giant 15-inch guns toward enemy targets, the turrets now aimed the world's largest radio telescope toward the universe.

Charles Husband's design was relatively straightforward, as the 250-foot dish was designed to have two motions (in the "altazimuth" plane). One was horizontal, around a circular railway track; the other up and down, to cover the sky between the horizon and the zenith. The circular track carried and rotated the great structure of girders supporting two great steel towers 180° apart. On top of each tower was mounted one of the second-hand ship's gun turrets. Each was designed to accommodate one of the enormous circular or cylindrical pivots, one on each side of the dish, enabling the dish to be elevated up and down. By means of these two movements, the 250-foot dish could be directed toward any part of the sky visible from the latitude of Manchester.

26.2 The Jodrell Bank Radio Telescope, Cheshire. (By kind permission of Professor Ian Morison.)

The interior of the dish consisted of a steel mesh, parabolic in shape, and from the centre of the dish rose a great steel pylon or receiver. As in Reber's first dish radio telescope of 1938, the radio waves were picked up by the dish and focused to the top of the pylon, where they were conveyed to the electronic recording station and laboratory on the ground.

This novel scientific instrument took five years to build, with some funding hold-ups along the way. As a proud Manchester native, and a young schoolboy fascinated with astronomy and science, I recall the kudos that Jodrell Bank brought to Manchester, and the celebrity status accorded to Bernard or, as he became in 1961, *Sir* Bernard Lovell. I yearned to behold this wonderful and unusual telescope with my own eyes! I still have vivid recollections of a summer Sunday evening drive, squeezed into our tiny Austin Seven car with mother, dad, and two grandparents, from our smoky industrial Pendlebury in north Manchester into the leafy Cheshire countryside to see the vast construction as it neared completion. The enormous engineering structure was a truly unforgettable sight, rising above the trees, with cows grazing peacefully in the nearby fields.

As with many great pioneering inventions, the construction costs for Jodrell Bank telescope overran original estimates. The government ploughed in yet more money, but much gratitude was owed to the funds provided by the Foundation created by William Morris, Lord Nuffield, the former Oxford bicycle repairer then multi-millionaire car manufacturer, who cared for scientific and medical research. In total, the great telescope cost some £700,000, a colossal sum of money at the time. In August 1957, the 250-foot dish was positioned to receive radio signals at 160 megahertz from the zenith, and began to collect data from the radio sky. It was adjusted and the data then analysed by what was then a powerful computer.[7]

The unique contributions that Sir Bernard's great radio telescope made to the space race blazed Jodrell Bank across both the Soviet East and the free West after 1957. As we shall see in Chapter 27, Russia and the United States competed neck-and-neck from the mid-1950s onwards to put a man on the moon. Before this could be achieved, many key intermediate steps had to be taken, including setting an artificial satellite in orbit around the earth. In the early days, Russia stole the lead, especially in the summer of 1957 when it successfully launched the satellite Sputnik into orbit around the earth. Lovell's great telescope proved an ideal instrument for tracking Sputnik, as it was for a succession of increasingly far-ranging spacecraft, both Russian and American, its great dish giving it enormous sensitivity when it came to detecting faint or complex radio signals.

Jodrell Bank's steerable dish and great radio sensitivity made it an invaluable instrument for both tracking and receiving faint signals from spacecraft, especially when conjoined with other radio telescopes, and also with optical telescopes, to operate as part of an "interferometer" investigation.

Radio astronomy was advancing rapidly from the 1950s onwards, and the 250-foot instrument – the Lovell telescope – would have its technology regularly upgraded, both in terms of its physical and mechanical properties and of its radio receiving and analytical functions. One significant development, as computing advanced, was to install digital control systems, which conferred powers and research capacities which no one could have imagined back in 1952.

The 250-foot Lovell telescope is not the only major radio telescope at Jodrell Bank, for in 1964 another steerable telescope, the Mark II, with a 25-metre-diameter radio receiver dish, came into operation. It was located some 1,394 feet away from the 250-foot dish. When both the Lovell (Mark I) and 25-metre (Mark II) telescopes were electronically conjoined, they formed an interferometer system, each capable of picking up a slightly different wavelength signal from the same radio source, and enabling the pinpointing of a radio source in deep space to within an angular area of about half an arc-minute. This would correspond to an extent of sky measuring about one-sixtieth of the diameter of the moon: a tiny angle. When the Jodrell telescopes were working in conjunction with other radio telescopes, in Cambridge, Europe, the USA, or Australasia, even smaller angles could be pinpointed. Thus it became possible to make clear identifications of optically visible star or galaxy sources with radio sources. This further enabled astronomers to identify "radio stars", and perhaps even more puzzling at first, to pinpoint places in space that were sources of powerful radio emissions but which appeared totally black and empty to an optical telescope.

As Einstein, Hubble, Eddington, Chandrasekhar, Lemaître, and their colleagues back in the 1930s had shown, the visible light universe is a bizarre kind of place, and the radio astronomers after 1950 would show the invisible universe to be an even more bizarre one.

Other great radio telescopes

Lovell's 250-foot steerable telescope at Jodrell Bank shifted the entire gear of radio astronomy, and since the 1950s it has shown what a fascinating place the radio universe is. But it was by no means the only major game-changing big telescope by the 1960s.

Sir Martin Ryle, at Cambridge, was soon going beyond his early array of wooden poles and chicken wire to design a great one-mile telescope, and further interferometer arrays. Instead of one big dish telescope, as at Jodrell, Sir Martin developed arrays of large dishes, capable of being moved on rails across more than a mile of the Cambridgeshire countryside. In Ryle's great Cambridge array, dishes of around 18 metres in diameter and moveable along an east-west track up to 3 miles long, which could achieve the sensitivity of a single aperture of prodigious size, enabled radio astronomers to do at least two things. Firstly, it allowed them to survey data from a wide angle of sky, and secondly, to pinpoint radio sources with incredible precision, to identify radio needles in cosmological haystacks. This work would win the 1974 Nobel Prize for Physics for Sir Martin and his colleague Anthony Hewish.

This principle was extended even further in 1973, with the building of the "Very Large Array" of no less than twenty-seven 82-foot-diameter receiver dishes, near Socorro, New Mexico, USA. The dishes could be moved along a "Y"-shaped configuration of tracks totalling 13 miles, and the VLA added yet more sensitivity and precision to our study of the radio universe. It is known as the "Karl G. Jansky" array, in honour of the first radio astronomer. But in 1963, a single radio dish of no less than 930 feet was built at Arecibo, Puerto Rico. This utilized a vast natural depression in the earth's surface, across which a metallic radio-wave-reflecting mesh was suspended, capable of being tensioned until it formed a parabolic dish. Three gigantic steel pylons, each several hundred feet high, were then set around the great dish, some 120 degrees apart. Suspended from the top of these were steel cables supporting the receiving apparatus. This was placed at the exact radio-focus of the 930-foot dish, to pick up the signals from space. The Arecibo instrument contained all the usual parts of a radio telescope – a basic configuration shared with the early telescopes of Jansky, Reber, and Lovell's 250-foot dish – a parabolic or other metallic dish to pick up and focus the signals, combined with a central antenna to collect the focused signal and then direct it to the astronomers' analytical laboratory. The Arecibo was not far north of the equator, ideally placed for receiving both northern and southern hemisphere radio signals, especially as it pointed permanently upwards, at the zenith.[8]

As one might expect from the rapid development of electronic and engineering technology since 1950, the Lovell telescope of 1957 is no longer the world's largest steerable radio telescope. The Green Bank, West Virginia, USA, and the Effelsberg, Germany, telescopes now exceed it. Yet radio telescopes, unlike their optical brethren, are not obliged to work as a single unit, for while big lenses and mirrors produce singular images, radio telescopes can be linked up to produce vast interferometer arrays, giving them greater accuracy in pinpointing radio objects in the sky. They can, however, be subject to their own particular type of sky pollution, as can optical telescopes, especially if they are located in suburban areas. We saw above how Reber did most of his radio astronomical listening between midnight and 6 a.m., because there was minimal domestic interference.

Sir Bernard Lovell encountered similar problems in the early days of Jodrell Bank. Police, ambulance, and taxi radio "noise" could be an unpredictable nuisance, although once identified it could be filtered out. But the Jodrell Bank radio astronomers were sometimes pestered by other problems: an interference or new noise would come and go at given times of the day or night, although at times which accorded with the solar or civil day rather than the stellar day. As the telescope was in a rural area, in dairy farming country, it was eventually realized that certain noises came and went as the local farmers' time switches activated the electric fences. Other devices could also cause mischief, such as electrical switches controlling various forms of agricultural and local village machinery. Proud of Jodrell Bank and its worldwide status though people were, Sir Bernard sometimes needed to use his diplomatic skills to ensure that local farmers, publicans, and others let him know when various electrical devices were likely to be activated.

Agricultural electric fences were hardly radio sources, but sudden changes in electrical potential in a domestic circuit could register on the recording equipment. Did the unexpected click herald a major discovery, or was it just the cows coming home? Teething troubles apart, radio astronomy would rapidly exert a profound and transformative effect upon our knowledge of the universe.

The achievement of radio astronomy

Modern cosmological knowledge would not be where it is today but for the discovery of the "radio universe" around the middle of the twentieth century. This includes not just straightforward radio waves from space, but also the detection by specialist telescopes of X-rays and alpha, beta, and gamma particles. One such discovery, as we saw in Chapter 25, was the cosmic background radiation coming out of the big bang about 13.7 billion years ago. This optically invisible aftershock of energy suffusing space provides radio-observed evidence for the big bang in a similar way to that in which airborne echoes continue to resonate following a violent terrestrial explosion. Yet without radio instrumentation, we would have remained ignorant of this radiation.

Radio telescopes are also used to sweep the sky in its entirety, in much the same way as William Herschel had swept it for the optically visible star-fields, nebulae, and clusters of the Milky Way. Consequently, we now possess a highly detailed radio map of the sky, covering both hemispheres, amplifying and improving the first radio map built up by Reber from Illinois. These surveys, of the sky began to reveal certain regions radiating intense localized energy sources, which turned out to be what would be called "supernova remnants", the remains of stars that exploded due to internal atomic changes. Several have exploded during well-recorded astronomical history, although no one at the time knew what had happened, except that an occasional new optically visible star would suddenly flare up to emit intense light, sometimes appearing even brighter than Venus, at least for a short time.

In 1054, for example, Chinese Mandarin astronomers recorded the appearance of a new or guest star in what we call the constellation of Taurus (the Chinese used different names for their star groups). The new star could even be seen in broad daylight. Then in November 1572, the Danish astronomer Tycho Brahe and others across Europe saw a similar stellar flare-up in Cassiopeia. After a radio sweep of the sky, the remnants of these long-extinct supernovae were still found to be intensely active. While giving out little or no visible light any more, they still remained extremely powerful radio transmitters.

In 1968, only a few months after Jocelyn Bell Burnell's Cambridge discovery of the first pulsar,[9] David H. Staelin, Edward Reifenstein III, and other astronomers at the Green Bank, West Virginia radio telescope detected two pulsars in region of the Crab Nebula. This was very significant because, knowing from Chinese astronomical records that the Crab supernova had blown up in 1054, twentieth century astronomers could accurately trace how its remnants had evolved across 900 years. Within the remnants' expanding gas and dust cloud were some visible light-emitting stars. Could one or both of them be pulsars? Once we had established the location of a light-emitting pulsar, we could also obtain its spectra and a host of chemical and physical details that might provide a vital key to understanding the physics of pulsars.

The pulsar – wherever it was in the Crab Nebula remnants – rotated thirty-three times every second, being a collapsed neutron star spinning at incredible speed, and sending a flashing beam of energy into space as it did so: which could be picked up as a series of rapid pulses with a terrestrial radio telescope. Knowing the precise flash period of the pulsar, it was possible for optical astronomers to devise a shutter mechanism tuned to operate in exact synchronicity with the pulsar's signals.[10] Let us assume that all the pulsating energy coming from the pulsar contained its visible light as well as its radio beam. If the spinning camera shutter were synchronized to be closed each time a pulse came down to the optical telescope (thus preventing light from the pulsar reaching the photographic plate), it would never register on the resulting photograph. All of the non-pulsating stars, however, which emitted a continuous stream of light, would register on the photograph. By noting which star was missing from the Crab Nebula when the camera shutter was closed, therefore, it would be possible to identify the pulsar and make it available for detailed study.

Gravitational lensing also became possible when radio and optical telescopes worked together. A gravitational lens might be a region of space where a stream of optical and radio energy from far away in or beyond our own galaxy passed by an object that was nearer to us yet which possessed an intense gravitational field. The object's gravitational field would bend the energy-stream from behind in a similar way to which a glass lens bends the light passing through it. And just as a glass lens enables us to see things beyond itself, so a gravitational lens can reveal immensely distant yet otherwise invisible galaxies beyond itself.

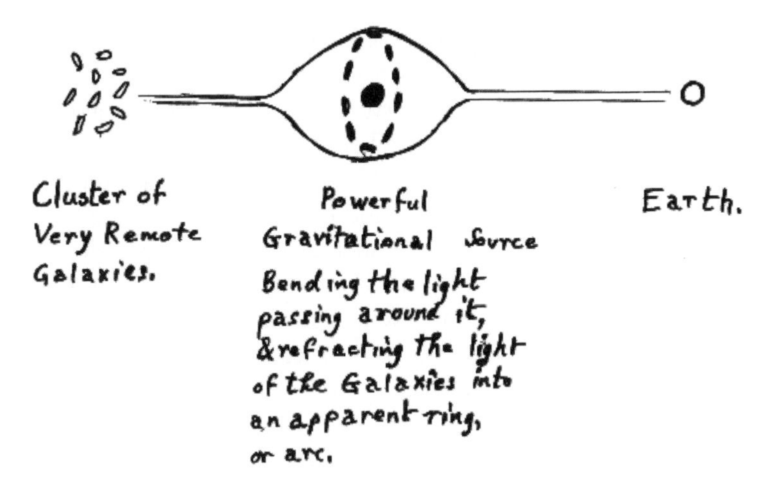

26.3 Gravitational lensing. Direct light from very distant galaxies is obscured by a closer, intensely powerful gravitational source. Yet this source can sometimes take the light and energy approaching it from behind (from the distant galaxies), and bend it by means of its own gravitational force. This can result in a curious-looking clump of faint galaxies, which modern astronomers came to realize were distant objects that had been gravitationally lensed by an intervening gravitational source. (Drawing by Allan Chapman.)

This was first done with the Jodrell Bank telescopes in the 1970s, where the great 1957 Lovell telescope and the smaller Mark II receiver located a radio field, later correlated optically with a pair of dim blue stars. Since the 1970s, gravitational lensing has become a valuable new technique for the deep-space astronomer, giving access to remote regions invisible by more straightforward radio or optical methods.

Radio astronomy, therefore, especially working in conjunction with the optical branches of the science, has revealed a vast wealth of information about the deep-space universe unimagined in 1950, and proved useful closer to home. The planet Jupiter, for example, has also shown itself to be a powerful radio source, enabling optical astronomers to better understand the complex and often turbulent belt systems of the planet: one almost sufficiently large, at 88,890 miles in diameter, to become a star in its own right. (Jupiter is one-tenth the size of the sun, and rotates in an incredible 9.8 hours.) Likewise, our knowledge of the sun has been greatly amplified by radio astronomy, and improved our knowledge of meteor storms and the aurora borealis.

The early twentieth century saw prodigious advances in optically based astronomy, matched by fundamental theoretical developments. In the wake of Karl Jansky, Grote Reber, Sir Bernard Lovell, Sir Martin Ryle, and others, radio astronomy took the astronomical sciences to heights that Harlow Shapley and even Sir Arthur Eddington would have found beyond imagination. The inevitable question now was "Where next?", especially when we factored in data obtained from telescopes and probes orbiting the earth and moving through space.

Sir Bernard Lovell: a recollection

I first saw Sir Bernard Lovell as a teenager, somewhere in the early or mid-1960s, when he delivered a public lecture on Jodrell Bank and the new radio astronomy to a capacity audience in the Great Hall of Bolton Technical College, Lancashire. He held the audience spellbound, as he opened up the wonders of the radio universe, and of the great telescope which he directed. I saw him, as did my parents, at a distance of 20 or 30 feet, as one of the crowd. In later years, I enjoyed the privilege of conversing with him at Royal Astronomical Society meetings and exchanging comments about astronomical history. Nor was astronomy our only shared interest:

Sir Bernard was also organist at St Peter's Church, Swettenham, the Cheshire parish in which he lived. I was an aspiring organist – though never as accomplished as he was. I always enjoyed reading about his ecclesiastical musical activities in the local Manchester press, and my parents sent me cuttings after I went to university in 1969, for Sir Bernard's activities were always news in Manchester.

26.4 Sir Alfred Charles Bernard Lovell. (By kind permission of the Royal Astronomical Society.)

On 14 March 1997, he made what I understand was his last visit to a meeting of the Royal Astronomical Society, at Burlington House, Piccadilly, London. Following the meeting, a group of Fellows of the RAS went off, as usual, to dine together at the Athenaeum Club in Pall Mall. I chatted to Bernard, and then, after the formal dinner was over, he, Patrick Moore, and George Cole suggested that we go for a drink in the large Drawing Room upstairs. Late-night train and bus timetables back to Oxford were forgotten, as I said "Yes." How could I pass up such an opportunity? The eighty-five-year-old Sir Bernard was in no especial hurry, as he was staying the night at the Athenaeum anyway, so the four of us went up and talked. I was fascinated by what he told me about the early history of Jodrell Bank, the space race, and Britain's very weighty contributions to radio astronomy. We even talked about our mutual love of church organs, and especially, the music of Sir William Herschel. Sir

Bernard was a powerful presence in every sense. Yet he was also a man of humility: courteous, kindly, and with a gentle sense of humour. I will never forget the occasion.

Chapter 27

"Fly Me to the Moon": The Birth of the Space Age

The 1950s and 1960s witnessed the achievement of one of humanity's most ancient dreams: to leave the earth's surface, and not merely fly in the air – which had become commonplace – but rise up into space itself, and at least potentially, up to the moon and beyond. Many factors would play their part in that monumental achievement, including the space race between the mutually hostile Soviet Union and the USA, representing as they did two fundamentally different views of the human condition in the twentieth century. It has been argued that the conquest of space played a significant role in preventing overt warfare between these two atomic bomb-possessing superpowers, as the glory of space conquest became a bloodless way of contending for prestige. Yet within the quarter-century following the end of World War II in 1945, rival Russians and Americans would go through all the trials, experiments, and engineering triumphs of getting into space, culminating, on 20 July 1969, with the first human beings, Neil Armstrong and Edwin Eugene – "Buzz" – Aldrin, making a safe landing on the lunar Sea of Tranquillity, and returning safely home in triumph to a jubilant USA and Western world. That, and future space expeditions and probes, both manned and unmanned, over the following decades would teach us more about the fine detail of the solar system than any earth-based telescope could ever do.

Rockets into space

Gunpowder was one of the great inventions of medieval China, perhaps dating from the tenth century. The Chinese appear to have used black powder to devise rockets, which they used for both military and celebratory firework displays. One legend, however, has it that a Chinese Mandarin named Wan Hu tried to fly in the air in some sort of vehicle powered by rockets attached to a chair. It was in Western Europe, however, that gunpowder came to be used for all manner of things after 1267, when Friar Roger Bacon of Oxford was said to have devised his own mixture of saltpetre, sulphur, and charcoal, as described in his *Opus Majus*. Later, another ingenious monk in Germany, legendarily said to have been one Berthold Schwartz (Black Berthold), had the bright idea of packing gunpowder into a tube to produce the first gun, although Chinese sources record bamboo guns being used in the Far East before that.

Although the troops of the King of Mysore in India had used rockets against the British in the 1790s, it was an English military engineer and experimentalist at the Woolwich Arsenal, south London, who turned the gunpowder rocket into a truly deadly device during the Napoleonic Wars. This was Colonel Sir William Congreve FRS, who, after 1804, tried to perfect a pointed iron tube nearly 4 inches in diameter, filled with an evenly packed gunpowder propellant that would carry a warhead nearly 2 miles, to explode or cause fires on impact. Congreve's biggest rockets were deadly, though their great weakness lay in their inaccuracy and the rocketeer's inability to control their paths once launched. Congreve transformed the rocket from being a firework into a powerful, payload-carrying piece of technology with a future potential.[1] Even before his time, "Firemasters" developed rockets for entertainment or for signalling purposes, and devised multiple-stage rockets, so contrived as to rise to great heights before exploding in displays of colour.

Congreve's military rockets began to excite the curiosity of ingenious civilian experimentalists who glimpsed their scientific potential. In France in the early nineteenth century, for example, Claude Ruggieri launched rats, mice, and other animals high into the air with powerful rockets, to test how rapid acceleration and suddenly diminishing air pressure affected them. By the 1840s, it was even being shown that the rocket did not need its time-honoured stick, as newly attached fins provided stability and could improve the flight path by making the rocket rotate, like a shell fired from a rifled cannon.

As we saw in Chapter 11, when Jules Verne wrote his adventure novel *From the Earth to the Moon* in 1865, it was a giant cannon-shell that carried his three bold adventurers into space, and not a rocket. For once away from the earth and in space, what would a rocket's thrust push against? It would slow down and eventually drift, well before reaching the moon. The man who, perhaps more than any other individual, would talk seriously of space travel, in 1903, and propose the rocket as a means of propulsion, was a largely self-taught Russian engineer and physicist who had been stone deaf from the age of nine: Konstantin Eduardovich Tsiolkovsky.

The rocket men

In 1903, Tsiolkovsky was forty-five years old. He had already produced experimental designs for airships and war devices, but after his *Exploration of Cosmic Space by Means of Reaction Devices* (1903) he would be acknowledged as establishing several key principles in scientific rocketry and space travel. His term "reaction devices" supplies a major clue to his thinking, because he recognized a rocket motor worked in accordance with Newtonian and other physical principles, perhaps the most important being that every *action* produced an equal and opposite *reaction.* A rocket motor did not in fact need something to "push" against – such as the earth, or even terrestrial gravity – because a powerful release of energy in one direction, such as the rocket's blast, would automatically release energy in the opposite direction, to propel it forward. So even if a rocket motor fired in airless outer space, it would still push the spacecraft with equal force in the opposite direction, making it possible to travel through space.

This meant that the fiery energy discharge could take place in the vacuum of space, so long as the rocket fuel carried its own oxygen. In his various rocket publications, however, Tsiolkovsky realized that solid fuels, such as gunpowder, were inadequate for space travel. It was, for instance, largely impossible to stop gunpowder from burning once it started. Rocketry required the ability to switch the spacecraft engines on and off, as flight details required. It was unnecessary for the rocket motors to keep burning once the vehicle was launched into space, for Newtonian physics showed that once free from earthly gravity, the craft would continue at its initial velocity indefinitely, without requiring a single extra rocket blast. Additional bursts of the engines were needed only to change flight direction, or ensure a controlled landing on the moon, a planet, or mother earth. Tsiolkovsky then realized that a viable space vehicle had to be powered by liquid fuel stored in on-board tanks, which could be fired to produce thrust or be shut off as circumstances required. Sadly, he failed to attract financial support, and consequently was never able to build experimental rockets. Even so, his publications laid down much of the theoretical groundwork that would underlie twentieth-century space rocketry, as late Czarist Russia slid increasingly into chaos, then the bloodbath of World War I, before the mayhem of the 1917 Communist Revolution occurred.

Things were very different in the United States, which prospered, at least until the Wall Street financial crash of 1929. In that land of entrepreneurs and ingenious engineers, rocketry attracted the attention of the academic physicist and engineer, Robert Hutchings Goddard. In 1919, he published a famous monograph entitled *A Method of Reaching Extreme Altitudes*.[2] He had already demonstrated the rocket's suitability for space travel, in 1915, when he successfully fired an early rocket in a vacuum chamber, and found that – in accordance with Newton's laws – it worked perfectly. While Tsiolkovsky was largely a theoretician, Goddard was an experimentalist, and having got the Smithsonian Institution and the pioneer aviator Charles Lindbergh on his side, Goddard succeeded in obtaining a handsome annual research grant of $15,000 from the Guggenheim Foundation. In 1926, he built and then successfully flew the world's first liquid fuel rocket. With small tanks for petroleum and liquid oxygen respectively, and a firing chamber in which both fuel products could be combined and combusted, he was ready to launch his pioneering liquid fuel rocket on 16 March 1926. As he had no formal research field station at that date, however, the historic launch took place at his aunt's farm at Auburn, Massachusetts.

The 4-foot rocket roared aloft for a mere two seconds, and achieved a 40-foot ascent before its fuel ran out. It was insignificant compared to what a conventional gunpowder entertainment rocket could achieve by that date, but a new technological "baby" had been born which would go on to change the whole scale of scientific knowledge.[3] Funded by the Guggenheim Foundation, Goddard was able to set up a proper rocket research station at Roswell, New Mexico, where by 1935 he had sent a liquid fuel rocket to an altitude of 12.5 miles, at a supersonic speed. Rocket propulsion was not his only concern, for he also developed gyroscopic devices to maintain flight stability, along with other refinements. Goddard was also sufficiently astute as to patent his rocket innovations, about 100 in all, and following his death in 1945 his widow received 1 million dollars from the US Government for their usage in rocketry.

Working in parallel with Goddard, and inadvertently co-discovering some of his principles, was the Romanian (later German) medical student turned rocket pioneer Hermann Julius Oberth. In 1922 Oberth submitted a doctoral thesis to Heidelberg University on how one might reach "escape velocity" and break free from gravity by means of rockets. The thesis was rejected, but he went ahead in 1923 to publish his text under the title *Die Rakete zu den Planetenräumen*, or "The Rocket into Interplanetary Space". For some time, he was unaware of replicating work already done by Tsiolkovsky and Goddard, though he later acknowledged their priority. Oberth made it clear from subsequent remarks that his childhood reading of Jules Vernes's *From the Earth to the Moon* inspired him, as it did many others (myself included). He almost succeeded in building a rocket as a spin-off for Fritz Lang's film *Frau im Mond* ("the Woman in the Moon") in 1929, an early space travel film for which he was scientific advisor. By this time, Oberth was the driving force behind the German *Verein für Raumschiffahrt*, or Society for Space Travel, though following an accident when an experimental mixture of petroleum and liquid air exploded, aggravated by various disappointments and depression, Oberth simply vanished from the scene.

When World War II broke out in 1939, Germany soon made use of the available expertise to assemble a team of physicists and engineers who would produce the deadly V1 and V2 military rockets that would play havoc with London's air defences. Under the Versailles Treaty following World War I, Germany had been forbidden to devote research resources to conventional military technology, such as guns and explosives. While Oberth and other members of the Society for Space Travel were inspired in the early 1930s by the rocket as a vehicle for space travel, such researches inevitably led to the accumulation of technical expertise that made possible the Peenemünde rocket facility on the Baltic coast, from which the military rockets would be launched against London by 1944.

By 1931 the Berlin rocketeers had developed an aluminium engine weighing a mere three ounces, yet capable of a 70-pound thrust, and early rockets (not infrequently assisted by access to Goddard's own legally accessible patents), some of which were reaching 1.5 miles altitude by 1932, were being brought safely home by parachutes.[4] Then in 1937, the 25-year-old Wernher Magnus Maximilian Freiherr von Braun, from a high-status German family and an active member of *Verein für Raumschiffahrt*, was appointed civilian head of the German government's rocket programme.

Although he worked for the Reich throughout World War II, and masterminded the "Vengeance Weapon 2" or V2 which bombarded London from across the North Sea, von Braun was not a Nazi. Nor did he commit specific war crimes. When the Nazi Reich collapsed in 1945, von Braun and his team of scientists and engineers made a strategic surrender to the American forces. Other German rocket scientists fell into the hands of the invading Russians. This, ironically, supplied the expertise and personnel which would lay the foundations for both the USA and Soviet space programmes.

Von Braun had a distinguished career in America, becoming Director of the Marshall Space Flight Center of NASA. From here, his team developed military ballistic rockets, such as the "Jupiter C", which put Explorer 1, the first US artificial satellite, into orbit on 31 January 1958, several months after Sputnik. The mighty "Saturn V" rockets, standing as high on the launch pad as a cathedral spire, which propelled the Apollo spacecraft to the moon, came from von Braun's rocket stable. Perhaps more than any other scientist, he became the technical architect and driving force behind the American space race. He died in 1977, aged sixty-five.

The first space flights

The development of rocket physics and technology was one of the great achievements of human ingenuity, on a level with the medieval clock, Newtonian physics, the steam engine, electricity, cars, aeroplanes, and modern medicine in its wider cultural impact. Like them, rocketry was an iconic statement of the creative power of the human intellect, requiring technical skill and sophisticated social organization. In the two centuries since Congreve, the rocket has been used as an instrument of scientific discovery and of destruction, and in the digital age, a vehicle with which to launch telecommunications satellites and facilitate the internet.

By the 1940s, rockets were sufficiently powerful to reach the 100 miles upwards that would just put them into space. It was not only the destructive V1 and V2 rocket technology as developed by Nazi Germany that was important. Manchester University's Alan Turing's fundamental mathematical work would go a long way toward developing the first computers. Rockets operate and fly in orbits defined by Newtonian gravitation theory, but a successful space flight requires not only launching the rocket into a given curved path, but also being able to tweak and adjust its reaction engines and flight path once the vehicle is in space. This control demands fast-moving mathematics of the highest order, and not even an army of trained women and men equipped with slide rules and logarithmic tables could hope to keep up with the fluid higher mathematics involved in a space journey. Hence the need for enormous number-crunching computers, equipped with valves and later with transistors, which could respond to a rocket in orbit and control its engines and flight path. As rocketry and space travel became ever-more sophisticated, computers would evolve, doing such things as transmitting electronic pictures to us from space.

In this Cold War era, the USA and the USSR were, as we have seen, competing neck to neck, for "firsts" and the global prestige that always accrues to winners. The Russians seized the first victories, after 4 October 1957, when Sputnik 1 was successfully launched into an orbit around the earth.[5] I remember, as a boy, standing outside on a chill autumn evening to see the artificial satellite move slowly across the sky. It appeared as a bright star, reflecting the sunlight still shining at high altitude, before suddenly disappearing upon entering the earth's shadow. Just a month later, the Russians did it again, when they launched the dog Laika into space. The poor little animal was doomed from the start, as she would burn up when re-entering the earth's atmosphere after the various sensors attached to her body inside the capsule had monitored the physiological reactions of the first living creature to enter space. The dog-loving British in particular took pity on Laika, and there were comments about "heartless Commies" (Communists). By 1960, however, the Russian spacecraft Sputnik 5 had successfully launched animals into space, and brought them safely back to earth.

The ideological race was partly squared when the American Explorer 1 satellite detected the energy from the Van Allen belts around the earth, although Sir Bernard Lovell kept the British flag proudly flying as his newly commissioned Jodrell Bank radio telescope collected and relayed data impartially from both US and Soviet probes. The Russians then won another triumph in October 1959 when Luna 3 sent back pictures from the far (or dark) side of the moon that can never be seen from earth. Even these blurred images compelled astronomers to form a new understanding of lunar history, for the far side was so different from the side which is gravitationally locked toward the earth. There were no great lava-plain "seas" which are such a conspicuous feature of the earth-facing side. Instead, with the exception of a couple of small, round "seas", there was just one vast silvery land mass, peppered with craters and pits. How had this come about?

Nor was everyone happy about the new "Russian moon". Was it not just another devious Commie plot to hoodwink the West? Some "experts" even claimed that the far side photograph was not of the moon, but of a made-up painting upon which brush marks could be discerned. In fact, the far side photographs were entirely genuine, as later American probes confirmed.

Yuri Gagarin (1934–68), the first space man, 1961

In many ways, the dream of the ages of mankind breaking loose from the terrestrial realm was finally realized on 12 April 1961, when Yuri Gagarin of the Soviet Air Force was launched into space inside a Vostock spacecraft, orbited the globe, and made a safe landing home after a flight of one hour and 48 minutes. Gagarin flew at an altitude of only 91 miles, but that placed him above the atmosphere in space, and made him feel weightless. He became not only a Soviet hero, but a global one as well, for no one could deny his monumental achievement and sheer courage. Nor his charm: in an era of hard-faced, hard-line Soviet ideologues, here was a Russian with a big smile who, on his celebrity tours beyond the Soviet Union, chatted with English schoolchildren, joked with factory workers, warmly received civic honours bestowed upon him, and even became the subject of a large and popular tableau at the Blackpool Illuminations, which I remember well. Yuri Gagarin was so much more than a "performing monkey in a tin can" as some mockers claimed. He was the herald of *Homo sapiens* in space, and a daring trailblazer. It was all the more tragic when he was killed in an air crash in 1968.

The Apollo missions

The world's admiration for Gagarin was in no way diminished by the fact that the Americans were far from happy with appearing to play second fiddle in the 1950s and early 1960s space race. This led to NASA starting its series of Apollo space missions, and its programme was galvanized on 25 May 1961, when the newly inaugurated US President John F. Kennedy made a historic pledge before Congress. He pledged that America would land a man upon the moon before the end of the 1960s decade.

First came the rigorous testing of the technology, as the evolving dynasty of rockets, Saturn I, IB, and then the mighty Saturn V were put through their paces between 1961 and 1968. What made all of this possible was that prodigious power of industrial organization which the United States developed over the late nineteenth and twentieth centuries. One must see the space programme in context, alongside the railroads, electrical technology, Henry Ford's transformation of mass motor-car manufacture after 1912, the production-line manufacture of "liberty ships" during World War II, and the burgeoning aircraft industry. While the Soviet Union's vast capacity for human regimentation and centralized control all helped to launch Sputnik and Gagarin into space in the 1950s and 60s, it was the much more diverse ingenuity- and profit-driven US economy that was sufficiently flexible and responsive to land two men safely in the lunar Sea of Tranquillity in 1969, and bring them and their command module captain safely home.

The moon-landing was eventually accomplished only after several years of meticulous preparation, equipment trials in space, and rehearsals spanning the 1960s. In addition to the preparation of the hardware, the men selected to be astronauts perfected skills hitherto unrequired of human beings: how to leave the safety of the spacecraft, and go for a space-walk, and how to assemble and manipulate pieces of equipment and tools in the bizarre environment of gravity-free space. Accidentally let go of a spanner, and off it will float into space, to drift for all eternity. More frightening, do not snag or cut the umbilical cord that connects the walking spaceman to the safety of the command module, otherwise he will drift, dead, into the airless depths of cosmic eternity, his colleagues, only a few feet away, incapable of helping him. Speaking of any astronaut, whether American, Russian, or any other nationality as just "men in tin cans" obeying orders, as some ignorant mockers have, is to wholly misunderstand what was involved. These men, and later women, needed a calmness of mind, intelligence, quick thinking, and sheer courage that were truly out of this world.

The Apollo space flights had a remarkable safety record, although tragedy struck in January 1967 when Virgil "Gus" Grissom, Edward H. White II, and Roger B. Chaffee died on the launch pad when the Apollo 1 command module burst into flames. The following Apollo flights were successful, each one building in a planned sequence upon its predecessor. Between Apollo 7 in October 1968 and Apollo 10 in May 1969, a whole range of technologies were perfected, extending from sending astronauts around the moon to the sophisticated handling of the lunar module, a trial run for the lunar landing.

Touchdown: the Sea of Tranquillity, 20 July 1969

The world was holding its breath on the evening of Sunday 20 July 1969 (British Summer Time) as we awaited the landing of Neil Armstrong and Buzz Aldrin in Apollo 11 on the surface of the moon. In the early hours of Monday 21st, mankind's dream of the ages came to fruition, as two human beings made a soft, safe landing on the moon. Armstrong and Aldrin conducted scientific experiments, walked, jumped, left footprints in the lunar dust, took photographs, collected specimens of moon rock, and planted a US flag: all recorded by TV cameras to be broadcast worldwide.

Three men had travelled to the moon in Apollo 11, but only two of them, Armstrong and Aldrin, actually touched down. The third man was Michael Collins, whose job it was to stay behind in the moon-orbiting module, where he managed the wider enterprise, and ensured the safe docking of his friends when the time came for them to be blasted up from the lunar surface to begin their journey home to earth.[6] The now assassinated President Kennedy's 1961 pledge had been dramatically fulfilled. Up to Apollo 17 in 1972, a succession of further missions would tell us much more about space.

The parallels between the early Apollo missions, especially Apollo 8, which carried Frank F. Borman, James A. Lovell, and William A. Anders in orbit around the moon, and Jules Verne's fictional adventure of 1865, were striking. Both space expeditions carried three men, Verne's men being launched into space from Florida, USA, and the Apollo astronauts from Cape Canaveral/ Kennedy, in the same state. Just like the real-life Borman, Lovell, and Anders, so Messrs Ardan, Barbican, and Nicholl, their frock-coat tails flapping in weightless space, scrutinized the opposite side of the moon from their projectile window. Then upon arrival at a pre-calculated point in space whose lunar and terrestrial attraction (they hoped) might be counterpoised, Verne's intrepid adventurers fired gunpowder signal rockets to kick them toward earth. Both sets of astronauts, a century apart, made safe splashdowns, to be picked up and brought to safety and heroes' welcomes, by US ships: the fictional *Susquehanna* and the real-life USS *Yorktown*.

The Book of Genesis goes to the moon: Christmas 1968

Looking at "earthrise" from the Apollo 8 spacecraft as it orbited the moon would have been breathtaking under any circumstances, yet something occurred in that spacecraft which made the American Atheists' Society hit the roof, and instigate legal action. The three astronauts, Borman, Lovell, and Anders, each took turns to read sections of the creation story, the first ten verses of Genesis, to the waiting world, and then rounded it off by wishing us all a "Merry Christmas". Wisely, the US Supreme Court threw out the Atheists' challenge, which under any circumstances appears rather small-minded, seen within the wider context of the three men's monumental courage and achievement. Nor was this all. When Buzz Aldrin was on the surface of the moon with the Apollo 11 mission in July 1969, he took Communion with consecrated elements supplied by his home church. A copy of the Bible was also left on the moon. As we have seen in the above chapters, astronomy and religious faith have always enjoyed a close connection, and it is hardly surprising that the first humans to cross the void of space to see, and even to touch, another world should have been especially inspired.[7]

The end of manned missions

Although there have been no manned missions to the solar system since the 1970s, space vehicles have routinely conveyed people to the International Space Station, orbiting the earth. Here, Americans, Russians, and the Englishman Sir Tim Peake have performed a whole array of experiments some 220 miles in orbit above the earth's surface. The study of cosmic rays and gravity, and observations on the human body in a weightless environment have produced a rich harvest of new knowledge across two decades, in spite of the tragic loss of seven astronauts' lives in the returning Space Shuttle in 2003. But why were no more human-carrying missions sent further afield?

The answer may lie in developments in computer technology and robotics which few could even have imagined in 1965. Unlike human beings, a sophisticated robotic probe can go to sleep for months or years when not in use, then on awakening collect and relay vast quantities of data back to earth. It never gets tired, will not have a nervous breakdown, and leaves no grieving loved ones behind when it fades away or comes to a sticky end. In spite of the immense skill and courage displayed by intrepid astronauts of various nations, the unmanned probes to distant and hostile locations have taught us the most about the solar system over the last fifty years.

The unmanned space probes

From the 1960s onwards, a succession of probes have been designed and launched. These included the solar system Voyagers 1 and 2, launched in August 1977, and the Spirit and Opportunity Mars Rovers after 2003. The two Voyager probes were launched at a fortuitous time, when a grand alignment of the planets of the outer solar system was due to take place, in the hope that the spacecraft could fly past them to collect detailed physical and chemical data that could never be obtained from a terrestrial observatory. The whole Voyager enterprise only became possible after a great raft of sophisticated technologies had become available. These involved complex, controlled rocketry, solar cell electrical power, elaborate orbital manipulations, and a whole array of on-board analytical equipment, computers, and transmitters to send the data back to earth. And the Jet Propulsion Laboratory (JPL), California, and NASA had to be able to receive, store, analyse, interpret, and publish the billions of encrypted pixel signals coming down to earth.

The first problem was to design a Newtonian-gravitation-based orbit that would carry the probes into the outer solar system, so that Voyager 2 would fly past Jupiter, Saturn, Uranus, and Neptune, neither crashing into them nor straying wholly off target. This was the Grand Tour of the solar system. Voyager 1 was sent on a different orbit to gather data about Saturn's satellite Titan, which was believed to have an atmosphere, and was subsequently found to possess oceans of liquid methane. As they flew, over nearly three decades, precise instructions to the Voyagers' navigation systems enabled booster rockets to fire so as to tweak the orbits to obtain the correct angles and positions for the on-board cameras and instruments to do their work to optimum effect.

By the time that Voyager 2 was flying past Saturn, Uranus, and Neptune, the distances were so vast that not only was planet earth no more than what Carl Sagan styled a "pale blue dot", but the sun itself was little more than a very bright star. This meant that it could take a command signal from earth – travelling at 186,000 miles per second – up to four hours to reach the spacecraft and activate its machinery as it approached Neptune. A similar time was needed for the spacecraft's data to reach us. Meticulous planning and timing were essential, therefore, if everything were to go to plan. But it did, time and again. None of this would have been possible without the post-war revolution in computer technology and servo-control systems, as well as rocketry, as one branch of science and technology after another all came together to make the Voyager project yield its fresh fruits of knowledge.[8]

The cornucopia of scientific results that poured in from the two Voyager missions, and later the Cassini–Huygens, 2004–17 (Saturn), and New Horizons, launched in 2006 (Pluto), was amazing. There were the beautiful close-up pictures of the planets and their satellites, along with data about Jupiter's complex and turbulent meteorology and atmospheric chemistry, the structure of Saturn's rings, and the magnetic fields of the "gas giants" of the solar system. Neptune was found to have its own, albeit much less pronounced, set of rings similar to those of Saturn, as vast swathes of rocky lumps and particles, captured by the planet's powerful gravitational field, had come to arrange themselves into a plane around Neptune. The New Horizons probe, in 2015, produced splendid images of Pluto and its satellite Charon, and has now exited the solar system in the hope of imaging objects in the Kuiper Belt. Then there were the planetary satellite families, as the frozen oceans of Jupiter's satellites Ganymede, Europa, and Callisto gave rise to speculations about the possibility of liquid water beneath the ice, and whether, warmed by internal heat, they just might host some kind of life-forms. Saturn's moon Enceladus might also have life-related chemical processes going on in geothermally heated water below its icy mantle. Conversely, Jupiter's moon Io was discovered to be peppered with active hot gas, sulphur, and volcanic "lava" vents. Katherine de Kleer of the Pacific Gemini Observatory successfully imaged violent volcanic outbursts on Io in 2013 and thereafter, as the satellite spewed hot material some 200 miles above its surface. Likewise, Neptune's Triton had been found in 1989 to have active hot nitrogen vents, the ejected gas from which freezes in space, falls back down, and gives the satellite a smooth, snow-like covering.

Not even in the wildest imaginings of a space-travel enthusiast could this amount of data have been collected by manned craft. Robotics clearly were, and I believe still are, the way forward in space exploration. Both Voyager 1 and 2 are still transmitting useful space data back to us, including pictures of the whole solar system, which they have now left, but their return to earth is clearly impossible – as was fully understood when they were designed in the 1970s. Yet after forty years of active life, as they head for interstellar space, the Voyager spacecraft leave no tearful loved ones behind. Nor do the Cassini–Huygens or New Horizons missions, or the Mars planetary probes. Yet before returning to Mars, let us say something about an unmanned earth-orbiting probe which has greatly augmented our knowledge of deep space.[9]

The Hubble Space Telescope

In spite of the size and power of great optical telescopes, all earth-based astronomy is limited by the air above our heads, which even from a Californian mountaintop varies in atmospheric pressure, humidity, and light-scatter. By the mid-1980s, therefore, the space technology industry was discussing putting a large optical telescope into orbit around the earth. Here, it would be outside the atmosphere, and would enjoy perfect seeing conditions whenever the sun was not in its field of view. Named after Edwin Hubble, who first revealed the true "extra-galactic" nature of the one-time nebulae, the large lorry-sized optical laboratory, with its 8-foot-diameter primary mirror, was launched in April 1990. There were teething troubles, for the first pictures transmitted back to earth were blurred, and a spaceman-engineer had to be sent up 353 miles to make adjustments. From then onwards, however, the images of deep-space galactic structures which it sent back were simply wonderful. The dust clouds within the Eagle Nebula, in the Serpens constellation, became global icons, although Hubble's real value lay in the hundreds of high-definition images it sent to scientists of hitherto invisible galactic structures. What is more, successive advances in digital imaging technology since 1990 have enabled us to extract details which would have been impossible twenty years before, and all this has confirmed and refined our knowledge of the universe's expansion.

But now, let us return to Mars.

EXPLORING THE SURFACE OF MARS

Although the Soviet Union trailblazed much early space exploration during the 1950s and early 1960s, that lead would soon be lost to the United States, which would carry space exploration into the twenty-first century. How far this was due to the difference between centralized and free-market developments in technology is a debatable point, though I suspect that the industrial organizational lead which the USA had been developing from the 1880s played a part. But the basic fact is that by the 1960s Russian rockets were blowing up or crashing, while American ones were consistently delivering payloads into space.

As with the moon-landing preparations, so Martian studies began with fly-past and fly-around expeditions with the Mariner spacecraft. Detailed surface maps were constructed, as well as spectroscopic studies of the Martian surface chemistry and CO_2 atmosphere, which was found to have a pressure of no more than 5 to 10 millibars on the surface. (The earth's atmospheric pressure is just over 1,000 millibars, or 29.92 inches.) This was during the late 1960s, and Mariner 9 not only detected dry riverbed type features on Mars, but also obtained close-up pictures of the satellites Phobos and Deimos.

All of this led up to the Viking expeditions of 1975, which possessed two research components: an orbiter which encircled the planet and made an even more detailed map, and a Lander vehicle, a small portable laboratory, which was brought safely down upon the Martian surface. Its purpose was to study the rock and sand in its vicinity, and especially to search for possible micro-organisms and other signs of life. Its findings are still being debated today.

Gradually, and systematically, our knowledge of the Martian surface became increasingly detailed with each new probe, such as the Mars Global Surveyor of 1996. Great volcanoes such as Olympus Mons had been found, and immense fissures in the crust which made the Grand Canyon look like a mere scratch. All of these features were generally much bigger than their equivalents on earth, yet they were on a world only half the size (at 4,200 miles in diameter). Mars's smaller size and mass also led to the realization that its once mighty volcanoes might well have blasted rock fragments into space when erupting. Rocky debris collected by geologists and experts in meteorites from the earth's surface in certain stable desert terrains have been found to contain chemical structures closer to the spectroscopic studies of the Martian surface rocks than to those of terrestrial rocks. So has the earth's greater gravity captured rock fragments blasted into space by volcanoes on Mars, whose lower gravity meant that they never fell down again, but flew off into space?

By 2003 things had gone far enough to consider landing two unmanned, computer-controlled probes on to the Martian surface, on sites carefully selected for their geological interest by previous satellite surveys. In June and July 2003, first Spirit and then Opportunity were sent from earth, each, after a six-month flight, making a successful landing, the wheels, solar batteries, cameras, probes, and transmitters all unfolding on schedule. Then in 2012, they were followed by the even more sophisticated Curiosity, which deposited a car-sized laboratory on wheels on Mars. By 2017, this was not merely still operational, but had travelled over nearly 10 miles of Martian terrain. Spirit, Opportunity, and Curiosity do all sorts of things, such as bore holes into Martian rocks to analyse their chemistry and physics.

Between the early Mariner fly-past photographic surveys and the fruits of the three mobile probes after 2003, Mars had helped create a new branch of astronomy: planetary science. One of the most fascinating things has been the physical changes through which it had passed, whose traces could be clearly read and interpreted by terrestrial astronomers in their laboratories. The Martian surface, for example, seems once to have possessed sufficient water as to leave river-valley-type erosion and flow patterns. Then there were the craters, canyons, mountains, great volcanoes, and other features. Had there once been a much denser atmosphere, perhaps containing oxygen, which the lower gravitational pressure had allowed to leak away into space? Atmospheres are turbulent, as they heat and cool, and we on earth have been fortunate to have a gravitational pull sufficiently strong as to keep the oxygen, nitrogen, and carbon dioxide atoms firmly locked in and enveloping our planet. Little Mars was not so lucky. So had Mars once hosted some kind of life? Not Martians as the advanced beings envisaged by H. G. Wells or Hollywood, but simple structures such as bacteria or simple botanical forms?

It was the possibility of discovering such life-forms, or traces of life-forms, that inspired the late Professor Colin Pillinger's ill-fated Beagle 2 expedition of 2003. Named in honour of HMS *Beagle*, the ship that carried the young Charles Darwin around the world in the 1830s and played a part in shaping his later evolutionary ideas, Beagle 2 was a project of the European Space Agency (ESA). Sadly, photographs from January 2015 indicate that its folding energy and transmission panels were damaged upon landing, and it never sent a single signal back to earth. Landing a laboratory vehicle on to the Martian surface inevitably involves a good deal of luck. After the vehicle is released from its orbiting command module, it depends on parachutes in the thin Martian air to slow its descent. Then gas bags will inflate to soften the impact of hitting the Martian surface, although almost inevitably this will lead to the whole craft bouncing for some distance, which can easily make things go wrong – as seems to have happened to Beagle 2.[10]

Beyond doubt, modern robotics have transformed our knowledge of the solar system in a single lifetime. And while human beings have died in space accidents nearer to home, the planetary expeditions to Mars and beyond have not cost a single human life.

Terra-forming Mars

We know far more about Mars than about any other object in the solar system, making it, in the eyes of some people, a possible second home for the human race. Robotic probes have taught us much about Martian soil chemistry, and I know scientists, including some in Oxford, who have synthesized Martian soil in the laboratory. Such soil syntheses enable scientists to experiment with possible plants, gases, and atmospheres which might lead to the release of oxygen in large greenhouse-type buildings on Mars in the future. These just might provide environments within which humans might be able to live, and even cultivate foodstuffs. So, in the generations to come, could we perhaps "terra-form" Mars to generate living conditions suitable for humans? As things stand at the moment, it is a very long shot, although the prodigious advance of planetary science since 1950 has demonstrated that it is unwise to shut the doors upon what might be achieved a century or two hence. Can we travel beyond Mars, or even beyond the solar system? We will return to this in the last chapter.

Chapter 28

A Universe for the People: Sir Patrick Moore and the New Amateur Astronomy

Astronomy has had a large popular following since the eighteenth century, as we have seen. Several popular civic astronomical societies had come into being after 1859, while the national, and international, British Astronomical Association had been actively promoting astronomy after 1890, along with the American Astronomical Society of the Pacific after 1889. So what happened after 1950 to greatly enhance astronomical interest among the wider public, in Britain, Europe, the USA, and Japan? As we shall explore in this chapter, global astronomical events helped to stimulate this interest, along with mass-produced, high-quality telescopes and accessories, and the rise of several charismatic popularizers. So much hinged upon the communications revolution after 1950, and the rapid spread of television ownership in most Western countries.

Astronomy has, however, always suffered from a significant disadvantage not affecting most other sciences: it is rarely taught in schools as a science in its own right, complete with instruments and observing techniques. Physics, chemistry, and biology can be taught anywhere and at any time, so long as the correct laboratory facilities are available for teachers and pupils to work with, but astronomy is associated with the night, when school pupils are doing homework or tucked up in bed. However, sunspots may be observed in the daytime, in complete safety, by using a telescope to project an image of the solar surface on to a white screen.

In Britain, astronomy might best be taught at the public or boarding schools, because the pupils are there at night, and many of these institutions have fine observatories. Some of them date from the nineteenth century, with big refracting telescopes, so that science teachers can easily demonstrate the night sky to their charges. Marlborough College, Stonyhurst College, and Rugby School possess classic observatories, while as we saw in Chapter 17, Stonyhurst also played a major role in early astrophysical research. Yet it is surprising how many day-school pupils, and teachers, are wonder-struck when shown the moon, Venus, or Jupiter through an even modest astronomical telescope. This is always brought home to me whenever I, or friends in local astronomical societies, demonstrate telescopes on school science days or to the public at large. "Are they *really* the craters?" they will say when looking at the moon at fifty times magnification; or, "It looks like a new moon" when beholding the crescent Venus, or "What are these little stars near Jupiter?"

Popular astronomical fallacies

I am always puzzled by how many people, even in our own time, fear or mistrust science, as I know from ample personal experience. "Chemicals [with complicated names] are dangerous, because, unlike plants, they are not natural" is a common one. Likewise, "Physics is about dangerous rays" or "Some biologists in the laboratories are trying to do funny things to us, such as creating babies with two heads." Genetics is especially mistrusted, while genetically modified plants are sometimes referred to as "Frankenstein" foods.

Astronomy has aroused its own fears in certain quarters. As a boy, I recall an elderly relative warning my mother that the moonbeams entering my eye through my simple home-made spectacle-lens telescope would "soften my brain", and I could end up being "mooned" (rather like the "moon calf" Caliban in Shakespeare's *The Tempest*). I also have a long-standing, albeit eccentric, friend with a first-class degree in English Literature and a pride in his ignorance of science who hates black holes. "They are *evil*," he affirms. The facts that black holes possess no moral agency and are billions of miles away in the depths of space fail to convince him.

Then there are those folks who are into aliens, of whom I have met many over the years. They believe they have been abducted by beneficent beings from somewhere near Orion: beings who are deeply concerned about humanity's murderous tendencies, and want to save us from ourselves. I have even been given detailed descriptions of the interiors of their spaceships by people who have supposedly been for a galactic spin inside one. Why do these well-meaning wonderful beings only reveal themselves to a certain type of idealistic, suggestible, or not entirely sober person, in lonely places late at night? Why, after crossing the galaxy at twice the speed of light, do these creatures never come down at mid-day in London's Trafalgar Square, or on the front lawn of the White House in Washington, to deliver their messages of hope?

We looked at the historical fascination with intelligent life on other worlds in Chapter 11, and in spite of a resounding lack of evidence for any type of such life, blind faith in its existence continues unabated in people who would swear by their commitment to rationality and who abominate all forms of superstition. More will be said about this in Chapter 29.

Then there are the astrologers. I looked at the serious, and even academic, roots of astrology in my book *Stargazers*,[1] and how the growing evidence-based case for the Copernican and then the infinite universe had caused it to die away by 1700, at least in scientifically informed circles. Like a belief in aliens, astrology survives despite evidence to the contrary, even in Europe and the USA. The Victorian astrologer Zadkiel was given a good comic beating by Mr Punch in *Punch* magazine in December 1859, while *Old Moore's Almanack*, claiming a seventeenth-century lineage dating back to Theophilus Moore of Dublin in 1697, is still published. It is now, in 2017, in colour and an updated format.

28.1 The astrologer Zadkiel being beaten by Mr Punch. (*Punch* magazine, December 1859.)

There are probably more millions of people into astrology in the West today, given our greater population size, than there were in 1700. Popular magazines, newspapers, and the internet all recognize that predicting the future is a profitable business. People make a good living as professional astrologers, either writing astrological horoscope columns in the press or else undertaking private consultations. While it is true that they are making a good living out of peddling false hopes to people who don't do science, on one level I have a sneaking regard for them. They are simply playing the market, and doing very well, having mastered the art of what my own dear late mother Lily called "getting your living without working for it". As the Great American Showman Phineas Taylor Barnum legendarily put it 150 years ago, "There's one born every minute."

Modern astrologers do fully accept on one level that the earth goes around the sun, and that there are galaxies out there –

maybe inhabited by kind-hearted green, grey, or transparent folk, worried about the human race. So it has been necessary for them to re-engineer the traditional geocentric cosmos, and discover that powerful astrological forces have some sort of modern-science-related physical basis. Galactic magnetism, relativity, gravity, and atomic radiation have all been held responsible for why people born under Gemini the Twins are double-minded, or Taureans strong and bullish. I have even met modern astrologers who call upon fanciful notions of modern genetics as an explanation of why the horoscope "works": "Galactic radiation affects the DNA, you know"! This is fine on one level, for as the early twentieth-century uranium and radium researchers discovered, big doses of X-ray and other radiation types often resulted in leukaemia, as cell replication was affected. Yet no one has quite explained to me how particular kinds of space radiation give you the specific characteristics of a Scorpio or a Virgo in accordance with classical astrological criteria.

I also tend to find that many modern people into astrology have a penchant for alternative medical therapies and the powers of crystals, and a tendency to believe that indigenous tribes living in the Amazonian rain forest or elsewhere possess a profound wisdom that we technologized Westerners have lost, and should go back to. Yet as I have pointed out elsewhere, when it comes to predicting an election or referendum result, or how the economy is likely to perform if certain things take place, then our modern-day number-crunchers and pollsters are no more reliable. Renaissance monarchs had astrologers armed with brass astrolabes to help fathom the future. Their modern equivalents have experts with supercomputers. The results are often about the same.[2]

Television and astronomy's new popular audience

No human group in the history of our planet has been offered so many choices as post-war Europeans, Americans, Australasians, and perhaps the Japanese. All of this has come via a proliferation of multi-media technologies. On 2 June 1953, Her Majesty Queen Elizabeth II was crowned in Westminster Abbey. This coronation differed from all others, in the fact that ordinary folk in Britain could enjoy a grandstand seat to watch the event. Even if you did not belong to that privileged set who had seats in the great Abbey, and were not one of the jolly crowd standing in the pouring rain to watch the Golden Coach go down the Mall, you could watch the Coronation live in comfort at home. That is, so long as you had saved up around £100 to buy a heavy, 2-foot-cube wooden box enabling you to tune in to a 9-inch-diameter cathode ray tube in a darkened room, to gaze upon a flickering black and white image.

In 1953, in hard-up post-war Britain, £100 was a lot of money, equivalent to ten or twelve weeks' wages for the average industrial worker. Yet hundreds of thousands of ordinary British folk scraped together the necessary sum to invest in a telly for the Big Day, and for home entertainment. My maternal grandparents were among that number, and were joined that day by parents, aunts, uncles, and other family members perched on any available seat to watch, and cheer, when the Archbishop of Canterbury lowered the Crown of State upon the beautiful young Queen's head.

At the time, I was a little boy who had just celebrated his seventh birthday, though watching that historic event in the crowded front room of a small terraced house in industrial Lancashire supplied me with memories I will never forget, and only strengthened my burgeoning love of science, history, religion, monarchy, and public ceremonial. Yet what, one might ask, has all of this to do with astronomy? Nothing in itself, I admit, but the televising of the 1953 Coronation unleashed a new and transformative force into British society – as it had already begun to do in the USA – the power of television, which had opened up a Pandora's box of wonders, for good and ill. Astronomy, along with many other subjects, would benefit from it.

Sir Patrick Alfred Caldwell-Moore and *The Sky at Night*, 1957–2012

By the mid-1950s, many factors were exciting public interest in astronomy. There was the space race, as we have seen, which was going full tilt by 1957, while as we saw in Chapter 26, radio astronomy was opening up a vast new domain for astronomical research. The Mount Palomar 200-inch telescope was revealing deep-space wonders that provoked many questions for thinking people: including Oxford dons, coal-miners, teachers, and housewives.

The BBC realized that the time was now ripe for some kind of TV programme dealing with astronomy. This would be astronomy on all levels, from discussion of the big discoveries to practical help for hands-on amateur astronomers, many of whom had first been captivated by the clear night skies during the public lighting blackouts of the war.

28.2 Modern amateur astronomers at work, 2010. (By kind permission of the Mexborough and Swinton Astronomical Society.)

Several potential presenters for such a programme were considered by the BBC. None seemed quite right. What was needed was not a high-powered university professor, nor a popular entertainer who did not quite understand his script. It needed someone in between. They wisely settled upon a thirty-four-year-old former RAF officer and preparatory schoolmaster, who would, in 1957, present the first episode of a programme that he would continue to host down to his death in 2012. He would present every monthly instalment, as well as specials, with one single exception, when he became ill after eating a tainted duck egg. The programme was *The Sky at Night*, first broadcast on 24 April 1957, and the presenter, Patrick Moore, soon became a household name. His opening words, "Good evening, and welcome to *The Sky at Night*" would almost become a catchphrase. He was knighted in 2001, and won a BAFTA award for his services to television. Patrick also received the Jackson-Gwilt Medal of the RAS for his services to astronomy, and since 2012, the RAS has each year awarded a Patrick Moore Medal for schoolteachers who have made a noteworthy contribution to astronomy or geophysics.

Yet as Sir Patrick told me on several occasions – for I had the honour to get to know him, and to appear with him on eleven (I think) episodes of *The Sky at Night* – he never had a formal contract with the BBC. Instead, he had what he described as a "Gentlemen's Agreement", adding: "You know, there were *gentlemen* in the BBC in those days." Patrick turned out to be the perfect presenter for *The Sky at Night*. As a young man, he stood well over 6 feet tall, was powerfully built, eccentric, charismatic, and highly articulate. Though never university-trained (he said that his wartime service in the RAF made university impossible), he was an accomplished practical amateur astronomer from his pre-teenage years in the early 1930s, and the youngest ever member of the British Astronomical Association.[3]

28.3 Sir Patrick Alfred Caldwell-Moore. (With thanks to the photographer, Peter Louwman.)

Moonstruck: amateur astronomy and the moon after 1950

Our satellite, the moon, was the first celestial body toward which Thomas Harriot and Galileo first quite independently turned their telescopes in 1609, and the moon has remained the favourite object for amateur astronomers, both beginners and more advanced observers, until large-aperture "go-to" computer-controlled high-quality telescopes became both widely available and affordable during the 1980s. At a quarter-million miles distance, and with all the complex detail of its broken and rugged surface, the moon was ideally placed for modest-aperture telescopes. As we saw above, even well-educated people often gasp when they first see the moon through even a small telescope.

My own first sight of the telescopic moon was through a *c.* ×20 magnification penny-in-the-slot instrument on Torquay promenade when I was nine or ten years old, on holiday with my parents. I was transfixed, and could not scrounge enough pennies from my parents to enable me to make sufficient repeat observations. How I wanted a telescope of my own! Over the years, I have never ceased to be delighted when I have given both schoolchildren and adults their first telescopic glimpse of the moon – free of charge.

Since the mid-seventeenth century there have been numerous published atlases of the moon, including those made by photography from the 1870s onwards. In fact, amateurs became the foremost "selenographers", or moon-mappers, and in 1946 Hugh Percy Wilkins produced and published a splendid chart some 300 inches across. Wilkins and the young Patrick Moore would come to work together, while the British Astronomical Association Lunar Section was always popular among Association members.[4] Making an accurate drawing of the moon, however, was not easy, because under varying conditions of illumination, as the phase constantly changed, features seemed to metamorphose in shape and size. To become a serious selenographer, therefore, it was first necessary to know the moon like the back of your hand. This demanded a serious commitment of time and energy, to go to the telescope eyepiece, at any hour from dusk to dawn, and observe what was visible. It helped, naturally, if you did not have a job which tired you out, and if you could function on three or four hours' sleep in every twenty-four. Most Grand Amateur astronomers had leisure and money, and did not need to be at work, bright and perky, at 8 a.m.

Once a lunar observer had become visually familiar with the lunar surface, there was the cartographic problem of how to understand, then draw, a sphere on a flat piece of paper. Inevitably, the shape of lunar features becomes distorted the closer they are to the edge, and as we on earth can never place ourselves so as to see round the edge of the moon, you need a good sense of geometrical perspective to do lunar mapping. On the other hand, the moon's surface appears only in stark, monochrome black and white, with grey for the "seas", and there are no colour tints to worry about, unlike when drawing Mars, Jupiter, or Saturn. A good black pencil and a sheet of white paper are the traditional ways to depict the lunar surface. Because the moon is wholly airless, moreover, the light is never softened. Stark white is seen when we observe features in direct sunlight, with relief and moulding provided by the inky-black shadows. Lunar draughtsmanship is a skill, therefore, that develops only after many hours at the telescope eyepiece, matched by trying to capture on paper what you think your brain is telling you. It is a true combination of art and science.

Yet if a selenographer is aspiring to a wholly accurate moon chart, he or she needs to use an eyepiece micrometer, with an adjustable pair of eyepiece cross-wires, which can be used to measure the shapes, sizes, and relationships of objects in exact angular terms. This was the way in which selenographers aiming at the highest possible cartographic accuracy worked before photography, and in accordance with which many modern serious amateurs still work today. One might reasonably ask, however, "Why map the moon by hand now that we can do it photographically?" In addition to the sheer delight in drawing and telescope manipulation that many serious modern amateur astronomers experience, there is another factor. When we look at the moon or any other celestial body through a telescope, we inevitably look through the dense and turbulent gases of our terrestrial atmosphere. This "boiling" effect of the air often momentarily destroys critical telescope clarity, resulting in photographs that often have blurred parts on a lunar feature. As the human eye can work at a slower, more concentrated rate than the exposure of a photograph, it can also absorb detail, filter out blurs, and build up a remarkably coherent overall image which can be drawn with enhanced clarity. Thus, a skilled lunar draughtsman can often capture more fine surface detail than a camera. Nowadays, as for so long in the past, amateurs have perfected and developed these skills.

Only fairly recently has it become possible to develop truly high-definition astronomical cameras working in conjunction with personal computers. In such cases, a lunar photographer may expose several dozen digital images of the same lunar feature over a period of several minutes. During this time atmospheric disturbances will come and go, even on a clear night, slightly blurring some parts of the field of view while leaving others crystal clear. Thus, with several images to play with, the astronomical photographer – usually an amateur working in their back garden – can compile a stack of images of, let us say, the same lunar crater. Then, using the computer, all the best bits of each image can be selected and collated, to result in an overall photograph of astonishing clarity. In this way, an amateur astronomer armed with a good 10-inch-aperture reflecting telescope, a digital camera, and a computer with the right software can take a garden photograph of the lunar surface, as good as the Mount Palomar 200-inch telescope could have done in 1960 when using conventional gelatine photographic plates.

Yet is the moon's surface *always* starkly monochromatic?

TRANSIENT LUNAR PHENOMENA, OR "TLP"S

There has been much discussion in the lunar community, both professional and amateur, about the possible present-day continuity of volcanic activity on our satellite. Academic lunar geologists are generally of the opinion that the moon has been geologically "dead" for millions of years, yet amateur astronomers sometimes report having seen patches of colour, usually reddish in hue, near certain lunar craters, suggesting, perhaps, the escape of gas (not flames) from the moon's interior, and occasional bright spots on its unilluminated parts. (Sir William Herschel reported one such lunar bright spot, or "volcano", to the Royal Society in 1787,[5] although no one nowadays believes it to have been an erupting volcano.) Though not a regular lunar observer myself, I once saw an interesting redness on the moon.

On 18–19 May 1992, my wife Rachel and I were staying as house-guests of Sir Patrick Moore at "Farthings" ("Far-Things"), Selsey, Hampshire. Rachel had gone to bed, and Patrick and I were talking together in his drawing-room. About 2 a.m. on the 19th, Patrick said "The moon will soon be up", and we went out to his great 15-inch-aperture reflecting telescope mounted in one of the domes in his garden. Patrick opened up, and under a clear, pre-dawn sky, trained the large instrument upon the rising moon, directing it, as I would soon discover, to the crater Aristarchus with an eyepiece giving ×270 magnification. Unusually unforthcoming, as if not to give anything away, he invited me to look through the eyepiece. I saw what appeared to be a distinctly reddish haze above Aristarchus: stark in the otherwise monochromatic moonscape. I described it to Patrick. His reply was "*Exactly.*" It was a "TLP".

We both continued to observe for about 45 minutes, until the reddish patch had gradually dissipated, and I turned in at 4 a.m. When we came down to breakfast in the morning, Patrick showed me his official report of the observation which he had prepared for the British Astronomical Association Lunar Section, where I found myself logged as Patrick's co-observer of the TLP. Transient lunar phenomena are relatively scarce and, as he said, it was very unusual for a non-regular lunar observer to see one. Was there some kind of gaseous emission from Aristarchus, which was refracting the light to the red end of the spectrum, before it faded away? As to its true nature, I suspend judgment, but as for the colour patch itself, I have no doubt of its reality.

Good telescopes for all

It was the relatively cheap, often homemade, silver-on-glass reflecting telescope popularized by the Revd Thomas William

Webb in the 1860s that gave an especial boost to popular amateur astronomy during the Victorian age. By the 1950s and 1960s, however, scores of new amateur astronomical societies were coming into being, and the cheapest way to acquire a decent astronomical telescope was to make one. As we have seen, glass mirror blanks could be purchased from firms such as Pilkington's in St Helens, Lancashire, while the necessary pieces of kit for making a 6-inch-glass mirror (the favourite beginner's telescope) could be purchased from the local hardware shop, or by mail order. All this was taking place on a scale that would have warmed the heart of the Revd Webb.

Having got his (telescope makers were usually men) glass mirror blanks, carborundum abrasive, pitch, and other items, the aspiring astronomer faced many hours of grinding his mirror, first to a spherical curvature, then "parabolizing" the curve to ensure that all the reflected light came to one single focus. "Foucault tests" would be applied to check focused image quality, then after much tinkering with and perfecting the optics, the astronomer would send the mirror away to be commercially "aluminized" (no longer "silvered") to make its optical surface highly reflective. The only bought-in ready-made pieces would be the 45° "flat" mirror and the eyepieces.

Building a stable mount for one's lovingly fashioned optics was often a triumph of ingenuity. Old car axles with their ball-bearing wheel hub attachments were sometimes pressed into service to provide a beautifully smooth rotating telescope mount. Large plastic sections of (new) drainpipes were used as telescope tubes, and old door fittings, bolts, and bars were adapted to create catches and adjusters. It was a world which I knew well by the time I was twenty, as did many of my astronomical friends. Every one of the burgeoning astronomical societies in the mid-1960s had its own inner club of amateur telescope makers, keen to teach any novice who came along.

By the late 1960s, however, amateur telescope-making was gradually becoming redundant as the post-war Japanese optical industry took off. By 1968, I was the proud owner of two very good Japanese telescopes: a Newtonian reflector and a refractor, both supplied complete with mounting stands and folding tripods. They gave surprisingly good-quality images, and were much cheaper than anything of comparable specification by British or European manufacturers. This trend has intensified over the ensuing decades. Mirrors and object glasses have got bigger and better, while prices have stayed remarkably affordable. At the present time, one can equip a modest private observatory, with telescope mount, computer controls, and even a strong, commercially produced PVC plastic observatory building, for the price of a modest second-hand car. One no longer needs to own a brewery or have a generous rich spouse to do serious Grand Amateur astronomy, as did the Victorians. Office workers, train drivers, teachers, and even a professional children's entertainer whom I know are all involved in serious amateur astronomy. Nor does a modern amateur with commercially manufactured equipment need to master all the details of the sky to find a given nebula or dim planet. Modern "Go-to" telescopes contain a programmed microchip memory of the positions of Messier and NGC (New General Catalogue) galaxies, and even the current locations of the moon and planets. All you need to do is switch on the electrics, key the code for the "Andromeda galaxy" into the computer, and stand back. The instrument makes a whirring noise, the telescope moves all on its own, then stops. Apply your eye to the eyepiece, and lo and behold, the Andromeda Galaxy awaits your gaze!

Should one next attach a digital camera to the eyepiece, a set of images can be easily recorded, and then processed on a personal computer. In this way, an ingenious bus driver working from a back garden in Northampton can have a detailed picture of the Andromeda galaxy which would have astonished Edwin Hubble at Mount Wilson in 1930.

The researches of modern amateur astronomers

By the early twenty-first century, serious amateur astronomy had become an organized pursuit across the Western world. Britain,

Europe, the USA, Australasia, South Africa, and Japan had become the "dynamo locations". The Japanese had not only initiated the global revolution in cheap, high-quality telescopes and ancillary equipment, but also enjoyed a huge advantage in geographical location. Positioned as it is on the extreme eastern edge of the Asian land mass, and with several thousands of miles of sea separating it from the American land mass, Japan catches the eastern rising of the sky before anyone else does. Comet Ikeya-Seki, discovered by the amateur astronomers Kaoru Ikeya and Tsutomu Seki in 1965, inspired many fellow-amateurs across the globe, as did that discovered by Yuji Hyakutake in January 1996, which, with its very long tail, was a fine sight.

Where modern amateur astronomers fulfil a valuable role in modern astronomical research is in monitoring the heavens for a variety of phenomena, such as comets. Amateur research also includes monitoring the periodicity of variable and coloured stars, both within our own and in other galaxies, along with watching for unexpected phenomena on our own sun, such as solar storms, faculae, and unusual prominences seen in hydrogen alpha light. Comet-sweeping is a major area where amateurs do valuable work, as well as meteor and aurora borealis and aurora australis studies. Regarding aurora studies and solar storms, I know serious amateurs who have built sensitive magnetometers and electroscopes capable of responding to changes in the sun's electromagnetic behaviour. One has built an instrument with a sensitive weightless pointer. When a narrow beam of light is shone upon a tiny mirror inside the instrument, it is reflected 20 feet or so across the room to fall upon a fixed scale. This weightless pointer makes possible a high level of sensitivity for the magnetometer. Another friend in Edinburgh regularly photographs the northern sky to record the aurora borealis.

All this is only a part of what modern serious amateur astronomers are now doing. Equipped, as many now are, with very substantial telescopes, video-recording cameras, and a whole host of computer technology, they regularly produce splendid images of Jupiter's changing belt systems, Saturn's rings, and the surface of Mars, and monitor TLPs on the moon.

The post-1950 amateur astronomy movement

Several further factors have played a significant role in the post-1950 rise of amateur astronomy. Shorter working hours, or at least more flexible hours, have helped, as the majority of people these days, at least in Britain, no longer have to go marching through the factory gates at 7:30 a.m. The great rise in the self-employed sector of the economy has also given independence and flexibility of working time to millions, especially when self-employed people earn their living working on a computer. An enthusiastic amateur might choose to observe the heavens until 2 a.m., roll out of bed at 9:30 a.m., and gradually come round hugging a large mug of coffee as the computer warms up for the day's work.

Personal computers and mobile telephones can put amateurs with similar interests in contact with each other, anywhere in the world. Observations, discoveries, and beautiful digital images of distant nebulae taken in one's garden can now flash around the world in the twinkling of an eye. In addition to digital networks, most amateur astronomers, at least in Britain, have their own societies where they can meet face to face, listen to lectures, compare telescopes and observations, and combine science with friendship and relaxation. While the amateur astronomical society may have been a Victorian invention, scores of new ones have come into being over the last century, and especially since the 1950s, when the space race turned the attention of millions of people skywards. Astronomical journalism plays a major role as well, with magazines like the British *Astronomy Now* and the American *Sky and Telescope* providing authoritative articles, advice, and beautifully reproduced illustrations.

Many astronomical societies own fine private observatories which members can use, some dating back to the 1890s, with others stretching across the twentieth century. Yet modern urban and suburban environments, with their sky-glow light pollution, are not good observation sites, even if one does have access to a magnificent 8-inch-aperture Thomas Cooke refractor dating back to Queen Victoria's time. Twenty-first-century amateur astronomers, however, like most other folks, probably drive a car.

Modern astronomical telescopes, with their strong plastic tubes, fittings, and accessories, are compact and lightweight. So if serious amateurs find that streetlights now ruin their garden sky, they can easily pack up their 5-inch-aperture computer-controlled refractor, or a reflector equivalent, into the back of their car, and off they go. Should their best local dark sky be found in the Kentish Weald, Wales, the Pennines, or the St Croix River an hour's drive from busy Minneapolis St Paul, Minnesota, all well and good. They might meet up with their astronomy club friends, each with their own portable observatory, taken out of the car boot. Such out-of-town, car-accessible locations can also host "star parties" of astronomical camping and caravanning friends, who might combine observing with a barbecue.

These are all aspects of modern amateur astronomy which no Victorian could have imagined – but this is how modern-day amateur astronomy operates and thrives.[6] One is rarely in the company of British amateur astronomers for long before the name "Patrick Moore" comes up. Yet before looking at him, we must mention the work of a very different popular astronomical writer.

Carl Edward Sagan and *Cosmos*, 1980

Carl Sagan's thirteen-part *Cosmos* TV series, shown first in the USA and then syndicated globally, followed by the book of the series, caused a popular sensation at the height of the Cold War in1980. Sagan's magnum opus did not just deal with astronomy, but also with its historical and philosophical aspects, and in particular, it articulated his firmly held view that space must be richly inhabited, not only by intelligent life, but also by high civilizations. Sagan was an initiator and staunch supporter of SETI, the Search for Extra-Terrestrial Intelligence, after 1984. His poetic description of planet earth as a "pale blue dot" in space, in the wake of the Voyager 1 photograph in 1990, would become one of the iconic phrases of the age.[7]

Yet Carl Sagan and Patrick Moore had very different agendas. Sagan, the academically trained scientist, was interested in making people think and philosophize, whereas Moore, the self-taught amateur, wanted people to get outside, look at the heavens through telescopes, and do hands-on practical astronomy, and was sceptical about flying saucers and unproven grand ideas. In many respects, they appealed to different types of minds; yet both Sagan, who died of cancer complications in 1996 aged sixty-two, and Moore, who died at eighty-nine in 2012, exerted a profound influence upon astronomy and astronomically related thinking.

SIR PATRICK MOORE: THE MAN AND THE ASTRONOMER

Patrick Moore made no claim to being an academic cosmologist or understanding the intricacies of relativity theory. He offered no opinions about possible beginnings and endings, about space being warped, bent, or evolving. But as a young man, he had ground mirrors, and built an excellent 12½-inch Newtonian reflecting telescope which, until his death on 9 December 2012, still stood in honourable retirement in its home-made "run-off" shed in his garden at Selsey. With this instrument, the younger Patrick had spent countless hours surveying, drawing, and remapping the moon. Patrick never ceased to claim, "I am a moon man, you know."

Yet in addition to the moon, Patrick had, through decades of practical experience stretching back to his boyhood, built up a meticulous knowledge of the night sky. The constellations, star clusters, and nebulae, and where each of the planets happened to be for any given night of the year, were all firmly noted and arranged in his photographic mind. He could tell you exactly what was where in the sky for pretty well any night of the year. This is a skill rarely possessed by modern academic professionals, but which he shared not only with contemporary amateurs, but with the whole "Grand Amateur" tradition of astronomy extending back through the centuries. This practical mastery of astronomy, plus his engaging fluency of speech and lucidity as an author, made him an ideal communicator of the science to the growing legions of British amateur astronomers.

Patrick Moore's *Guide to the Moon* (1953) and his iconic *The Amateur Astronomer* (1957) remain models of clear and engaging technical prose-writing, going through numerous revised editions over the years, and constantly being brought up to date. *The Amateur Astronomer* especially taught the aspiring amateur how to judge and to use a telescope, what was what in the sky, and where the constellations and planets were, and included the stories behind some famous discoveries; as well as lots of practical tips. His books were carefully and abundantly illustrated, and while Patrick always claimed that he was bereft of artistic skills, he was, even so, a fine lunar and planetary draughtsman. And when fame came to him in a big way with *The Sky at Night*, as we saw above, he displayed no interest whatsoever in being a "celebrity", for he still drove around the country in his old Ford car – the "Ark" – to lecture to countless amateur astronomical societies, and always declined payment of any kind. He was a man of extraordinary generosity: both of his own time, and out of his own pocket.

One incident that really brought home Patrick's easy manner, willingness to impart knowledge, and generosity took place when I was with him at the British Astronomical Association Centenary Conference in Liverpool in 1990. A poorly dressed and clearly hard-up boy came up to Patrick requesting his autograph on a crumpled piece of paper. Instead of signing it, Patrick took the lad over to the big commercial bookstand, and told him to choose whichever book he wished, irrespective of the author. Patrick then paid for the book, signed it, and presented it to the boy.

On top of all this was his voluminous letter-writing, battered out with two spatulate fingers on his antique Woodstock typewriter. Everyone who wrote to Patrick – be it the Astronomer Royal, Mr X of Taunton struggling to "parabolize" his newly ground mirror, the person who had seen a flying saucer hovering above his garden shed, the egregious "Mrs T. Pott", or the schoolchild seeking advice about how to make a telescope from a pair of granny's old spectacles – received a full, personal reply. There are many eminent professional astronomers today who freely admit that Patrick was the first to inspire them to study astronomy, as others claim for Carl Sagan.

But what sort of a man was Patrick? As I had the honour to get to know him quite well, and to broadcast with him between the late 1980s (I think) and 2012, I offer some personal observations. Most notably, he was jolly good company. He made no bones about his views on political correctness: he was a British patriot, a monarchist, and quite destitute of any left-wing credentials. In terms of religion, however, we differed. He had been an unbeliever, but following the death of his beloved mother Gertrude Lilian at a great age in 1981, he edged toward a sort of spiritualism, though personal belief was an area where his otherwise abundant loquacity always dried up. Should one be walking down the street or be on a train with him, heads would turn, and people would point, smile, and wave. I vividly remember the journey with Patrick, travelling standard class between Liverpool and Oxford, after the above-mentioned 1990 BAA meeting. At every station, people came up to the carriage window to wave. It was a bit like being on a Royal Progress!

When his old chum since wartime days, the astronomer Colin Ronan, was present, the sparks of wit and humour flew. In many ways Colin was the antithesis of Patrick: always impeccably dressed and groomed, and looking every inch the retired Major that he was, though with a sense of mischief that was never far away. Young Patrick, by contrast, always appeared sartorially shambolic, wearing a baggy blue suit looking, as my old geography teacher once suggested, "as though it had been cut with a circular saw", and a rather skewed RAF tie, and with unruly hair. Yet Colin and Patrick got on like a house on fire. Before Colin's sudden death in 1995, we often dined together at the Athenaeum Club following Royal Astronomical Society meetings, and if our friend Andrew Murray, retired from the Royal Greenwich Observatory, was also dining, we were in for an entertaining evening. As an astronomical historian, I later wrote down the stories, snippets, and jokes that passed around the table. (The mischievous Colin Ronan would sometimes make Patrick growl by referring to his Selsey home, "Farthings", as "Fartings".)

Patrick's personal appeal extended well beyond astronomy, encompassing all manner of unlikely people. He was firmly against all blood sports and anything harmful to animals, and, like Edwin Hubble (and I), was a devoted cat-lover. On one occasion, an elegant young lady I used to know in Oxford came up to me and said "You know Sir Patrick Moore, don't you? I have *adored* him since being a child. He is a *great English gentleman*!" The young lady in question had no interest in astronomy. What is more, she rode, shot, and was into country sports. Yet what captivated her was Patrick the "great English gentleman". I also knew a young taxi driver, who probably could not tell the difference between Jupiter and Venus, yet when he knew that the Selsey address to which he was driving me to film was the home of Sir Patrick, the car almost veered off the road. "Sir Winston Churchill and Sir Patrick Moore are my heroes!", he said. When, upon my arriving and explaining, Patrick generously invited him to join us and the *Sky at Night* team for lunch, he was over the moon. "Wait till I tell all my mates at the George and Dragon that I had lunch with Sir Patrick Moore!" he exclaimed. As in the case of the "county" lady, the appeal of Patrick to my driver lay in his standing as a "great English gentleman".

How many people he inspired to do astronomy, during his fifty-five years of *The Sky at Night* broadcasting and lecturing, is probably incalculable. His appeal was multi-faceted, and much of it had to do with his personality, eccentricity, engaging fluency, encyclopaedic knowledge, and complete accessibility. British astronomy's premier celebrity, by 1990, had no minders, no personal assistants, and no promotional company, and would stop and chat to anyone. It is hard to imagine any other individual who, single-handedly, did more to promote popular and amateur astronomy across the English-speaking world, in an unbroken public career spanning fifty-five years between the ages of thirty-four and eighty-nine.[8]

Sir Patrick passed away peacefully on the morning of Sunday 9 December, 2012. I was invited to his closely guarded secret private funeral in Chichester soon after. As I said of Edmond Halley at the end of Chapter 3, I look forward someday to catching up with Sir Patrick and other friends "in a Brighter Light and a Greater Glory".

Chapter 29

Postscript: Creation Revisited: Where Do We Stand Today?

In 1850, everyone knew what astronomy was, and what astronomers did. They studied the heavenly bodies through powerful telescopes, mapped the moon, and measured the tiny angles between the stars. All of this was done at night, in observatories, using the human eye and delicately adjusted precision instruments. By 1910, however, an astronomer seemed a less clear-cut personage. Most cutting-edge research was now being done photographically. One might work from stellar spectra or galactic plates in the measuring laboratories of Harvard or Potsdam, or even be a chemist producing pure chemical reagents to act as standards against which to compare stellar spectra. Such deep-space astrophysicists and astro-chemists need never set foot in an observatory at night, but could work office hours in the laboratory.

By 2000, the astronomical community contained geophysicists, planetary physicists, and geologists, radio and radiation specialists, statisticians, computer engineers, and software and telecommunications experts. All were devoted to advancing our knowledge of the cosmos. The only people doing traditional "eye at the telescope" observing at night were the amateur astronomers. Even the more serious amateurs were engaged in professional-astronomy-related research, such as patrolling the night sky to search for supernovae. I know amateurs whose computer-controlled telescopes will regularly patrol parts of the sky where deep-space supernovae are likely to be found while their owners are asleep in bed. In the morning, a new supernova might be found to have registered its exact position on the computer. This can then be reported to the professional research teams who will ascertain its spectral, radio, and other characteristics, to add to our growing corpus of cosmological knowledge.

Considering the size of the international community worldwide, I suspect that there are more "eye at the telescope" astronomers now than there were in 1850. Many amateur astronomers are not out to make discoveries, but simply love beholding the glory of the starry firmament, through binoculars or simple telescopes with an ×100 magnification. Modern astronomy is not a broad so much as a vast "church", with professional stellar and planetary scientists at one end and contented back-garden sky-lovers at the other.

Life on other worlds and space travel, twenty-first-century style

Selenites and little green men have long since receded into history as serious candidates for celestial beings. Yet what fascinates me is how the "search for life in space" has almost become a credo for many scientists, funding agencies, American presidents, and media people. We *must* find some sort of "life" out there – whether it exists or not! The message sometimes comes over like the creedal statement of committed believers. Exoplanets of the right size, and rotating within the habitable zone of their parent star, are avidly searched for, with eligible exoplanet discoveries hitting the global media at regular intervals. All one needs is the right gravitational conditions and the possibility of an atmosphere, and lo, a potentially habitable world hits our TV screens. Why are we so concerned with not being alone, and why should the discovery of even a hint of oxygen or an amino acid, combined with evolution, set cosmically lonely hearts racing on Carl Sagan's "pale blue dot" which we call earth?

Before addressing this desire, though, let us first consider some of the practical problems involved in finding life on other worlds and being able to do something meaningful about it. Most of the people interested in life on other worlds that I have encountered tend to be primarily concerned with *intelligent* life, beings somehow capable of relating to us intellectually. Yet even if galactic space abounded with intelligent life, there are major limiting factors involved before we can hope for any meaningful dialogue. Firstly, that life would need to be at a developmental stage which parallels that of where Homo sapiens happens to be today. If the inhabitants of "exoplanet X" happened to be only as far advanced as Neanderthals, or had they tried to send us electronic messages when we ourselves were but Neanderthals, then no ideas could have been exchanged.

Secondly, had highly intelligent beings been sending us signals in 1850, or at any time before Karl Jansky and Grote Reber built the first radio telescopes in the 1930s, then once again the post office would have been firmly shut. Similarly, what chance do we stand of having a meaningful exchange by sending encrypted radio signals of goodwill to beings whose technological development is 500 years ahead of ours? In short, any meaningful communication with "aliens" must predicate a synchronicity of both intellect and communications technology. Otherwise, doors, ears, and minds will remain closed against each other.

Thirdly, we must remember that when considering interstellar space communication of any kind, we are dealing with inconceivably vast distances and transmission times of years, centuries, or even millennia. As both light and radio waves, being part of the same electromagnetic spectrum, travel at 186,000 miles per second, these physical factors must inevitably lay constraints upon communication. "Local" radio signals within the solar system would take 8 minutes to reach the sun, and 4 hours to get to Neptune. Then beyond the solar system transmission times escalate: over 4 years to the nearest star, in the Alpha and Proxima Centauri star system; over 11 years to 61 Cygni; 30,000 years from our solar system to the centre of our Milky Way galaxy; and another 100,000 or so years across the galaxy. Should we receive a message saying "Greetings from the Andromeda galaxy!" (the nearest extra-Milky-Way galaxy to us), the message has already been on its way for at least 2.24 million years, and our reply would take an equal time to reach them: a "return of post" time of nearly 5 million years in all.

When all of these factors are considered, one might begin to appreciate why I remain unconvinced that any realistic dialogue with intelligent beings on other worlds is remotely possible. Some argue that all we need to do is learn how to accelerate the speed of radio signals, while turning a committed fantasist's blind eye to the basic laws of physics, such as Einstein's theory of relativity. But we have to be careful never to conflate science *fact*, measured, tried, and tested, with science *fiction*. Hollywood and other branches of the media entertainment industry have created a whole new "alternative reality" for many people worldwide, and *Star Trek*, *Star Wars*, *Doctor Who*, and similar space fantasy creations have come to be elevated to religious cult dimensions. In this never-never world, characters may hop from star to star, or galaxy to galaxy: may materialize, dematerialize, and cross infinities of time and space at the flick of a switch; but let us remember that this is *not* physics. Nor should we lull ourselves into thinking that what we might summon up with our imagination stands the remotest chance of becoming physical reality.

I fully admit that no one knows what we will discover in 100 or 500 years' time, but what has been a sheet anchor to astronomy over the centuries has been an emerging and beautifully integrated body of ideas called the laws of physics. Think of Roger Bacon and the optical physics of the rainbow in *c.* 1260, William Gilbert and terrestrial magnetism in 1600, Kepler's Laws of Planetary Motion, 1609–18, Newtonian gravitation, Faraday's electromagnetism of 1831, Clerk Maxwell's energy equations, the electron and neutron parts of the atom, Einstein's relativity, quantum, Lemaître's expanding universe, and radio waves. Each stands upon its predecessors not merely in a purely intellectual, but also in an observational, experimental and substantive tradition, and can be tested and verified by anybody, of any nationality, in the international market-place of science. Between them, they form the "laws of science", and within their rich explanatory structure I can see no room for the space fantasists' world being anything other than a fantasy. Therefore, I fail to see how we can ever galaxy-hop, or send a quick birthday greeting to an alien friend in the Orion Nebula.

With the possible exception of a "terraformed" Mars at some date centuries hence, I also remain sceptical that humans will ever, in the words of the TV space saga *Star Trek*, "boldly go" to voyage across the galaxy. The distances, quite simply, are too vast, and the laws of physics, space, and time are stacked against such adventures. Nor can we conceive of the psychological pressures that might act upon human beings trapped for years, decades, or a lifetime in a spaceship from which there can be no escape or relief. It would be nothing like what Renaissance sailors had to put up with during the months of a long-haul sea voyage, for there would be no sunshine, blue skies, sweet breezes, scenic variety, or normal food and drink, and no joyous homecoming to look forward to for people heading for a galactic exoplanet. A life sentence in Dartmoor Prison would be like a holiday camp by comparison. In Dartmoor, at least you could enjoy the passing seasons and the sunshine, have visitors, meet new faces, and cherish the hope of escaping. How would people cope with the eternal boredom of a long-haul space voyage? Would they go mad, kill, and perhaps even eat each other? When we try to turn Hollywood science fiction into science fact, we must add in all the down-to-earth human components as well as the visionary.

Why are so many people unhappy with the human race being "alone" in the universe, and willing to invest their intellectual and emotional energies in all manner of future possibilities which go quite beyond the laws of physics as they have been built up, tried, and tested, over the centuries? To indulge in what might be called a "blind faith" in the future? For millennia, we humans have enjoyed an intimate and ultimately a deeply comforting relationship with beings that are somehow "up there", beyond the starry firmament: angels, ministering spirits, and God Himself. Could it be that people who, for whatever reason, abandon faith in a spiritual dimension, and perhaps call themselves rationalists or materialists, experience a bleakness at the thought of being masters of their own fate, and, like their theistic ancestors, return to the heavenly realms for comfort, to aliens, or a futuristic human colonization of space to give us a genetic immortality across time, space, and eternity? It interests me how people who spurn religion as "unreasonable" can sometimes indulge in such staggering flights of fancy!

Creation, cosmology, and the mind of God

There is the oft-mentioned scenario of the couple looking up at the starry heavens on a glorious clear night. One says to the other, "It makes you feel so small and insignificant, doesn't it?" I am the first to admit that what was once referred to as the "starry firmament" inspires me with a profound sense of wonder and awe. On the other hand, by all the available evidence, we mere humans are the only ones doing the seeing, asking the questions, and, over the last couple of millennia and more, gradually fathoming it all out. We insignificant humans, on the star-spangled shores of the Aegean, around 650 BC, first began to realize that not only did the heavens display the most beautiful and precise mathematical characteristics, but that we could also understand them and predict their behaviour. From there, we worked out the spherical shape of the earth, then as the centuries rolled by, we discovered what rotated around what, established the invisible laws that governed their movements, and, after 1860, began to crack the chemistry and physics of the sun and stars. In the twentieth century, we discovered that we lived in an expanding universe governed not only by gravity, but also by strange relativistic laws, while in the 1930s two American radio enthusiasts discovered a wholly new window into the cosmos: radio astronomy.

Next, we humans sent splendidly engineered spacecraft not only to the moon, but also to explore the outer planets, while the photographs sent back by the Hubble Space Telescope enchanted people across the world. In addition, we also ask "Big Questions", such as "Where did it all come from?", "What will eventually happen to it?", and, most mystifying of all, "Why?" Why is it here, and why are we even bothering to ask these questions, when, I suspect, rabbits, butterflies, and even chimpanzees do not? We humans are not just interested in straightforward processes and sequences, such as "How many days to the new moon?" We want to understand the *causes*. "What started it?" "Why is it this way rather than another?" "How can I test and verify it?" We are relentlessly curious, even about things which have no conceivable bearing on the daily struggle for existence, such as which elements formed first after the big bang, and even whether anything could have existed before "creation".

I would suggest there is what might be considered an even greater wonder: why is the universe coherent, on so many levels? On one level, for example, it coheres because seemingly disparate bits fit elegantly together into a bigger picture: such as the way in which the sodium "D" lines in the spectrum can be activated both by turning a spectroscope toward the sun and by burning a pinch of common salt (sodium chloride) in a flame. Chemical elements, physical laws, light, radio waves, and atomic radiation, all fit elegantly together.

Perhaps most wonderful of all, our eyes, brains, and nervous systems can make sense of it all, of what looks like a stunning display of organized and connected coherence that embraces the micro and macro worlds, extending from the cells and synapses within our brains to the physics of black holes and the big bang. Even if some people, as far back as the Roman philosopher Lucretius in *De Rerum Natura* ("On the Nature of Things") in *c.* 60 BC, have been trying to tell us that all this seeming coherence is an illusion,[1] one is still left asking "Why?" In purely Darwinian terms, it is hard to comprehend what kind of evolutionary survival advantage such a profoundly delusional understanding of the nature of reality could serve.

So could it just be that we "insignificant" humans might possess something that is special after all, and that this specialness enables us to gain insight into the very nature of creation? And by extension, that when we gaze into the starry sky, we might just be getting a glimpse into the mind of God?

Appendix

The Cock Lane Ghost, or the "Ghost Catch"

The Ghost will be heard in Cock-lane to-night
The Ghost, The Ghost
The Ghost will be heard in Cock-lane to-night
Our Children and Parsons and Lords to affright
Our Children and Parsons, our Children and Parsons, our Children and Parsons and Lords to affright
Hark, a Knock, List, a Scratch, Hark, Hark, List, List, Hark, List
I'll Swear, I'll Swear, I'll Swear it is so.
Do… do… diddle diddle, diddle diddle…

(This is followed by several more lines of "dos", "diddles", and recapitulations, in the jolly rollicking manner of an eighteenth-century "Catch".)

Notes

Full bibliographical details for books and articles cited in the notes may be found in the Further Reading section under the relevant chapter heading.

Chapter 1

1. From a sermon preached by Revd John Paton, former Precentor of Christ Church Cathedral, Oxford.
2. Proctor, *Flowers of the Sky* (1903 edn), pp. 265–67.
3. Chapman, *Gods in the Sky*, chs. 4–5.
4. Chapman, *Stargazers*, chs. 4–5.
5. Chapman, "A new perceived reality: Thomas Harriot's moon maps".
6. W. Herschel, "Astronomical Observations… tending to investigate… the arrangement of the celestial bodies in space".

Chapter 2

1. Cook, *Edmond Halley*, pp. 132–34.
2. Cook, *Edmond Halley*, pp. 132–52.
3. Thrower, *The Three Voyages of Edmond Halley*, pp. 162–63.
4. Flamsteed to Sharp, 18 Dec. 1703: *The Correspondence of John Flamsteed* vol. 3, p. 47.
5. Halley, "Consideration of the change of Latitude of some of the principal fix't stars".

Chapter 3

1. Hevelius, *Cometographia* (1668), cited in Yeomans, *Comets*, pp. 86–87.
2. Hooke, *Lampas*, p. 47; *Cometa*, p. 51, etc.
3. Halley, *A Synopsis of the Astronomy of Comets*, p. 20.
4. Hooke, "Earthquake Discourses", *Posthumous Works*, pp. 210–450; also in Tan Drake, pp. 153–371.

5. Chapman, "Edmond Halley's use of historical evidence", p. 180.
6. Thrower, *The Three Voyages of Edmond Halley*, p. 174; Cook, *Edmond Halley*, p. 278.

Chapter 4

1. Chapman, *Physicians, Plagues and Progress*, chs. 22, 26.
2. Hadley, "An Account of a Catadioptrick Telescope".
3. "James Short", *Oxford DNB*; Bryden, *James Short and his Telescopes*, p. 32.

Chapter 5

1. Michell, "On the Means of Discovering the Distance, Magnitude, & C … of the Fixed Stars".
2. Lubbock, *The Herschel Chronicle*, pp. 8–9.
3. Lubbock, *The Herschel Chronicle*, p. 8.
4. The Cock Lane Ghost: see Appendix.
5. C. Herschel, *Memoir and Correspondence*, p. 36.
6. Lubbock, *The Herschel Chronicle*, p. 43.
7. Lubbock, *The Herschel Chronicle*, p. 60.
8. C. Herschel, *Memoir and Correspondence*, p. 35.
9. C. Herschel, *Memoir and Correspondence*, p. 35.
10. C. Herschel, *Memoir and Correspondence*, p. 39; Lubbock, *The Herschel Chronicle*, p. 78.
11. Lubbock, *The Herschel Chronicle*, pp. 79–85.
12. For full MS details see Chapman, "Out of the meridian: John Bird's equatorial sector", pp. 441–42.

Chapter 6

1. Lubbock, *The Herschel Chronicle*, p. 79.
2. W. Herschel, *The Scientific Papers*, vol. 1, l, li.
3. W. Herschel, "Account of Some Observations Tending to Investigate the Construction of the Heavens". See also Herschel, "Construction" papers 1785, 1811, 1814.
4. *The Life and Letters of the Reverend Adam Sedgwick*, vol. 2, p. 107.
5. De Saint-Fond, *A Journey through England and Scotland to the Hebrides in 1784*, pp. 61–62.

6. Hanham and Hoskin, "The Herschel Knighthoods: fact and fiction".

Chapter 7

1. Chapman, *Stargazers*, pp. 292–312.
2. Woolf, *The Transits of Venus*, p. 57 and plates 8 and 10.
3. Woolf, *The Transits of Venus*, pp. 148, 193, 195.
4. Woolf, *The Transits of Venus*, p. 193.
5. Woolf, *The Transits of Venus*, pp. 136–38, 182–87.
6. Andrews, *The Quest for the Longitude*, pp. 168–254; Beaglehole, *The Life of Captain James Cook*.

Chapter 8

1. Chapman, *Dividing the Circle*, p. 102.
2. Chapman, "The pit and the pendulum: G. B. Airy and the determination of gravity".
3. Chapman, *Stargazers*, pp. 398–409.
4. Chapman, *Dividing the Circle*, p. 104.
5. Maskelyne, "An Account of Observations made on Schiehallion".
6. Chapman, *Dividing the Circle*, p. 106.
7. Cavendish, "Experiments to determine the Density of the Earth"; Powers, *The Honourable Henry Cavendish*, pp. 24–30.

Chapter 9

1. W. Herschel, "Observations tending to investigate the Sun" (Further Reading, ch. 16).
2. W. Herschel, "Investigation of the powers of Prismatic Colours".
3. Caroline Herschel to Sir Joseph Banks, 17 Aug. 1797, in C. Herschel, *Memoir and Correspondence*, pp. 94–95 (Further Reading, ch. 5).

Chapter 10

1. John Herschel to Charles Babbage, MS. letter, 18 Dec. 1815.
2. W. Herschel, "Astronomical Observations… tending to investigate… the arrangement of the celestial bodies in space".
3. John Herschel to Margaret Herschel, letter, date uncertain, 10 Aug.

1841 or 1843. Cited in Schaaf, "The poetry of light". Further details in Chapman, *The Victorian Amateur Astronomer*, pp. 323–24, n. 17.
4. J. Herschel, "Of the Aberrations of Compound lenses and Object Glasses".
5. J. Herschel, "Account of some observations made with a 20-foot reflecting telescope".
6. M. Herschel, *Lady Herschel. Letters from the Cape*, p. 27.
7. J. Herschel, MS. "Diary", 29 Aug. 1839.
8. Ch. XL, 364 (World's Classics, 1993).
9. Desmond and Moore, *Darwin*, pp. 90–91.

Chapter 11

1. Dick, *Celestial Scenery*, pp. 365–66.
2. Baum, "The man who found a city on the moon".
3. Locke, "Great astronomical discoveries lately made by Sir John Herschel… at the Cape of Good Hope".
4. *The War of the Worlds*, Book II, ch. 8, p. 168 (Penguin edn, 2005).

Chapter 12

1. Bodleian Library, Somerville Papers Dep. c. 355 MSAU-3, p. 5; McMillan, *Queen of Science*, p. 9.
2. Somerville, *Personal Recollections*, pp. 121–22; McMillan, *Queen of Science*, p. 99.
3. Somerville, "On the Magnetizing power of the more refrangible solar rays", p. 139. Also Parkin, "Mary Somerville (1780–1872): her correspondence and work on chemistry".
4. *On the Mechanism of the Heavens*, "Preliminary Dissertation", p. xl.
5. *Mechanism of the Heavens*, "Preliminary Dissertation", p. lxvi.
6. Somerville, *Personal Recollections*; McMillan, *Queen of Science*, pp. 139–41.
7. Somerville, *On the Connexion of the Physical Sciences*, Section XXXVI, "meteors": 3rd edn (1836), pp. 404–06; 9th edn (1858), p. 421.
8. Somerville, *Personal Recollections*, p. 290; McMillan, *Queen of Science*, p. 235.
9. Somerville, *Physical Geography*, ch. 1, p. 44, etc.
10. Somerville, *Connexion of the Physical Sciences*, final sentence.
11. Somerville, *Personal Recollections*, p. 373; McMillan, *Queen of Science*, p. 299.

Chapter 13

1. Howse, *Nevil Maskelyne*, pp. 201–02.
2. Airy, *Autobiography*, p. 29.
3. Airy, *Autobiography*, pp. 56–57.
4. Trollope, *Barchester Towers* (1857), ch. 31.
5. Airy, *Autobiography*, p. 112.
6. Airy, *Autobiography*, p. 356.
7. Genesis 47; Courtney, "The historical meridian".
8. Chapman, "Standard Time for all."
9. Airy, "Astronomer Royal's Journal"; Chapman, "Private research and public duty".
10. Chapman, "Airy's Greenwich staff" and "Porters, watchmen, and labourers".

Chapter 14

1. Sprat, *History of the Royal Society*, p. 67.
2. William Lassell to Lord Rosse, 6 May 1852, Birr Castle Archives, Ireland, K10.1; see Chapman, *The Victorian Amateur Astronomer*, p. 347 n. 74.
3. William Lassell to Wilhelm Struve, 7 Sept. 1850, Lassell MSS. 8:2, fol. 39, RAS Library; see Chapman, *Astronomical Instruments and Their Users* XVII, p. 346.
4. See Mollan, *William Parsons, Third Earl of Rosse.*
5. See Hartwell House "Albums" and "Scrapbooks" in Gunther MSS, Museum of the History of Science, Oxford. Full bibliographical references in Chapman, *The Victorian Amateur Astronomer*, pp. 333–40.

Chapter 15

1. Daguerre (transl. Memes), *The History and Practice of Photogenic Drawing.*
2. Schaaf, *Out of the Shadows*, and "The poetry of light", pp. 3–15.
3. Currently preserved in store in the Museum of the History of Science, Oxford.
4. Roberts, *Photographs of Stars, Star Clusters and Nebulae.*

Chapter 16

1. Comte, *Cours de Philosophie Positive* II, "Astronomie", pp. 8–9.
2. Merle, "Weather Diary".
3. Schwabe, "Sonnen-Beobachtungen"; "Extract of a Letter from H. Schwabe to Mr Carrington".
4. Clark, *The Sun Kings*, p. 73.
5. *James Nasmyth Engineer*, pp. 341, 370, 378.

Chapter 17

1. Gingerich, "Unlocking the chemical secrets of the cosmos".
2. Mills and Brooke, *A Sketch of the Life of Sir William Huggins*, p. 21.
3. Huggins, "On the Spectra of some of the Chemical Elements".
4. Huggins and Miller, "On the Spectra of some of the Nebulae".
5. Sources cited in Chapman, *The Victorian Amateur Astronomer*, p. 348, n. 79.
6. Elliot and Mollan, *William E. Wilson*.

Chapter 18

1. Chapman, *The Victorian Amateur Astronomer*, p. 226.
2. Tobin, *The Life and Science of Léon Foucault*, pp. 207–14.
3. Marriott, "Webb's telescopes".
4. Berthon, "Romsey Observatory".
5. Samuel Cooper's Venus transit photographs are included in Jenkins, "A plea for the reflecting telescope".
6. Smiles, *Men of Invention and Industry*, pp. 361–69.
7. Evans, *Seryddiaeth a Serydddwyr*, p. 273.
8. See Chapman, *The Victorian Amateur Astronomer*, pp. 378–80, nn. 14–31, for John Jones primary sources, including private communications with relatives.
9. Somerville, *Personal Recollections*, p. 100; McMillan, *Queen of Science*, p. 82; Chapman, *Mary Somerville*, pp. 24, 148, n. 24 (Further Reading, ch. 12).
10. *The Life of Roger Langdon*, pp. 73, 76.
11. Smiles, *Men of Invention and Industry*, p. 333.

12. John Leach: MS. sources in Chapman, *The Victorian Amateur Astronomer*, pp. 188–200, 374–75, nn. 24–37.
13. Thomas Bush: MS. sources in Chapman, *The Victorian Amateur Astronomer*, pp. 200–03, 375–77, nn. 43–63.
14. "Johnson Jex. A study for the Million", *Norfolk Advertiser*, 17 Jan. 1852; cutting in Hartwell House "Album", Museum of the History of Science, Oxford, MS. Gunther 37, pp. 82–83. See Chapman, *The Victorian Amateur Astronomer*, p. 372, nn. 6–9.

Chapter 19

1. Wright, *Sweeper of the Sky. The Life of Maria Mitchell.*
2. Hutchins, *British University Observatories*, pp. 71, 132, 297; Chapman, *The Victorian Amateur Astronomer* pp. 157, 366, nn. 73, 74.
3. Brück, *Women in Early British and Irish Astronomy*, p. 157.
4. See Brück, *Agnes Mary Clerke and the Rise of Astrophysics.*
5. For Florence Taylor primary sources, see Chapman, *The Victorian Amateur Astronomer*, pp. 392, n. 64; 396, n. 5; 399, n. 29; 402, n. 79; etc.
6. Elizabeth Brown, "Programmes of the Directors of the Observing Sections. Solar Section", p. 59.
7. Chapman, *The Victorian Amateur Astronomer*, p. 281.
8. For lady eclipse chasers see Creese, "Elizabeth Brown", p. 195; McKim, "Women in the British Astronomical Association" (*A&G*); Brück, *Women in Early British and Irish Astronomy*, pp. 221–33.
9. Bailey, "100 years of Fellowship"(*A&G*).
10. Bowler, "Persistent pioneers" and Masters, "Women of the future in the RAS" (both *A&G*).

Chapter 20

1. Solly, *Working Men*, pp. 8–9.
2. Hawes, *Henry Brougham*, p. 229.
3. Wordsworth, *The Major Works*, ed. S. Gill (1984), pp. 322–23.
4. Mayhew, *London Labour and the London Poor*, vol. 3, "The Telescope Exhibitor", pp. 79–83.
5. Mayhew, *London Labour and the London Poor*, vol. 3, p. 80.
6. Airy, "Popular Astronomy: A Series of Lectures delivered at Ipswich"; Airy, *Autobiography*, pp. 195–96.

7. Mayhew, *London Labour and the London Poor*, vol. 3, pp. 81–88.
8. "London Lecturers – The Astronomical Mania", Hartwell House "Album", Museum of the History of Science, Oxford, MS. Gunther 36, pp. 1–3; Chapman, *The Victorian Amateur Astronomer*, ch. 9, p. 167.
9. Jabez Inwards, *The Comet. Is the World in Danger?*, handbill, in Hartwell House "Album", Museum of the History of Science, Oxford, MS. Gunther 37, p. 111; Chapman, *The Victorian Amateur Astronomer*, p. 369, n. 28.
10. Yeomans, *Comets*, pp. 178–79, for "Miller's Comet".
11. "Richard Anthony Proctor", Hutchins, *Oxford DNB*.
12. Crowe, *The Extraterrestrial Life Debate*, p. 548.
13. *Reminiscences and Letters of Sir Robert Ball*, pp. 188–89.
14. Manchester Astronomical Society "Minute Book, 1903–1909", 23 Sept. 1908. See Chapman, *The Victorian Amateur Astronomer*, pp. 394–95, nn. 97, 107 for full references.

Chapter 21

1. Krisciunas, *Astronomical Centers of the World*, pp. 124–40: 125.
2. Krisciunas, *Astronomical Centers of the World*, pp. 131–32.
3. Haley, "Williamina Fleming and the Harvard College Observatory".
4. For a catalogue of Clark telescopes, see Warner, *Alvan Clark & Sons*.
5. These telescopes are discussed in King, *History of the Telescope*, pp. 282–415.

Chapter 22

1. Ball, *The Story of the Heavens*, pp. 319–20.
2. Halley, "On the Infinity of the sphere of the fix'd Stars".
3. Ridpath, "Michelson Morley Experiment".
4. DeVorkin, "Stellar evolution and the origin of the Hertzsprung–Russell Diagram", pp. 97–98.
5. DeVorkin, "Stellar evolution and the origin of the Hertzsprung–Russell Diagram".
6. *Phillips Astronomy Encyclopedia*, "White Dwarf", pp. 436–37 (Further Reading, ch. 21).
7. Leavitt, "1777 Variables in the Magellanic Clouds".
8. Johnson, *Miss Leavitt's Stars*.

9. Slipher, "Radial velocity observations of the spiral nebulae".
10. Nussbaumer and Bieri, *Discovering the Expanding Universe*, p. 48.
11. Trimble, "The 1920 Shapley–Curtis discussion".

Chapter 23

1. "Molecule", *Nelson's Encyclopaedia.*
2. See Pasachoff, *Marie Curie and the Science of Radioactivity.*
3. "Ernest Rutherford", *Chambers Biographical Encyclopaedia of Scientists.*
4. Bohr, *On the Constitution of Atoms and Molecules.*
5. Baum and Sheehan, *In Search of Planet Vulcan*; Ley, *Watchers of the Skies*, pp. 189–202.
6. Cited by Gingerich in "Albert Einstein: a laboratory in the mind", p. 282.
7. Eddington and de la Crommelin, "A new theory of the universe".
8. Isaacson, *Einstein*, p. 262.
9. Ollerenshaw, *To Talk of Many Things*, p. 43.

Chapter 24

1. Sandage, "Edwin Hubble, 1889–1953".
2. Smart, *The Riddle of the Universe*, p. 176.
3. Hubble, "Extra-galactic nebulae" and "A relation between distance and radial velocity among extra-galactic nebulae".
4. Ventrudo, "How a janitor at Mount Wilson Observatory measured the size of the universe".
5. Sandage, "Current problems in the extragalactic distance scale".
6. W. Herschel, "Observations and Experiments tending to investigate the local arrangement of the celestial bodies in space", pp. 302–03.

Chapter 25

1. Lemaître, "The expansion of the universe. A homogeneous universe of constant mass and increasing radius as accounting for the radial velocity of extra-galactic nebulae".
2. Holder and Mitton, *Georges Lemaître.*
3. De Sitter, *The Observatory*, 53 pp. 37–39; Chapman, *The Observatory*, 130, pp. 198–200.

4. Nussbaumer and Bieri, *Discovering the Expanding Universe*, pp. 63–87.
5. Mitton, "Fred Hoyle's birth centennial".
6. Eddington, *Stars and Atoms*, p. 50.
7. Chandrasekhar, "The highly collapsed configurations of stellar mass" and "The maximum mass of ideal White Dwarfs".
8. Turek, *Georges Lemaître and the Pontifical Academy of Sciences.*

Chapter 26

1. Pfeiffer, *The Changing Universe*, p. 9.
2. Jansky, "Radio waves from outside the solar system".
3. Pfeiffer, *The Changing Universe*, p. 15.
4. Pfeiffer, *The Changing Universe*, p. 25.
5. Lovell, "The effects of defence science on the advancement of astronomy".
6. Lovell, *Astronomer by Chance.*
7. Lovell, "The Jodrell Bank Radio Telescope".
8. See Sullivan III, *Cosmic Noise and the History of Early Radio Astronomy.*
9. Hewish, Bell, et al., "Observations of a rapidly rotating pulsating radio source".
10. Staelin and Reifenstein III, "Pulsating radio sources near the Crab Nebula".

Chapter 27

1. Congreve, *The Details of the Rocket System*, see "Rocket Ammunition". Also Aldwinckle, "The development of gunpowder-based military pyrotechnics in the period *c.* 1790–1846".
2. See Further Reading for full details.
3. Springer, "Early experimental programs of the American Rocket Society, 1930–41".
4. See Freedman, *2000 Years of Space Travel*, p. 229.
5. Siddiqi, *Sputnik and the Soviet Space Challenge.*
6. Turnill, *The Moonlandings*, chs 15–16.
7. Chaikin, *A Man on the Moon*, pp. 204ff.
8. See Bell, *The Interstellar Age.*
9. Cooper, "From here to eternity. 40 years [and counting] of the Voyagers".
10. Pillinger, *Space is a Funny Place.*

Chapter 28

1. Chapman, *Stargazers*, ch. 23.
2. Chapman, letter, *The Spectator*, 19 Nov. 2016.
3. Moore, *80 Not Out*, pp. 7–9.
4. See Moore, *Guide to the Moon.*
5. William Herschel, "An Account of three Volcanoes on the Moon".
6. Lankford, "Amateurs: astronomy's enduring resource".
7. Sagan, *Pale Blue Dot.*
8. See Mobberley, *It Came from Outer Space Wearing an RAF Blazer!*

Chapter 29

1. T. Lucretius Carus, *De Rerum Natura* (transl. Stallings), Book 1.

List of In-text Illustrations

Chapter 1

Chapter 2

Chapter 3

Chapter 4

Chapter 5

Chapter 6

Chapter 7

Chapter 8

Chapter 9

Chapter 10

Chapter 12

Chapter 13

Chapter 14

Chapter 15

Chapter 16

Chapter 17

Chapter 18

Chapter 19

Chapter 20

Chapter 21

Chapter 22

Chapter 23

Chapter 24

Chapter 25

Chapter 26

Chapter 28

Appendix

Further Reading

ABBREVIATIONS

AA	*The Antiquarian Astronomer,* journal of the Society for the History of Astronomy
AJ	*Astrophysical Journal*
AN	*Astronomy Now*
Annals	*Annals of Science*
A&G	*Astronomy & Geophysics*, journal of the Royal Astronomical Society
BJHS	*The British Journal for the History of Science*
JAHH	*The Journal of Astronomical History and Heritage*
JBAA	*Journal of the British Astronomical Association*
JHA	*Journal for the History of Astronomy*
MNRAS	*Monthly Notices of the Royal Astronomical Society*
Notes and Records	*Notes and Records of the Royal Society of London*
Oxford DNB	*Oxford Dictionary of National Biography*
Phil. Trans.	*Philosophical Transactions of the Royal Society of London*
PASP	*Publications of the Astronomical Society of the Pacific*
Vistas	*Vistas in Astronomy*

General Reading

John D. Barrow, Cosmic Imagery. *Key Images in the History of Science* (The Bodley Head, London, 2008).

Hermann A. Brück, *The Story of Astronomy in Edinburgh from its Beginnings Until 1975* (Edinburgh Univ. Press, 1983).

Robert Bud and Deborah Jean Warner (eds), *Instruments of Science: An Historical Encyclopedia* (Garland, New York and London, 1998).

Allan Chapman, *Dividing the Circle: The Development of Critical Angular Measurement in Astronomy, 1500–1850* (Wiley-Praxis, Chichester, 1990, 1995).

Allan Chapman, *Astronomical Instruments and Their Users: Tycho Brahe to William Lassell* (Variorum, Aldershot, 1996).

Allan Chapman, *The Victorian Amateur Astronomer: Independent Astronomical*

Research in Britain, 1820–1920 (Wiley-Praxis, Chichester, 1998; revised edn, Gracewing, 2017).

Norman Cohn, *Cosmos, Chaos, and the World to Come: The Ancient Roots of Apocalyptic Faith* (Yale Univ. Press, 1993).

Peter Doig, *A Concise History of Astronomy* (Chapman & Hall, London, 1950).

J. L. E. Dreyer and H. H. Turner (eds), *History of the Royal Astronomical Society: vol. 1, 1820–1920* (RAS, London, 1923).

John Fauvel, Raymond Flood, and Robin Wilson (eds), *Oxford Figures: Eight Centuries of the Mathematical Sciences* (Oxford Univ. Press, 2000; 2nd edn, 2013).

J. V. Field and Frank A. L. J. James (eds), *Renaissance and Revolution: Humanists, Scholars, Craftsmen and Natural Philosophers in Early Modern Europe* (Cambridge Univ. Press, 1993).

Galileo Galilei, *Discoveries and Opinions of Galileo, translated with an introduction and notes by Stillman Drake* (Doubleday Anchor Books, New York, 1957).

Owen Gingerich, *The Great Copernicus Chase and Other Adventures in Astronomical History* (Sky Publishing, Cambridge Mass., Cambridge Univ. Press, 1992).

Mark Girouard, *Life in the English Country House: A Social and Architectural History* (Yale Univ. Press, 1978).

I. S. Glass, *Victorian Telescope Makers: The Lives and Letters of Thomas and Howard Grubb* (Institute of Physics, Bristol and Philadelphia, 1997).

I. S. Glass, *Revolutionaries of the Cosmos: The Astro-Physicists* (OUP, 2005).

Edward Grant, *Planets, Stars, and Orbs: The Medieval Cosmos, 1200–1687* (CUP, 1994).

Robert Grant, *History of Physical Astronomy from the Earliest Ages to the Middle of the Nineteenth Century* (London, 1852).

Robert Hooke, *Micrographia* (1665), pp. 241–46.

Michael Hoskin, *Cambridge Illustrated History of Astronomy* (CUP, 1997).

Derek Howse, *Greenwich Observatory, 3: The Buildings and Instruments* (Taylor & Francis, London, 1975). (Seventeenth to twentieth centuries.)

Roger Hutchins, *British University Observatories 1772–1939* (Ashgate, Aldershot, 2008).

Jamie James, *The Music of the Spheres: Music, Science, and the Natural Order of the Universe* (Copernicus, Springer-Verlag, New York, 1993).

H. C. King, *The History of the Telescope* (Charles Griffin, London, 1955).

Kevin Krisciunas, *Astronomical Centers of the World* (CUP, 1988).

John Lankford (ed.), *History of Astronomy: An Encyclopedia* (Garland, New York and London, 1997).

Willy Ley, *Watchers of the Skies. An Informal History of Astronomy from Babylon to the Space Age* (Sidgwick & Jackson, London, 1963).
Andrew Liddle and Jon Loveday, *The Oxford Companion to Cosmology* (OUP, 2008).
Robert E. W. Maddison, *A Dictionary of Astronomy* (Hamlyn, London, 1980).
Paul Marston, *Great Astronomers in European History* (University of Central Lancashire, 2014).
John D. North, *The Fontana History of Astronomy and Cosmology* (Fontana Press, London, 1994).
A. Pannekoek, *A History of Astronomy* (Allen & Unwin, London, 1961).
Colin A. Ronan, *The Astronomers* (Evans Brothers, London, 1964).
Colin A. Ronan, *Their Majesties' Astronomers: A Survey of Astronomy in Britain Between the Two Elizabeths* (The Bodley Head, London, 1967).
Bruce Stephenson, *Marvin Bolt, and Anna Felicity Friedman, The Universe Unveiled: Instruments and Images through History* (CUP, 2000).
René Taton and Curtis Wilson (eds), *Planetary Astronomy from the Renaissance to the Rise of Astrophysics: Part A, Tycho Brahe to Newton* (CUP, 1989).
René Taton and Curtis Wilson (eds), *Planetary Astronomy from the Renaissance to the Rise of Astrophysics: Part B, The Eighteenth and Nineteenth Centuries* (CUP, 1995).
R. J. Tayler (ed.), *History of the Royal Astronomical Society: vol. 2, 1920–1980* (RAS and Blackwell, Oxford, 1987).
E. Wilfred Taylor and J. Simms Wilson, revised P. D. Scott Maxwell, *At the Sign of the Orrery: The Origins of the Firm of Cooke, Troughton, & Simms, Ltd.* (booklet, no publisher or date, but probably York, 1950).
Stuart Ross Taylor, *Destiny or Chance: Our Solar System and its Place in the Cosmos* (CUP, 1998).
Albert Van Helden, *The Invention of the Telescope* (American Philosophical Society, 1977).
Ewen Whitaker, *Mapping and Naming the Moon: A History of Lunar Cartography and Nomenclature* (CUP, 1999).
Peter Whitfield, *Mapping the Heavens* (British Library, 1995).
Robert Wilson, *Astronomy Through the Ages. The Story of the Human Attempt to Understand the Universe* (Taylor & Francis, London, 1997).

Chapter 1

Francesco Bianchini, *Observations Concerning the Planet Venus*, trans. Sally Beaumont and Peter Fay (Springer-Verlag, London, 1996).

Bill Bryson (ed.), *Seeing Further. The Story of Science and the Royal Society* (HarperPress, London, 2010).

Allan Chapman, *Gods in the Sky. Astronomy, Religion, and Culture from the Ancients to the Renaissance* (Channel 4 Books, Pan Macmillan, 2001).

Allan Chapman, "A new perceived reality: Thomas Harriot's moon maps", *A&G* 50 (Feb. 2009), pp. 27–33. (The moon maps are in the Petworth House Archive, HMC 241/9, fol. 20, West Sussex Record Office, Chichester.)

Allan Chapman, *Stargazers: Copernicus, Galileo, the Telescope and the Church* (Lion Hudson, 2014).

Louise Cochrane with Charles Burnett, ed. Peter Wallis, *Adelard of Bath: The First English Scientist* (Bath Royal Literary and Scientific Institution, 2013).

Dennis Danielson, *The First Copernican: Georg Joachim Rheticus and the Rise of the Copernican Revolution* (Walker and Co., New York, 2006).

John Fauvel, Raymond Flood, Michael Shortland, and Robin Wilson (eds), *Let Newton Be! A New Perspective on his Life and Works* (OUP, 1988).

Robert Fox (ed.), *Thomas Harriot. An Elizabethan Man of Science* (Ashgate, Aldershot, 2000).

James Hannam, *God's Philosophers. How the Medieval World Laid the Foundations of Modern Science* (Icon Books, London, 2009).

William Herschel, "Astronomical Observations… tending to investigate… the arrangement of the celestial bodies in space", *Phil. Trans.* 107 (1817), pp. 302–31. Reprinted in *The Scientific Papers of Sir William Herschel* vol. 2, ed. Dreyer (see under Chapter 5), p. 575.

Barry Hetherington, *A Chronicle of Pre-Telescopic Astronomy* (John Wiley, Chichester, 1996).

Arthur Koestler, *The Sleepwalkers. A History of Man's Changing Vision of the Universe* (Hutchinson, London, 1959).

Thomas Kuhn, *The Copernican Revolution. Planetary Astronomy in the Development of Western Thought* (Harvard Univ. Press, 1966).

Stephen C. McCluskey, *Astronomies and Cultures in Early Medieval Europe* (CUP, 1998).

Patrick Moore, *The Great Astronomical Revolution 1534–1687 and the Space Age Epilogue* (Albion Publishing, Chichester, 1994).

John North, *God's Clockmaker. Richard of Wallingford and the Invention of Time* (Hambledon, London and New York, 2005).

Richard A. Proctor, *Flowers of the Sky*, 15th edn (London, 1903).

David Sellers, *In Search of William Gascoigne, Seventeenth-Century Astronomer* (Springer, 2012).

John W. Shirley (ed.), *Thomas Harriot, Renaissance Scientist* (Clarendon Press, Oxford, 1974).

Bruce Stephenson, Marvin Bolt, and Anna Felicity Friedman, *The Universe Unveiled* (see General Reading).

E. G. R. Taylor, *The Mathematical Practitioners of Tudor and Stuart England* (CUP, 1968).

Hugh Thurston, *Early Astronomy* (Springer-Verlag, New York, 1994).

Christopher Walker (ed.), *Astronomy Before the Telescope* (British Museum Press, 1996).

Richard S. Westfall, *The Life of Isaac Newton* (CUP, 1993).

CHAPTER 2

Angus Armitage, *Edmond Halley* (London, 1966).

John L. Birks, *John Flamsteed. The First Astronomer Royal* (Avon, London, 1999).

Allan Chapman, "Edmond Halley's use of historical evidence in the advancement of science", *Notes and Records* 48, 2 (1994), pp. 167–91. (Triennial John Wilkins Lecture, Royal Society, 1994.)

Alan Cook, *Edmond Halley. Charting the Heavens and the Seas* (OUP, 1998).

The "Preface" to John Flamsteed's Historia Coelestis Britannica, edited and introduced by Allan Chapman, based on a translation by Alison Dione Johnson (National Maritime Museum Monographs no. 52, Greenwich, 1982).

John Flamsteed to Abraham Sharp, 18 Dec. 1703, in *The Correspondence of John Flamsteed, the First Astronomer Royal*, vol. 3, ed. Eric G. Forbes, Lesley Murdin, and Frances Willmoth (Institute of Physics, Bristol and Philadelphia, 2002), p. 47.

Eric G. Forbes, *Greenwich Observatory, 1: Origins and Early History* (Taylor & Francis, London, 1975).

Edmond Halley, "An Account of Several Nebulae or Lucid Spots like Clouds...", *Phil. Trans.* 29 (1716), pp. 390–92.

Edmond Halley, "An Account of the later surprizing Appearance of the Lights seen in the Air on the sixth of March last", *Phil. Trans.* 29 (1716), pp. 406–28.

Edmond Halley, "Consideration of the change of Latitude of some of the principal fix't stars", *Phil. Trans.* 30 (1718), pp. 736–38.

Edmond Halley, "Of the infinity of the Sphere of the Fix'd Stars", *Phil. Trans.* 31 (1720), pp. 22–23.

Edmond Halley, "Of the number, order and light of the Fix'd Stars", *Phil. Trans.* 31 (1720), pp. 24–26.

Brian Harpur, *The Official Halley's Comet Book* (Hodder and Stoughton, London, 1985, 1986).
Eugene MacPike (ed.), *Correspondence and Papers of Edmond Halley* (Clarendon Press, Oxford, 1932).
Colin Ronan, *Edmond Halley, Genius in Eclipse* (London, 1970).
Norman J. W. Thrower (ed.), *The Three Voyages of Edmond Halley in The Paramore, 1698–1701* (Hakluyt Society, Second Series no. 156, London, 1981). (See also Appendix of Halley's sea charts.)
Herbert Hall Turner, *Astronomical Discovery* (London, 1904).
Frances Willmoth (ed.), *Flamsteed's Stars. New Perspectives on the Life and Work of the First Astronomer Royal, 1646–1719* (National Maritime Museum, Greenwich, 1997).

CHAPTER 3

Allan Chapman, *England's Leonardo. Robert Hooke and the Seventeenth-Century Scientific Revolution* (Institute of Physics, Bristol and Philadelphia, 2005), pp. 81–83.
Allan Chapman, "Edmond Halley's use of historical evidence in the advancement of science" (see under Chapter 2).
Alan Cook, *Edmond Halley* (see under Chapter 2).
Ellen Tan Drake, *Restless Genius: Robert Hooke and his Earthly Thoughts* (OUP, Oxford and New York, 1996). (Reprints Hooke's geological "Earthquake Discourses" from *Posthumous Works* (1705).)
Edmond Halley, *A Synopsis of the Astronomy of Comets* (London, 1705).
Edmond Halley, "A Short Account of the Saltness of the Ocean… to discover the Age of the World", *Phil. Trans.* 29 (1715), pp. 296–300.
Edmond Halley, "Some Considerations about the Cause of the Universal Deluge laid before the Royal Society on 12th December 1694", *Phil. Trans.* 33 (1724), pp. 118–21.
Brian Harpur, *The Official Halley's Comet Book* (see under Chapter 2).
John Russell Hind, *Comets: A Descriptive Treatise upon those Bodies…* (London, 1852).
Robert Hooke, *Lampas, or, Descriptions of some Mechanical Improvements of Lamps & Waterpoises* (London, 1677).
Robert Hooke, *Cometa, or remarks about Comets* (London, 1678).
Robert Hooke, "Discourses of Earthquakes", in *The Posthumous Works of Robert Hooke*, ed. Richard Waller (1705), Section V, pp. 210–450. (See Tan Drake, above, for modern text of "Discourses".)

Norman Thrower (ed.), *The Three Voyages of Edmond Halley in The Paramore, 1698–1701* (see under Chapter 2).
Herbert Wendt, *Before the Deluge. The Story of Palaeontology*, transl. Richard and Clara Wilson (Gollancz, London, 1968).
William Whiston, *A New Theory of the Earth* (London, 1696).
Donald K. Yeomans, *Comets. A Chronological History of Observation, Science, Myth, and Folklore* (John Wiley, New York, Chichester, 1991).

CHAPTER 4

Allan Chapman, *Physicians, Plagues and Progress. A History of Western Medicine from Antiquity to Antibiotics* (Lion Hudson, Oxford, 2016).
J. A. Bennett, *Church, State, and Astronomy in Ireland: 200 Years of Armagh Observatory* (Armagh, Belfast, 1990).
Nicolas Bion, *The Construction and Principal Uses of Mathematical Instruments*, translated from the French by Edmond Stone (London, 1723).
David Bryden, *James Short and his Telescopes* (Edinburgh, 1968).
John Hadley, "An Account of a Catadioptrick Telescope… with the Description of a Machine Contriv'd by Him for the Applying it to Use", *Phil. Trans.* 32 (1723), pp. 303–12.
Thomas Keith, *A New Treatise on the Use of the Globes, or a Philosophical View of the Earth and Heavens* (London, 1805).
H. King, *History of the Telescope* (see General Reading).
John R. Millburn, *Benjamin Martin: Author, Instrument-Maker, and "Country Showman"* (*Science and History* 2, Leiden, 1976).
John R. Millburn, *Benjamin Martin Supplement* (London, 1986).
John R. Millburn, *Adams of Fleet Street. Instrument-Makers to King George III* (Ashgate, Aldershot, 2000).
Jonathan Peacock, *Jeremiah Dixon, Scientist, Surveyor, and Stargazer* (Bowes Museum, Durham; undated, post-2013).
Jonathan Powers, *The Philosopher Lecturing on the Orrery. The identity of the man in Joseph Wright's famous painting and the rediscovery of a half-forgotten Enlightenment figure* (iOpening Press, University of Derby, 2016).
"James Short", by Tristram Clarke, *Oxford DNB* (OUP, 2004).
Robert Smith, *A Compleat System of Opticks* (Cambridge, 1738).
E. G. R. Taylor, *The Mathematical Practitioners of Hanoverian England, 1714–1840* (CUP, 1966).
E. Wilfred Taylor and J. Simms Wilson, *At the Sign of the Orrery* (see General Reading).
Anthony Turner, *Early Scientific Instruments: Europe 1400–1800* (London, 1987).

CHAPTER 5

Angus Armitage, *William Herschel* (Thomas Nelson and Sons, London, 1962).
James A. Bennett, "'On the Power of Penetrating into Space', the telescopes of William Herschel", *JHA* 7 (1976), pp. 75–108.
Frank Brown, *William Herschel, Musician and Composer* (William Herschel Society, Bath, 1990).
Allan Chapman, "William Herschel and the measurement of space", in Chapman, *Astronomical Instruments and Their Users* (see General Reading), XII.
Allan Chapman, "Out of the meridian: John Bird's equatorial sector and the new technology of astronomical measurement", *Annals* 52 (1995), pp. 431–63.
Caroline Herschel, *Memoir and Correspondence of Caroline Herschel*, ed. Mrs John Herschel (John Murray, London, 1879).
Caroline Herschel, *Caroline Herschel's Autobiographies*, ed. Michael Hoskin (Science History Publications, Cambridge, 2003).
William Herschel, *The Scientific Papers of Sir William Herschel. Including Early Papers Hitherto Unpublished*, 2 vols., ed. J. L. E. Dreyer (Royal Society and Royal Astronomical Society, 1912).
Michael A. Hoskin, *William Herschel, Pioneer of Sidereal Astronomy* (London, 1959).
Michael Hoskin, "The cosmology of Thomas Wright of Durham", *JHA* 1 (1970), pp. 44–52.
Michael Hoskin, *The Herschel Partnership, as Viewed by Caroline* (Science History Publications, Cambridge, 2003).
Michael Hoskin, *The Herschels of Hanover* (Science History Publications, Cambridge, 2007).
Kenneth Glyn Jones, *The Search for the Nebulae* (Alpha Academic, Science History Publications Ltd., Chalfont St Giles, England, 1975).
R. V. Jones, "Through music to the stars. William Herschel, 1738–1822", *Notes and Records* 33, 1 (Aug. 1978), pp. 37–56.
Immanuel Kant, *Universal Natural History and Theory of the Heavens*, transl. S. L. Jaki (Scottish Academic Press, Edinburgh, 1981).
Constance A. Lubbock (ed.), *The Herschel Chronicle. The Life-Story of William Herschel and his Sister Caroline Herschel* (CUP, 1933).
John Michell, "On the Means of Discovering the Distance, Magnitude, & C … of the Fixed Stars, in Consequence of the Diminution of the Velocity of their Light…" *Phil. Trans.* 74 (1784) pp. 35–57.
Robert Smith, *A Compleat System of Opticks* (see under Chapter 4).

Robert Smith, *Harmonics, or, The Philosophy of Musical Sounds* (London, 1749; 2nd edn, 1759).
A. J. Turner, *Science and Music in Eighteenth-Century Bath* (Bath, 1977).

CHAPTER 6

Andrew Hanham and Michael Hoskin, "The Herschel Knighthoods: fact and fiction", *JHA* 44 (2013), pp. 149–64.
Caroline Herschel, *Memoir and Correspondence* (see under Chapter 5), p. 36 for "glees and catches".
William Herschel, "Account of some observations tending to investigate the Construction of the Heavens", *Phil. Trans.* 74 (1784), pp. 437–51. Also Herschel's further "Construction" papers, *Phil. Trans.* 75 (1785), 1811, 1814.
William Herschel, "Catalogue of one thousand new nebulae and clusters of stars", *Phil. Trans.* 76 (1786), pp. 457–99.
William Herschel, *The Scientific Papers of Sir William Herschel* (see under Chapter 5).
Michael Hoskin, assisted by David Dewhirst, *William Herschel and the Construction of the Heavens* (Oldbourne, London, 1963). Reprints sections from Herschel's published papers.
Michael Hoskin, *Stellar Astronomy: Historical Studies* (Science History Publications, 1982).
Michael Hoskin, "Alexander Herschel: the forgotten partner", *JHA* 35 (2004), pp. 387–420.
Michael Hoskin, *Discoverers of the Universe. William and Caroline Herschel* (Princeton Univ. Press, 2011).
Constance A. Lubbock, *The Herschel Chronicle* (see under Chapter 5).
Andreas Maurer and E. G. Forbes, "William Herschel's astronomical telescope", *JBAA* 81 (1971), pp. 284–91.
John Michell, "An inquiry into the probable Parallax and Magnitude of the fixed stars", *Phil. Trans.* 57 (1767), pp. 234–64.
John Michell, "On the Means of Discovering the Distance, Magnitude, & C.…of the Fixed Stars" (see under Chapter 5).
Barthélemy Faujas de Saint-Fond, *A Journey Through England and Scotland to the Hebrides in 1784* (Paris, 1797), ed. and transl. Archibald Geike (Glasgow, 1907).
The Life and Letters of the Reverend Adam Sedgwick, LL.D., D.C.L., FRS, vol. 2, ed. J. W. Clark and T. M. Hughes (CUP, 1890): p. 107 for "sweeping the cobwebs out of the sky".

Thomas Wright, *An Original Theory or New Hypothesis of the Universe* (1750).

CHAPTER 7

William J. H. Andrewes, *The Quest for the Longitude. The Proceedings of the Longitude Symposium, Harvard University, 1993* (Collection of Astronomical Scientific Instruments, Harvard Univ., 1996).
J. C. Beaglehole, *The Life of Captain James Cook* (Hakluyt Society, 1974).
Allan Chapman, *Stargazers* (see under Chapter 1).
Michael Chauvin, *Hōkūloa. The British 1874 Transit of Venus Expedition to Hawaii* (Bishop Museum Press, Honolulu, 2004).
Rupert T. Gould RN, *John Harrison and his Timekeepers* (National Maritime Museum, 1958).
Derek Howse, *Nevil Maskelyne. The Seaman's Astronomer* (CUP, 1989).
Derek Howse and Beresford Hutchinson, *The Clocks and Watches of Captain James Cook 1769–1969* (Antiquarian Horology Society, 1969).
Eli Maor, *June 8, 2004. Venus in Transit* (Princeton University Press, 2000). (Venus transits from the seventeenth to the twenty-first century.)
Jonathan Peacock, *Jeremiah Dixon* (see Chapter 4). (Mason–Dixon Line.)
Albert Van Helden, "Measuring the solar parallax: the Venus transits of 1761 and 1769 and their nineteenth-century sequels", in Taton and Wilson (eds.), *Planetary Astronomy, Part B* (see General Reading), pp. 153–68.
Harry Woolf, *The Transits of Venus. A Study of Eighteenth-Century Science* (Princeton Univ. Press, 1959).
Man is not Lost. A record of two hundred years of astronomical navigation with the Nautical Almanac (HMSO, London, 1968).

CHAPTER 8

Henry Cavendish, "Experiments to determine the Density of the Earth", *Phil. Trans.* 88 (1798), pp. 469–526.
Seymour L. Chapin, "The shape of the earth", in Taton and Wilson (eds.), *Planetary Astronomy, Part B* (see under General Reading), pp. 22–34.
Allan Chapman, *Dividing the Circle* (see General Reading), pp. 102–06.
Allan Chapman, "The pit and the pendulum: G. B. Airy and the determination of gravity", in Denys Vaughan (ed.), *The Royal Society and the Fourth Dimension. The History of Timekeeping* (see under Chapter 13), pp. 70–78.
I. S. Glass, *Nicolas-Louis De La Caille. Astronomer and Geodesist* (OUP, 2013).
Nevil Maskelyne, "An Account of Observations made on Schiehallion for finding its Attraction", *Phil. Trans.* 65 (1775), p. 500.

Pierre L. M. Maupertuis, "Le figure de la terre...", *Mémoires de l'Académie Royale des Sciences* (Paris, 1737), pp. 386–466.
Pierre L. M. Maupertuis, "Observations astronomiques et physiques faites en le Caïenne", pp. 233–326. Chapter X, "De la longueur du pendule à secondes de temps", p. 320. In *Mémoires de l'Académie Royale des Sciences. Depuis 1666 jusqu'à 1699.* Tome VII, Part 1 (Paris, 1729).
Jean Picard et al., *The Measure of the Earth... by Diverse Members of the Academy of Sciences in Paris*, transl. Richard Waller (London, 1687). (Bodleian Library, Oxford, copy bound up in Claude Perrault's *Natural History of Animals* (London, 1687).)
Jonathan Powers, *The Honourable Henry Cavendish, FRS, FSA. The Man who Weighed the Earth* (iOpening Books, Derby Univ., 2012).

Chapter 9

"Friedrich Wilhelm Bessel", by W. Fricke, *Dictionary of Scientific Biography* ed. C. C. Gillespie, vol. 2 (Charles Scribner, New York, 1970), pp. 97–102.
Jeffrey Burley and Kristina Plenderleith (eds.), *A History of the Radcliffe Observatory, Oxford. A Biography of a Building* (Green College / Radcliffe Observatory, Oxford, 2005).
James Fergusson, *Astronomy Explained upon Sir Isaac Newton's Principles, and made easy for those who have not studied Mathematics* (London, 1799).
Mark Girouard, *Life in the English Country House* (see General Reading).
Pamela Gossin, *Thomas Hardy's Novel Universe. Astronomy, Cosmology, and Gender in the Post-Darwinian World* (Ashgate, Aldershot, 2007).
R. W. Harris, *Reason and Nature in 18th-Century Thought* (Blandford Press, London, 1968).
Caroline Herschel's Autobiographies, ed. Michael Hoskin (see under Chapter 5).
William Herschel, "Investigation of the Powers of Prismatic Colours to heat and illuminate objects...", *Phil. Trans.* 90 (1800), pp. 255–83.
Michael Hoskin, *The Herschel Partnership* (see under Chapter 5).
Patrick Moore, *Caroline Herschel. Reflected Glory* (William Herschel Society, Bath, 1988).
John Pringle Nichol, *Thoughts on Some Important Points Relating to the System of the World* (William Tait, Edinburgh, 1846).
Jonathan Powers, *John Whitehurst's Theory of the Earth* (iOpening Books, Derby Univ., 2013).
Jonathan Powers, *The Philosopher Lecturing on the Orrery* (see under Chapter 4).
Andrew Robinson, *The Last Man who Knew Everything: Thomas Young, the Anonymous Polymath* (Oneworld, Oxford, 2006).

Simon Schaffer, "Herschel in Bedlam: natural history and stellar astronomy", *BJHS* 11 (1980), p. 81.

CHAPTER 10

Gunther Buttman, *The Shadow of the Telescope. A Biography of John Herschel*, transl. B. F. J. Pagel and Davis S. Evans (Lutterworth, Guildford and London, 1974).

Allan Chapman, "An occupation for an independent gentleman: astronomy in the life of John Herschel", *Vistas* 1993, Elsevier Science, 1994, pp. 1–25. (Also in Chapman, *Astronomical Instruments and Their Users* (see General Reading), XIII.)

Agnes Clerke, *The Herschels and Modern Astronomy* (Cassell, London, 1895).

Adrian Desmond and James Moore, *Darwin* (1991; Penguin, 1992), pp. 90–91.

D. Evans, T. J. Deeming, Betty Hall Evans, and Stephen Goldfarb (eds), *Herschel at the Cape. Diaries and Correspondence of Sir John Herschel, 1834–1838* (Univ. of Texas, Austin and London, 1969).

John Fauvel, Raymond Flood, and Robin Wilson (eds), *Möbius and his Band. Mathematics and Astronomy in Nineteenth-Century Germany* (OUP, 1993).

John Herschel to Charles Babbage, letter, 18 Dec. 1815, Royal Society Manuscripts H.S. 20:30. (Further details in Chapman, *The Victorian Amateur Astronomer* (see General Reading), p. 323, n. 13.)

John Herschel, "Of the Aberrations of Compound lenses and Object Glasses", *Phil. Trans.* 111 (1821), pp. 222–67.

John Herschel, "Account of some observations made with a 20-foot reflecting telescope; Comprehending [item] (3) An Account of the Actual State of the Great Nebula in Orion…", *Memoirs of the RAS* II (1826), pp. 459–97.

John Herschel, "Observations of Nebulae and Clusters of stars made at Slough, with a Twenty-feet Reflecting Telescope between 1825 and 1833", *Phil. Trans.* 123 (1833), pp. 359–505.

John Herschel, *Results of Astronomical Observations Made During the Years 1834, 5, 6, 7, 8 at the Cape of Good Hope* (Smith, Elder & Co., London, 1847).

John Herschel, *Outlines of Astronomy* (London, 1849).

John Herschel's papers: *Catalogue of Scientific Papers (1800–1863) Compiled by the Royal Society of London*, III (1867), pp. 322–28.

John Herschel, "Diary", MS copy in Royal Society Library.

Margaret Herschel, *Lady Herschel. Letters from the Cape 1834–1838*, ed. Brian Warner (Friends of South African Library, Cape Town, 1991).

William Herschel, "Astronomical Observations..." (see under Chapter 1).
D. G. King-Hele (ed.), *John Herschel 1792–1871. A Bicentennial Commemoration* (Royal Society, London, 1992).
Patrick Moore, *John Herschel, Explorer of the Southern Skies* (William Herschel Society, Bath, 1991).
John Pringle Nichol, *Views of the Architecture of the Heavens* (William Tait, Edinburgh, 1858).
Larry J. Schaaf, "The poetry of light: Herschel, art and photography", in King-Hele (ed.), *John Herschel 1792–1871* (see above), pp. 3–15.
Eileen Shorland, *Sir John Herschel. The Forgotten Philosopher* (The Herschel Family Archive, 2016). (An excellent study based largely upon Herschel family archives.)

CHAPTER 11

Jim Al-Khalili, *Aliens: Science from the Other Side* (Profile Books, London, 2016).
William Astore, *Observing God: Thomas Dick, Evangelicalism, and Popular Science in Victorian Britain and America* (Ashgate, Farnham , 2001).
Richard Baum, "The man who found a city on the moon", *AN* Aug. 2017, pp. 68–69.
Michael J. Crowe, *The Extraterrestrial Life Debate, 1750–1900. The Plurality of Worlds from Kant to Lowell* (CUP, New York, 1986).
Thomas Dick, *The Christian Philosopher, or, The Connection of Science and Philosophy with Religion* (Longman, Hurst, Rees, Edinburgh, 1823).
Thomas Dick, *Celestial Scenery; Or, The Wonder of the Planetary Systems Displayed Illustrating The Perfections of the Deity* (Harper Brothers, New York, 1839, etc.). (See Ch. IX for "On the Doctrine of the Plurality of Worlds".)
David Gavine, "Thomas Dick, LL.D. (1774–1857)", *JBAA* 84 (1973–74), pp. 345–50.
Karl S. Guthke, *The Last Frontier. Imagining other Worlds from the Copernican Revolution to Modern Science Fiction*, transl. Helen Atkins (Cornell Univ. Press, Ithaca and London, 1983).
Richard A. Locke, "Great astronomical discoveries lately made by Sir John Herschel... at the Cape of Good Hope" (1835), reprinted in Faith K. Pizor and T. Allan Comp, *The Man in the Moon. An Anthology of Antique Science Fiction and Fantasy* (Sidgwick and Jackson, London, 1971), pp. 190–216.
Edward Walter Maunder, *Are the Planets Inhabited?* (Harper, New York, 1913).

Jules Verne, *From Earth to the Moon* (Paris, 1865) (many English translations after 1867).
Jules Verne, *Around the Moon* (Paris, 1870) (many English translations).
Herbert G. Wells, *The War of the Worlds* (Heinemann, London, 1898; Penguin, 2005). Bk. II, Ch. 8, p. 168 (Penguin) "But there are no bacteria on Mars, and directly these invaders arrived, directly they drank and fed, our microscopic allies began to work their overthrow."
H. G. Wells, *The First Men in the Moon* (G. Newnes, London, 1901).
John Willis, *All these Worlds are Yours: The Scientific Search for Alien Life* (Yale Univ. Press, 2016).

CHAPTER 12

Mary Brück, "Mary Somerville, mathematician and astronomer of underused talents", *JBAA* 106, 4 (1996), pp. 201–06.
Allan Chapman, *Mary Somerville and the World of Science* (Canopus, Bath, 2004; Springer, 2015).
Christina Colvin (ed.), *Maria Edgeworth. Letters from England 1813–1844* (Clarendon Press, Oxford, 1971).
Sarah Parkin, "Mary Somerville (1780–1872): her correspondence and work on chemistry" (Oxford Univ. MChem Thesis, 2001, Bodleian Library, Oxford).
Elizabeth A. Patterson, *Mary Somerville and the Cultivation of Science 1815–1840* (Martinus Nijhoff, Kluwer, Boston, The Hague, Lancaster, 1983).
Richard A. Proctor, "Mary Somerville", obituary, *MNRAS* 33 (1873), pp. 190–97.
Mary Somerville, *Personal Recollections from Early Life to Old Age of Mary Somerville*, by her daughter Martha Somerville (Roberts Brothers, Boston, 1874).
Queen of Science. Personal Recollections of Mary Somerville, ed. Dorothy McMillan (Canongate Classics 102, Edinburgh, 2001). (Re-edited and unexpurgated edition based on the original MS.)
Mary Somerville, "On the Magnetizing power of the more refrangible solar rays", *Phil. Trans.* 116, 2 (1826), pp. 132–39.
Mary Somerville, *On the Mechanism of the Heavens* (London, 1831). (See "Preliminary Dissertation".)
Mary Somerville, *On the Connexion of the Physical Sciences* (Murray, London, 1835, and later editions).
Mary Somerville, *Physical Geography* (2nd edn, London, 1849).
Mary Somerville, *Microscopic and Molecular Science* (London, 1869).

(Mary Somerville's surviving papers are in the Bodleian Library, Oxford, and the Royal Society Library, London.)

Chapter 13

Autobiography of Sir George Biddell Airy, KCB etc., ed. Wilfred Airy (CUP, 1896).

G. B. Airy, "Astronomer Royal's Journal", Cambridge Univ. Library MS. RGO 6.

Richard Baum and William Sheehan, *In Search of Planet Vulcan. The Ghost in Newton's Clockwork Universe* (Plenum Trade, New York and London, 1997).

Mary Brück, *The Peripatetic Astronomer. The Life of Charles Piazzi Smyth* (Adam Hilger and Institute of Physics, 1988).

Allan Chapman, "Sir George Airy (1801–1892) and the concept of international standards in science, timekeeping, and navigation", *Vistas* 28 (Pergamon Press, 1985), pp. 321–28. Also in Chapman, *Astronomical Instruments and Their Users* (see General Reading), XVI.

Allan Chapman, "Private research and public duty: George Biddell Airy and the search for Neptune", *JHA* 19 (1988), pp. 121–39. Also in Chapman, *Astronomical Instruments and Their Users* (see General Reading), XIV.

Allan Chapman, "Standard Time for all. The electric telegraph, Airy, and the Greenwich Time Service", in Frank A. J. L. James (ed.), *Semaphores to Short Waves* (Royal Society for Arts, Manufacturers, and Commerce, 1998), pp. 40–59.

Allan Chapman, "Porters, watchmen, and labourers, and the crime of William Sayers: the non-scientific staff of the Royal Observatory, Greenwich, in Victorian times", *JAHH* 6 (2003), pp. 27–36.

Allan Chapman, "Airy's Greenwich staff", *AA*, 6 January 2012, pp. 4–16.

Stephen Courtney, "The historical meridian: antiquity and Scripture in the public work of George Biddell Airy", *JHA* 49, 2 (2018), pp. 135-57.

Edwin Dunkin, *A far off vision. A Cornishman at Greenwich Observatory, being Dunkin's "auto-biographical notes"*, ed. Peter D. Hingley and T. C. Daniel (Royal Institution of Cornwall, 1999).

Bernard S. Finn, *Submarine Telegraphy. The Grand Victorian Technology* (HMSO, London, 1973).

Raymond Flood, Adrian Rice, and Robin Wilson (eds), *Mathematics in Victorian Britain* (OUP, 2013).

H. M. Harrison, *Voyager in Time and Space. The Life of John Couch Adams, Cambridge Astronomer* (Book Guild Ltd., Sussex, 1994).

Derek Howse, *Nevil Maskelyne. The Seaman's Astronomer* (CUP, 1989).

L. T. Macdonald, "Making Kew Observatory: the Royal Society, the British Association and the politics of early Victorian science" *BJHS*, vol. 48, part 3 (September, 2015), pp. 409–33.

Lee T. Macdonald, *Kew Observatory and the Evolution of Victorian Science, 1840-1910* (Pittsburg Univ. Press, 2018).

Lee T. Macdonald, "The origins and early years of the Magnetic and Meteorological Department at Greenwich Observatory, 1834–1848", *Annals*, July 2018 (online publication).

Edward Walter Maunder, *The Royal Observatory, Greenwich. A glance at its history and work* (Religious Tracts Society, 1900).

Anita McConnell, *Instrument Makers to the World. A History of Cooke, Troughton, & Simms* (William Sessions, York, 1992).

A. J. Meadows, *Greenwich Observatory, 2: Recent History 1836–1975* (Taylor & Francis, London, 1975).

Eleanor Mennim, *Transit Circle. The Story of William Simms 1793–1860* (William Sessions, York, undated, *c.* 1993).

Patrick Moore, *The Planet Neptune. A Historical Survey before Voyager* (Wiley, Chichester, 1988, 1996).

John Pringle Nichol, *The Planet Neptune, an Exposition and History* (Edinburgh, 1848).

H. H. Turner, *Astronomical Discovery* (London, 1904).

Denys Vaughan (ed.), *The Royal Society and the Fourth Dimension. The History of Timekeeping* (Proceedings of the Joint Royal Society and Antiquarian Horology Meeting, 25 June 1993, Antiquarian Horology Society, Sussex, autumn 1993).

CHAPTER 14

Hermann A. Brück, "Lord Crawford's Observatory at Dun Echt 1872–1892", *Vistas* 35 (1992), pp. 81–138.

Mary Brück, *Women in Early British and Irish Astronomy. Stars and Satellites* (Springer, Dordrecht, 2009).

Allan Chapman, "William Lassell (1799–1880): practitioner, patron, and "Grand Amateur" of Victorian astronomy", *Vistas* 32 (1989), pp. 341–70. Also in Chapman, *Astronomical Instruments and Their Users* (see General Reading), XVII.

Mike Edmunds, "Founders of the RAS: William Pearson" *A&G* 58, 5 (Oct. 2017), p. 12.

John Herschel, "Observations of Nebulae and Clusters of stars…" (see under Chapter 10).

Roger Hutchins, *British University Observatories 1772–1939* (see General Reading).
Jack Meadows, *The Victorian Scientist. The Growth of a Profession* (British Library, 2004).
Charles Mollan (ed.), *William Parsons, Third Earl of Rosse. Astronomy and the Castle in Nineteenth-Century Ireland* (Manchester Univ. Press, 2014).
J. Morrell and A. Thackray (eds.), *Gentlemen of Science. Early Correspondence* (Royal Historical Society, 1984).
James Nasmyth, *James Nasmyth Engineer: An Autobiography*, ed. Samuel Smiles (John Murray, London, 1889).
James Nasmyth and James Carpenter, *The Moon: Considered as a Planet, A World and a Satellite* (Murray, London, 1874).
William Parsons, Lord Rosse, "Observations of the nebulae", *Phil. Trans.* 140 (1850), pp. 499–515.
William Parsons, Earl of Rosse, "Of the construction of specula of six-feet aperture and a selection from the observations of nebulae made with them", *Phil. Trans.* 151 (1861), pp. 681–745.
George Peacock, *Life of Thomas Young, MD, FRS, &C.* (London, 1855).
W. Garrett Scaife, *From Galaxies to Turbines. Science, Technology, and the Parsons Family* (Institute of Physics, Bristol and Philadelphia, 2000).
Barbara Slater, *The Astronomer of Rousdon. Charles Grover 1842–1921* (Steam Mill Press, Norwich, 2005).
William H. Smyth, *Cycle of Celestial Objects*, and the *Bedford Catalogue* (Part II of the *Cycle*) & *Aedes Hartwelliana* (London, 1844 and 1860).
Thomas Sprat, *A History of the Royal Society of London* (London, 1667; 2nd edn 1702).
Thomas Woods, *The Monster Telescope* (1845; reprinted CUP, 2010). (On Lord Rosse's 72-inch telescope.)

CHAPTER 15

Trudy E. Bell, "History of astrophotography", *Astronomy* 4 (1979), pp. 66–79.
L. J. M. Daguerre (transl. J. S. Memes), *The History and Practice of Photogenic Drawing with a New Method of Dioramic Painting* (Smith, Elder & Co., London, 1839).
Warren De La Rue, "Mr De La Rue on celestial photography", *MNRAS* 19 (1859), pp. 353–58.
Henry Draper, "On a Reflecting Telescope for Celestial Photography

erected at Hastings, near New York", *Report of the British Association for the Advancement of Science, 1860* (London, 1861).

Henry Draper, "Photographs of the Orion Nebula", *The American Journal of Science and Arts*, 3rd series, 20 (1880), p. 433; and Draper, "Photographs of the Orion Nebula", *Philosophical Magazine*, 5th series, 10 (1880), p. 388.

John William Draper, "On the process of Daguerreotype, and its application to taking Portraits from the Life", *Philosophical Magazine and Journal of Science* 17 (1840), pp. 217–25.

Michel Frizot (ed.), *A New History of Photography*, transl. Susan Bennett *et al.* (Köneman, Köln, 1994, 1998).

John Lankford, "The impact of photography on astronomy", in Owen Gingerich (ed.), *The General History of Astronomy* (ed. Michael Hoskin), vol. 4 part A, *Astrophysics and Twentieth-Century Astronomy to 1950* (CUP, 1984), pp. 16–39.

David Le Conte, "Warren de la Rue – pioneer of astronomical photography", *AA* 5 (2011), pp. 14–35.

Joseph Norman Lockyer, *Stargazing, Past and Present* (London, 1878).

James Nasmyth and James Carpenter, *The Moon* (see under Chapter 14).

Beaumont Newhall, *The Daguerreotype in America* (3rd edn, Dover, New York, 1976).

Isaac Roberts, *Photographs of Stars, Star-Clusters and Nebulae: Together with Information Concerning the Instruments and Methods Employed in the Pursuit of Celestial Photography* (Knowledge Office, London, 1893, 1899).

Larry J. Schaaf, *Out of the Shadows. Herschel, Talbot and the Invention of Photography* (Yale Univ. Press, 1992).

Larry J. Schaaf, "The poetry of light: Herschel, art and photography" (see under Chapter 10).

CHAPTER 16

Richard C. Carrington, "On the Distribution of the solar spots in latitude since the beginning of the year 1854", *MNRAS* 19 (1858), p. 1.

Richard C. Carrington, "Description of a singular appearance seen in the Sun on September 1, 1859", *MNRAS* 20 (1860), p. 31.

Stuart Clark, *The Sun Kings. The Unexpected Tragedy of Richard Carrington and the Tale of How Modern Astronomy Began* (Princeton Univ. Press, 2007).

Auguste Comte, *Cours de Philosophie Positive II* (Paris, 1835), "Astronomie", pp. 8–9.

William Herschel, "Investigation of the Powers of Prismatic Colours..." (see under Chapter 9).
William Herschel, "Observations tending to investigate the Sun, in order to find the causes of symptoms of its variable emission of light and heat...", *Phil. Trans.* 91 (1801), p. 265.
Alexander von Humboldt, *Cosmos* [*Kosmos*]: *A Sketch of a Physical Description of the Universe*, transl. E. C. Otte (London, 1864).
Norman C. Keer, *The Life and Times of Richard Christopher Carrington, BA, FRS, FRAS (1826–1875)* (Norman C. Keer, private publication, 2000).
Joseph Norman Lockyer, *The Chemistry of the Sun* (London, 1887).
Lee T. Macdonald, "Solar spot mania: the origins and early years of solar research at Kew Observatory, 1852–1860", *JHA* 46, 4 (Nov. 2015), pp. 469–90.
Edward Walter Maunder, "Note on the Distribution of Sun-Spots in Heliographic Latitude", *MNRAS* 64 (1904), pp. 747–61.
A. J. Meadows, *Early Solar Physics* (Pergamon, Oxford, 1970).
William Merle, "Weather Diary", 1337–44, Bodleian Library, Oxford, MS.: cited in Robert Plot, *The Natural History of Oxfordshire* (London, 1705).
James Nasmyth Engineer (see under Chapter 14): pp. 341, 370, 378 for solar work.
Robert S. Richardson, "Sunspot problems old and new", *Popular Astronomy* 55 (1947), pp. 120–32.
Robert S. Richardson, "A century of sunspots", *PASP*, leaflet 213, vol. 5 (Nov. 1948).
Heinrich Schwabe, "Sonnen-Beobachtungen im Jahr 1843", *Astronomische Nachrichten* vol. 21, no. 495 (1844), pp. 233–36.
Heinrich Schwabe, "Extract of a Letter from H. Schwabe to Mr Carrington", *MNRAS* 17 (1856), p. 241.
Charles A. Young, *The Sun* (New York, 1881).

Chapter 17

Barbara J. Becker, *Unravelling Starlight. William and Margaret Huggins and the Rise of the New Astronomy* (CUP, 2011).
George Bishop, *Jesuit Pioneers of Modern Science and Mathematics* (Gujarat Sahitya Prakash, Gujarat, India, 2005).
Ian Elliott and Charles Mollan, *William E. Wilson (1851–1908). The Work and Family of a Westmeath Astronomer* (Charles Mollan, County Dublin, 2018).
Owen Gingerich (ed.), *Astrophysics and Twentieth Century Astronomy to 1950*,

vol. 4 part A of *The General History of Astronomy* ed. Michael Hoskin (CUP, 1984).

Owen Gingerich, "Unlocking the chemical secrets of the cosmos", in Gingerich, *The Great Copernicus Chase* (see General Reading), pp. 170–76.

J. B. Hearnshaw, *The Analysis of Starlight. One Hundred and Fifty Years of Stellar Spectroscopy* (CUP, 1986).

J. B. Hearnshaw, *The Measurement of Starlight. Two Centuries of Astronomical Photometry* (CUP, 1996).

Stephen P. Holmes and Charles Fitzgerald-Lombard, "The great Observatory at Downside, 1859–67", *AA* 11 (June 2017), pp. 33–44.

William Huggins, "On the Spectra of some of the Chemical Elements", *Phil. Trans.* 154 (1864), pp. 139–60.

William Huggins and W. Miller, "On the Spectra of some of the Nebulae", *Phil. Trans.* 154 (1864), pp. 437–44.

William Huggins, "Further Observations on the Spectra of some of the Stars and Nebulae with an attempt to determine… whether these Bodies are moving towards or from the Earth…", *Phil. Trans.* 158 (1866), pp. 529–64.

William Huggins, "The new astronomy: a personal retrospect", *Nineteenth Century* 41 (1897), pp. 907–29.

James Lequeux, *François Arago. A 19th-Century French Humanist and Pioneer in Astrophysics* (Springer, Dordrecht, Heidelberg, etc., 2016).

Joseph Norman Lockyer, *Stargazing, Past and Present* (see under Chapter 15). (A good treatment of spectroscopes and spectroscopy.)

Joseph Norman Lockyer, *The Chemistry of the Sun* (Macmillan, London, 1887).

Sabino Maffeo, SJ, *In the Service of Nine Popes. 100 Years of the Vatican Observatory* (Vatican Observatory and Pontifical Academy, Rome, 1991).

Charles E. Mills and C. F. Brooke, *A Sketch of the Life of Sir William Huggins* (London, 1936).

Robert S. Richardson, "A century of sunspots" (see under Chapter 16), pp. 103–10.

"Lewis Morris Rutherfurd", obituary, *MNRAS* 53 (Feb. 1893), pp. 229–31.

William Tobin, *The Life and Science of Léon Foucault. The Man who Proved the Earth Rotates* (CUP, 2003).

Stephen F. Tonkin (ed.), *Practical Amateur Spectroscopy* (Springer-Verlag, London, 2002).

CHAPTER 18

Edward Lyon Berthon, "Romsey Observatory", *The English Mechanic*, 13 Oct. 1871, p. 83.
Allan Chapman, *The Victorian Amateur Astronomer* (see General Reading).
Henry Draper, "On the Construction of a Silvered Glass Telescope Fifteen and a half inches in Aperture and its use for Celestial Photography", *Smithsonian Contributions to Knowledge* 14, 4 (1864), pp. 1–15.
John Silas Evans, *Seryddiaeth a Seryddwyr* ["Astronomy and Astronomers"] (Cardiff, 1923).
Owen Gingerich, "Unlocking the chemical secrets of the cosmos" (see under Chapter 17).
G. Parry Jenkins, "A plea for the reflecting telescope", *Journal of the RAS of Canada* V (1911), 59–75: p. 68. (Includes Samuel Cooper's Venus transit photographs.)
Roger Langdon, *The Life of Roger Langdon told by himself, with additions by his Daughter Ellen* (London, 1909).
Robert Marriott, "Webb's Telescopes", in Janet and Mark Robinson (eds), *The Stargazer of Hardwicke* (below), pp. 121–44.
Janet and Mark Robinson (eds), *The Stargazer of Hardwicke. The Life and Work of Thomas William Webb* (Gracewing, Hereford, 2006).
Samuel Smiles, *Men of Invention and Industry* (London, 1884).
William Tobin, *The Life and Science of Léon Foucault* (see under Chapter 17).
Keith Venables, "The real origins of amateur telescope making", *Sky & Telescope*, May 2017, pp. 22–27.
Thomas William Webb, *Celestial Objects for Common Telescopes* (Longmans, Green & Co., London, 4th edn 1881).
The History of the Manchester Astronomical Societies (no author given: UMIST, Manchester, 1992).

Chapter 19

Gabriella Bernardi, *The Unforgotten Sisters. Female Astronomers and Scientists before Caroline Herschel* (Springer, Praxis, Chichester, 2016).
Elizabeth Brown, "Programmes of the Directors of the Observing Sections. Solar Section" *JBAA* 1, 2 (1890), pp. 58–60.
"In Memoriam. Elizabeth Brown, F. R. Met. Soc.", obituary, *JBAA* 9, 5 (March 1899), pp. 214–15.
Mary Brück, *Agnes Mary Clerke and the Rise of Astrophysics* (CUP, 2002).
Mary Brück, *Women in Early British and Irish Astronomy* (see under Chapter 14).
Cambridge University Observatory Syndicate *Reports*, 1882, 1885.

J. R. Cole, *Fair Science. Women in the Scientific Community* (The Free Press, New York, 1979).

Mary Creese, "Elizabeth Brown (1830–1899), solar astronomer", *JBAA* 108, 4 (1998), pp. 193–97.

H. S. Davis, "Women Astronomers, 1750–1890", *Popular Astronomy* 6 (1890), pp. 211–16.

Roger Hutchins, *British University Observatories 1772–1939* (see General Reading).

Peggy Aldrich Kidwell, "Women in Astronomy", in *History of Astronomy: An Encyclopedia*, ed. John Lankford (Garland, New York, 1997), pp. 564–67.

John Lankford, "Women and women's work at Mt Wilson Observatory before World War II", in G. Good (ed.), *The Earth and the Heavens and the Carnegie Institution of Washington, History of Geophysics* 5 (1994), pp. 125–28.

Marilyn Bailey Ogilvie, *Women in Science: Antiquity through the Nineteenth Century. A biographical dictionary with annotated bibliography* (3rd edn, MIT Press, Cambridge Mass., 1986).

Olga S. Opfell, *The Lady Laureates: Women who have won the Nobel Prize* (Scarecrow Press, New Jersey and London, 1978).

Cecilia Payne-Gaposchkin, *An Autobiography and other Recollections* (CUP, 1984).

Dava Sobel, *The Glass Universe: The Hidden History of the Women who Took the Measure of the Stars* (4th Estate, Harper Collins, London, 2016).

H. Wright, *Sweeper of the Sky. The Life of Maria Mitchell, First Woman Astronomer in America* (Macmillan, New York, 1949).

In 2016, the RAS *Astronomy and Geophysics* journal (*A&G*) marked the centenary of women in the RAS Fellowship with a series of commemorative articles. They are:

Vol. 57, Issue 1, Feb. 2016: Mandy Bailey, "100 years of Fellowship", pp. 19–21; Michael Hoskin, "Caroline Herschel", pp. 22–25.

Issue 2, April 2016: Allan Chapman, "Mary Somerville", pp. 10–12; Barbara Becker, "Margaret Huggins", pp. 13–14.

Issue 3, June 2016: Mark Hurn, "Anne Sheepshanks", p. 11; Sue Nelson, "Williamina Fleming", pp. 12–13; Sue Bowler, "Annie Jump Cannon", pp. 14–15; Sara Russell, "Agnes Mary Clerke", pp. 16–17.

Issue 4, Aug. 2016: Allan Chapman, "Victorian lady amateurs", pp. 12–13; Richard McKim, "Women in the British Astronomical Association", pp. 14–17.

Issue 5, Oct. 2016: Jeremy Shears, "Mary Adela Blagg", pp. 17–18; Siân

Prosser, "Mary Proctor", pp. 19–20; Silvia Dalla and Lyndsay Fletcher, "Annie Maunder", pp. 21–23.
Issue 6, Dec. 2016: Sue Bowler, "Persistent pioneers: from 1916 to the present", pp. 14–18; Karen Masters, "Women of the future in the RAS", pp. 19–20.

Chapter 20

George Biddell Airy, "Popular Astronomy: A Series of Lectures delivered at Ipswich…",
The Astronomical Register (see "Peter Johnson" below).
Sir Robert Stawell Ball, *The Story of the Heavens* (Cassell, London, 1886).
Sir R. S. Ball, *In Starry Realms* (Isbister, London, 1893).
Sir R. S. Ball, *Star-Land. Being Talks with Young People about the Wonders of the Heavens* (Cassell, London, 1893).
Sir R. S. Ball, *The Story of the Sun* (Cassell, London, 1893, 1910).
Sir R. S. Ball, *In the High Heavens* (Cassell, London, 1894).
Sir R. S. Ball, *Great Astronomers* (London, 1895, 1906).
Sir R. S. Ball, *A Popular Guide to the Heavens* (Phillips, London, 1910).
Valentine Ball (ed.), *Reminiscences and Letters of Sir Robert Ball* (Cassell, London, 1915).
Peter J. Bowler, *Science for All: the Popularization of Science in Early Twentieth-Century Britain* (Univ. of Chicago Press, 2009).
Henry Brougham, *Paley's Natural Theology Illustrated. Preliminary Discourse by Lord Brougham* (London, 1835).
A. Chapman, *The Victorian Amateur Astronomer* (see General Reading).
Michael J. Crowe, *The Extraterrestrial Life Debate* (see under Chapter 11).
Edwin Dunkin, *The Midnight Sky. Familiar notes on the stars and planets* (Religious Tracts Society, 1891).
John Durant (ed.), *Museums and the Public Understanding of Science* (Science Museum, London, 1992).
The English Mechanic and World of Science, 1865–1926.
A. Fowler, *Popular Telescopic Astronomy* (Phillips, London, 1896).
Frances Hawes, *Henry Brougham* (Jonathan Cape, London, 1957).
Roger Hutchins, "Richard Anthony Proctor", *Oxford DNB* (2004).
James H. Jeans, *The Mysterious Universe* (CUP, 1930).
James Jeans, *Through Time and Space* (CUP, 1934).
Peter Johnson, "*The Astronomical Register* 1863–86", *JBAA* 100, 2 (1990), pp. 62–66.
Edward Walter Maunder, *Astronomy Without a Telescope* (London, 1904).

E. W. Maunder, *The Heavens and their Story* (London, 1910).
Henry Mayhew, *London Labour and the London Poor: A Cyclopaedia of the Conditions and Earnings of Those that Will Work…* (4 vols., London, 1851).
F.-A. Pouchet, *The Universe: or, The Infinitely Great and the Infinitely Little* (10th edn, English translation, Blackie, Glasgow and Edinburgh, 1883).
Henry Solly, *Working Men: A glance at some of their wants, with reasons and suggestions for helping them to help themselves* (London,1863).
William Wordsworth, "Star Gazers", 1807, in Wordsworth, *The Major Works* ed. Stephen Gill (OUP, 1984), pp. 322–23.
Donald K. Yeomans, *Comets* (see under Chapter 3).

Chapter 21

R. V. Bruce, *The Launching of Modern American Science, 1846–1876* (Knopf, New York, 1987).
Owen Gingerich (ed.), *Astrophysics and Twentieth-Century Astronomy to 1950* (see under Chapter 17). (Contains excellent chapters on the Harvard, US Naval, and Lick Observatories, along with big telescopes in the USA; also excellent articles on Greenwich, Paris, Potsdam, and other European observatories.)
M. Greene, *A Science not Earthbound. A Brief History of Astronomy at Carleton College* (at Northfield, Minnesota) (Carleton College, 1988).
George E. Hale, *The Study of Stellar Evolution: An Account of Some Recent Methods of Astrophysical Research* (Chicago Univ. Press, 1908).
Paul A. Haley, "Williamina Fleming and the Harvard College Observatory", *AA* 11 (June 2017), pp. 2–32.
Jeff Hecht, "The lesson of the Great Paris Telescope", *Sky and Telescope*, July 2017, pp. 28–33.
H. King, *History of the Telescope* (see General Reading).
Kevin Krisciunas, *Astronomical Centers of the World* (see General Reading).
John Lankford (assisted by Ricky L. Slavings), *American Astronomy: Community, Careers, and Power, 1859–1940* (Univ. of Chicago Press, 1997).
Elias Loomis, *The Recent Progress of Astronomy: Especially in the United States* (New York, 1851).
H. S. Miller, *Dollars for Research: Science and its Patrons in Nineteenth-Century America* (Washington Univ. Press, 1970).
Donald E. Osterbrock, *James E. Keeler, Pioneer American Astrophysicist, and the Early Development of American Astrophysics* (CUP, 1984).
Phillips Astronomy Encyclopedia, ed. Sir Patrick Moore (Phillips, London, 2002).

William Sheehan, *The Planet Mars. A History of Observation and Discovery* (Univ. of Arizona Press, 1996–97).
Dava Sobel, *The Glass Universe* (see under Chapter 19).
Deborah Jean Warner, *Alvan Clark & Sons. Artists in Optics* (Smithsonian, Washington, 1968).

CHAPTER 22

R. S. Ball, *The Story of the Heavens* (see under Chapter 20).
R. Berendzen, R. Hart, and D. Seeley, *Man Discovers the Galaxies* (Science History, New York, 1975).
Michael J. Crowe, *Modern Theories of the Universe from Herschel to Hubble* (Dover, New York, 1994).
David DeVorkin, "Stellar evolution and the origin of the Hertzsprung–Russell Diagram", in Gingerich (ed.), *Astrophysics and Twentieth-Century Astronomy to 1950* (see under Chapter 17), pp. 90–108.
George E. Hale, *The Study of Stellar Evolution* (see under Chapter 21).
Edmond Halley, "On the Infinity of the sphere of the fix'd Stars", *Phil. Trans.* 31, no. 364 (1720), pp. 22–24.
George Johnson, *Miss Leavitt's Stars. The Untold Story of the Woman who Discovered how to Measure the Universe* (Norton, New York, 2005).
Kenneth Lang and Owen Gingerich (eds), *A Source Book in Astronomy and Astrophysics 1900–1975* (Cambridge Mass., Harvard Univ. Press, 1979).
Henrietta S. Leavitt, "1777 Variables in the Magellanic Clouds", *Annals of Harvard College Observatory* LX (IV) (1908), pp. 87–110.
Harry Nussbaumer and Lydia Bieri, *Discovering the Expanding Universe* (CUP, 2009).
E. R. Paul, *The Milky Way Galaxy and Statistical Cosmology, 1890–1924* (CUP, 1993).
Ian Ridpath, "Michelson Morley Experiment", in *Collins Encyclopaedia of the Universe* (Harper Collins, London, 2001), pp. 86–87.
Harlow Shapley, "On the existence of external galaxies", *PASP* 31(1919), pp. 261–67.
Vesto M. Slipher, "Radial velocity observations of the spiral nebulae", *The Observatory* 40 (1917), pp. 304–06.
R. W. Smith, *The Expanding Universe: Astronomy's "Great Debate" 1900–1931* (CUP, 1982).
Virginia Trimble, "The 1920 Shapley–Curtis discussion: background, issues, and aftermath", *PASP* 107 (1995), p. 1133.

Chapter 23

R. Baum and W. Sheehan, *In Search of Planet Vulcan* (see under Chapter 13).
Niels Bohr, *On the Constitution of Atoms and Molecules*, Papers of 1913 reprinted from *The Philosophical Magazine*, with an introduction by L. Rosenfeld (Munksgaard Ltd, Copenhagen and New York, 1963).
Ronald W. Clark, *Einstein: The Life and Times* (Wing Books, New York, 1984).
A. S. Eddington and A. C. de la Crommelin, "A new theory of the universe", *The Illustrated London News*, 15 Nov. 1919, p. 766.
Albert Einstein, *The Meaning of Relativity* (Routledge, New York, 1945).
Owen Gingerich, "Albert Einstein: a laboratory in the mind", in *The Great Copernicus Chase* (see General Reading), pp. 282–92.
Otto Glasser, *Wilhelm Conrad Röntgen and the Early History of the Röntgen Rays* (John Bale, London, 1933).
Walter Isaacson, *Einstein: His Life and Universe* (Simon and Schuster, New York, 2007).
James H. Jeans, *Problems of Cosmogony and Stellar Dynamics* (CUP, 1919).
Willy Ley, *Watchers of the Skies* (see General Bibliography).Ray Mackintosh, Jim Al-Khalili, Bjorn Jonson, and Teresa Pena, *Nucleus. A Trip to the Heart of Matter* (Johns Hopkins Univ. Press and Canopus, Bristol, England, 2001).
"Molecule" in *Nelson's Encyclopaedia* XV (Thomas Nelson, London, *c.* 1910), pp. 391–93.
Dame Kathleen Ollerenshaw, *To Talk of Many Things* (Manchester Univ. Press, 2004). (Autobiography.)
Abraham Pais, *Subtle is the Lord. The Life and Science of Albert Einstein* (OUP, 1982, 2005).
Naomi Pasachoff, *Marie Curie and the Science of Radioactivity* (OUP, 1996).
Andrew Robinson, *Einstein: A Hundred Years of Relativity* (Palazzo, 2005).
Tony Rothman, *Everything's Relative: And other Fables from Science and Technology* (Wiley, New York, 2003).
Ernest Rutherford, *The Newer Alchemy* (CUP, 1937).
"Ernest Rutherford", *Chambers Biographical Encyclopaedia of Scientists*, eds John Daintith, Sarah Mitchell, and Elizabeth Tootill (Edinburgh, 1981), pp. 341–42.

Chapter 24

Richard Berendzen and Michael Hoskin, "Hubble's announcement of

Cepheids in spiral nebulae", *PASP*, leaflet 504, June 1971.
Gale Christianson, *Edwin Hubble: Mariner of the Nebulae* (Farrar, Straus, and Giroux, New York, 1995).
William Herschel, "Observations and Experiments tending to investigate the local arrangement of the celestial bodies in space", *Phil. Trans.* 107 (1817), pp. 302–31.
Norris S. Hetherington, *The Edwin Hubble Papers: Previously Unpublished Papers on the Extragalactic Nature of the Spiral Nebulae* (Pachart, Tucson, 1990).
Edwin Hubble, "Extra-galactic nebulae", *AJ* 64 (1926), pp. 321–69.
E. Hubble, "A relation between distance and radial velocity among extra-galactic nebulae", *Proceedings of the National Academy of Sciences* 15, 3 (1929), pp. 168–73.
E. Hubble, *The Realm of the Nebulae* (Yale Univ. Press, 1936).
E. Hubble, *The Observational Approach to Cosmology* (Clarendon Press, Oxford, 1937).
James H. Jeans, *Astronomy and Cosmogony* (CUP, 1928).
J. H. Jeans, *The Mysterious Universe* (see under Chapter 20).
Robert Naeye, *Through the Eyes of Hubble. The Birth, Life, and Violent Death of Stars* (Institute of Physics, Bristol and Philadelphia, 1998). (Images of objects through the Hubble Space Telescope.)
Allan Sandage, "Current problems in the extragalactic distance scale", *AJ* 127 (1958), pp. 513–26.
Allan Sandage, "Edwin Hubble, 1889–1953", *Royal Astronomical Society of Canada Journal* 83 (1989), pp. 351–62.
Alexander S. Sharov and Igor D. Novikov, *Edwin Hubble the Discoverer of the Big Bang Universe* (CUP, 1989).
W. M. Smart, *The Riddle of the Universe* (Longmans, London, 1968).
Brian Ventrudo, "How a janitor at Mount Wilson Observatory [Milton Humason] measured the size of the universe", *The Christian Science Monitor*, 19 May 2010.
Reginald L. Waterfield, *A Hundred Years of Astronomy* (Duckworth, London, 1938).
David O. Woodbury, *The Glass Giant of Palomar* (Dodd, Mead & Co., New York, 1939).

CHAPTER 25

A. L. Berger (ed.), *The Big Bang and Georges Lemaître: Proceedings of a Symposium in honour of G. Lemaître fifty years after his initiation of Big Bang*

Cosmology, Louvain-la-Neuve, Belgium, 10–13 Oct. 1983 (Springer and Reidel, Dordrecht, Netherlands, 1984).
Bruno Bertotti (ed.), *Modern Cosmology in Retrospect* (CUP, 1990).
Nigel Calder, *Violent Universe. An eye-witness account of the commotion in astronomy, 1968–69* (BBC, 1969).
S. Chandrasekhar, "The highly collapsed configurations of stellar mass", *MNRAS* 91 (1931), pp. 456–66.
S. Chandrasekhar, "The maximum mass of ideal White Dwarfs", *AJ* 74 (1931), pp. 81–82.
S. Chandrasekhar, *Oral History Interview* (American Institute of Physics, Center for the History of Physics, 1977).
Allan Chapman, Paper to RAS, *The Observatory* 130, no. 1217, Aug. 2010 (meeting of 8 Jan. 2010), pp. 198–200.
Willem de Sitter, Paper to RAS, *The Observatory* 53 (1930), pp. 37–39.
Arthur Stanley Eddington, *Stars and Atoms* (Clarendon Press, Oxford, 1927).
S. Eddington, *The Expanding Universe* (CUP, 1933).

Stephen Hawking, *A Brief History of Time. From the Big Bang to Black Holes* (Bantam Press, 1988).
Rodney Holder and Simon Mitton, *Georges Lemaître: Life, Science, and Legacy* (Astrophysics and Space Flight Library 395, Springer Verlag, Berlin, Heidelberg , 2013).
Helge Kragh, "A cosmologist's excursion into geophysics" [Fred Hoyle], *A&G* 56, 6 (Dec. 2015), pp. 15–17.
Georges Lemaître, "The expanding universe", *MNRAS* 91 (1931), pp. 490–501.
Georges Lemaître, "The expansion of the universe. A homogeneous universe of constant mass and increasing radius as accounting for the radial velocity of extra-galactic nebulae" (in French, Brussels, 1927; English translation, 1931), *MNRAS* 91 (1931), pp. 483–90.
Frank Levin, *Calibrating the Cosmos. How Cosmology Explains our Big Bang Universe* (Springer Verlag, Heidelberg, New York, 2007).
Mario Livio, *The Accelerating Universe. Infinite Expansion, the Cosmological Constant, and the Beauty of the Cosmos* (John Wiley & Sons Inc., Chichester, 2000).
Simon Mitton, *Conflict in the Cosmos: Fred Hoyle's Life in Science* (John Henry Press, 2005).
Simon Mitton, "Fred Hoyle's birth centennial", *A&G* 56, 6 (Dec. 2015), pp. 10–14.
H. Nussbaumer and L. Bieri, *Discovering the Expanding Universe* (see under Chapter 22).

T. Padmanabhan, *After the First Three Minutes. The Story of our Universe* (CUP, 1998).
Józef Turek, *Georges Lemaître and the Pontifical Academy of Sciences* (Specola Vaticana, Rome, 1989).
G. C. Vittlie, "Georges Lemaître", *Quarterly Journal of the RAS* 8 (1967), pp. 294–97.

CHAPTER 26

Jon Agar, *Science and Spectacle: The Work of Jodrell Bank in Post-War British Culture* (Harwood Academic, 1998).
D. O. Edge and M. J. Mulkay, *Astronomy Transformed: The Emergence of Radio Astronomy in Britain* (John Wiley, New York, 1976).
A. Hewish, Susan Jocelyn Bell, et al. "Observations of a rapidly rotating pulsating radio source" [pulsar], *Nature* 217 (1968), p. 709.
J. S. Hey, *The Evolution of Radio Astronomy* (Science History Publications, Neale Watson Academic, New York, 1973).
Karl G. Jansky, "Radio waves from outside the solar system", *Nature* 132, 3323 (1933), p. 66.
Bernard Lovell, "The Jodrell Bank Radio Telescope", *Nature* 180, 4576 (1957), pp. 60–62.
B. Lovell, *The Story of Jodrell Bank* (OUP, 1968).
B. Lovell, "The effects of defence science on the advance of astronomy", *JHA* 8 (1977), pp. 151–73.
B. Lovell, *The Jodrell Bank Telescopes* (OUP, 1985).
B. Lovell, *Astronomer by Chance* (Macmillan, London, 1990).
Ian Morison, "The Lovell Telescope: 60 years old and counting", *AN* Aug. 2017, pp. 40–43.
David P. D. Munns, *A Single Sky: How an International Community Forged the Science of Radio Astronomy* (Cambridge Mass., MIT Press, 2013).
Tim O'Brien and Theresa Anderson, "Recycling, rockets, and radio astronomy" [60 years of the Lovell Telescope], *A&G* 58, 5 (Oct. 2017), pp. 22–27.
John Pfeiffer, *The Changing Universe. The Story of the New Astronomy* (Gollancz, London, 1956).
David Staelin and Edward C. Reifenstein III, "Pulsating radio sources near the Crab Nebula", *Science* 162 (1968), pp. 1481–83.
Woodruff T. Sullivan III, "Early radio astronomy", in Gingerich (ed.), *Astrophysics and Twentieth-Century Astronomy to 1950* (see under Chapter 17), pp. 190–98.

Woodruff T. Sullivan III, *Cosmic Noise: A History of Early Radio Astronomy* (CUP, 2009).

CHAPTER 27

Mark Aldwinckle, "The development of gunpowder-based military pyrotechnics in the period *c.* 1790–1846, with particular regard to the development of the Congreve Rocket" (Oxford Univ. MChem Thesis,1989, Bodleian Library).

E. J. Beckland and D. Millard, *Congreve and His Works* (Science Museum, London, undated pamphlet, *c.* 1987).

Jim Bell, *The Interstellar Age: Inside the Forty-Year Voyager Mission* (Penguin Group, 2015).

Roger E. Bilstein, *Stages to Saturn: A Technological History of the Apollo/Saturn Launch Vehicles* (National Aeronautics and Space Administration, Washington DC, 1996).

Andrew Chaikin, *A Man on the Moon. The Voyages of the Apollo Astronauts* (Viking, 1994).

William Congreve, *The Details of the Rocket System* (London, 1814) (unpaginated).

Keith Cooper, "From here to eternity. 40 years [and counting] of the Voyagers", *AN* Aug. 2017, pp. 50–57.

Russell Freedman, *2000 Years of Space Travel* (Collins, London and Glasgow, 1965).

Robert H Goddard, *A Method of Reaching Extreme Altitudes* (Smithsonian Misc. Collection, vol. 71, no. 3 (Washington, 1919).

Patrick Moore, *Moore on Mercury. The Planet and the Missions* (Springer-Verlag, London, 2007).

Michael Perryman, *The Making of History's Greatest Star Map* (Springer, Heidelberg, Dordrecht, 2010).

Colin Pillinger, *Space is a Funny Place. Fifty years [and more] of space exploration seen through the eyes of cartoonists* (Barnstorm Productions, Cartoon Museum, London, 2007).

Faith K. Pizor and T. Allan Comp, *The Man in the Moon* (see under Chapter 11).

Asif A. Siddiqi, *Sputnik and the Soviet Space Challenge* (Gainsville Univ. Press, Florida, 2003).

Gurbir Singh, *Yuri Gagarin in London and Manchester. A Smile that Changed the World?* (Gurbir Singh, Astrotalkuk Publications, 2011).

Gurbir Singh, *The Indian Space Programme. India's incredible journey from the*

Third World to the First (Gurbir Singh, Astrotalkuk Publications, 2017).
R. W. Smith, *The Space Telescope: A Study of NASA. Science, Technology, and Politics* (CUP, 1989).
Anthony M. Springer, "Early experimental programs of the American Rocket Society, 1930–41", *Journal of Spacecraft and Rocketry* 40, 4 (July–Aug. 2003), pp. 475ff.
J. N. Tatarewicz, *Space Technology and Planetary Astronomy* (Indiana Univ. Press, 1990).
Reginald Turnill, *The Moonlandings. An Eyewitness Account* (CUP, 2003).
Tom Wolfe, *The Right Stuff* (Bantam, New York, 2001).

Chapter 28

Allan Chapman, *Stargazers* (see General Reading).
Allan Chapman, letter in *The Spectator*, 19 Nov. 2016, p 31.

Ruth Finnegan (ed.), *Participating in the Knowledge Society. Researchers Beyond the University Walls* (Palgrave, Macmillan, 2005).
William Herschel, "An Account of three Volcanoes on the Moon", *Phil. Trans.* 72 (1787), pp. 229–32.
John Lankford, "Amateurs: astronomy's enduring resource", *Sky & Telescope* 75 (1988), pp. 482–83.
Martin Mobberley, *It Came From Outer Space Wearing an RAF Blazer! A Fan's Biography of Sir Patrick Moore* (Springer, Heidelberg, 2013).
Patrick Moore, *Guide to the Moon* (Eyre and Spottiswood, London, 1953, etc.).
Patrick Moore, *The Amateur Astronomer* (CUP, 1957 (many editions)).
Patrick Moore, *Patrick Moore on the Moon* (Cassell, London, 2001).
Patrick Moore, *80 Not Out* (Contender Books, London, 2003).
Patrick Moore, *Stars of Destiny. A Scientific Look at Astrology* (Canopus, Bristol, 2005).
Patrick Moore, *Moore on Mercury* (see under Chapter 27).
Carl Sagan, *Cosmos* (Random House, New York, 1980).
Carl Sagan, *Pale Blue Dot. A Vision of the Human Future in Space*, (Random House, New York, 1994).
Stephen F. Tonkin (ed.), *Practical Amateur Spectroscopy* (see under Chapter 17).

CHAPTER 29

Jim Al-Khalili, *Aliens: Science from the Other Side* (see under Chapter 11).
Titus Lucretius Carus, *De Rerum Natura* ["On the Nature of Things"], *c.* 60 BC, transl. Alicia Stallings (Penguin Classics, 2007). (Many other translations.)
Edward Walter Maunder, *Astronomy and the Bible: An Elementary Commentary on the Holy Scripture* (New York, 1908; reprinted 2007).
Carl Sagan, *Pale Blue Dot* (see under Chapter 28).
Jon Willis, *All these Worlds are Yours: The Scientific Search for Alien Life* (see under Chapter 11).

Index[1]

[1] *f* refers to figures